Yarn and Sewing

Yarn and Sewing

Sandeep Roy

Yarn and Sewing

ISBN 978-93-5111-649-3

Published in 2015 in India by

RANDOM PUBLICATIONS

4376-A/4B, Gali Murari Lal, Ansari Road
New Delhi-110 002
Phone : +9111-43580356, 011-23289044, 011-43142548
e-mail: sales@randompublications.com,
info@randompublications.com, randomexports@gmail.com

Reprinted 2021

Type Setting by : Friends Media, Delhi-110089
Digitally Printed at: Replika Press Pvt. Ltd.

Preface

Sewing is the craft of fastening or attaching objects using stitches made with a needle and thread. Sewing is one of the oldest of the textile arts, arising in the Paleolithic era. Before the invention of spinning yarn or weaving fabric, archaeologists believe Stone Age people across Europe and Asia sewed fur and skin clothing using bone, antler or ivory needles and "thread" made of various animal body parts including sinew, catgut, and veins. For thousands of years, all sewing was done by hand. The invention of the sewing machine in the 19th century and the rise of computerization in the later 20th century led to mass production of sewn objects, but hand sewing is still practised around the world. Fine hand sewing is a characteristic of high-quality tailoring, haute couture fashion, and custom dressmaking, and is pursued by both textile artists and hobbyists as a means of creative expression.

The first known use of the word sewing was in the 14th century. Sewers working on a simple pattern need only a few sewing tools: measuring tape, needle, thread, cloth, and sewing shears. More complex patterns done on a sewing machine may only need a few more simple tools to get the job done, but there are an ever-growing variety of helpful sewing aids available, such as presser foot attachments for sewing ruffles, or hem repair glue. When the sewer has gathered the necessary tools to tackle a pattern, there are several elements of garment construction that are part of the process. Patterns will specify whether to cut on the grain or a bias cut. Construction stitches include edgestitching, understitching, staystitching and topstitching; seam types include the plain seam, zigzag seam, flat fell seam, French seam, and many others. Supporting materials, such as interfacing, interlining or lining, or fusing, may be used as well, to give the fabric a more rigid or durable shape. Volume can be added with elements such as pleats, or reduced with the use of darts. Sewing machines are now made for a broad range of specialised sewing purposes, such as quilting machines, computerized machines for embroidery, and various sergers for finishing raw edges of fabric.

I would like to thank my team for standing beside me throughout my career and writing this book. My special thanks go to "Random Publications" who have published the book.

–Sandeep Roy

Contents

1

Techniques

TACK (SEWING)

In sewing, to tack or baste is to make quick, temporary stitching intended to be removed. Tacking is used in a variety of ways:

- To temporarily hold a seam or trim in place until it can be permanently sewn, usually with a long running stitch made by hand or machine called a tacking stitch or basting stitch.
- X-shaped tacking stitches are also very common on vents (slits) on the back of men's suit jackets, or at the bottom of kick pleats on a woman's skirt. They are meant to hold the flaps in place during shipping and when on display in the store. They should be removed before being worn; however many buyers do not realize it. Brand labels loosely basted on the outer edges of the sleeves of suits as well as women's winter coat should also be removed after purchase. They are meant to help customers to easily identify the brands in the store without reaching into the collar.
- To temporarily attach a lace collar, ruffles, or other trim to clothing so that the attached article may be removed easily for cleaning or to be worn with a different garment. For this purpose, tacking stitches are sewn by hand in such a way that they are almost invisible from the outside of the garment.
- To transfer pattern markings to fabric, or to otherwise mark the point where two pieces of fabric are to be joined. A special loose looped stitch used for this purpose is called a tack or tailor's tack. This is often done through two opposing layers of the same fabric so that when the threads are snipped between the layers the stitches will be in exactly the same places for both layers thus saving time having to chalk and tack the other layer.
- A basting stitch is essentially a straight stitch, sewn with long stitches and unfinished ends. The basting stitch is used for temporarily holding sandwiched pieces of fabric in place. The stitch is removed after the piece is finished. Often used in quilting or embroidery.

CUT (CLOTHING)

Cut in clothing, sewing and tailoring, is the style or shape of a garment as opposed to its fabric or trimmings. The *cut* of a coat refers to the way the garment hangs on the body based on the shape of the fabric pieces used to construct it, the position of the fabric's grain line, and so on.

DARNING

Darning is a sewing technique for repairing holes or worn areas in fabric or knitting using needle and thread alone. It is often done by hand, but it is also possible to darn with a sewing machine. Hand darning employs the darning stitch, a simple running stitch in which the thread is "woven" in rows along the grain of the fabric, with the stitcher reversing direction at the end of each row, and then filling in the framework thus created, as if weaving. Darning is a traditional method for repairing fabric damage or holes that do not run along a seam, and where patching is impractical or would create discomfort for the wearer, such as on the heel of a sock. Darning also refers to any of several needlework techniques that are worked using darning stitches:

- Pattern darning is a type of embroidery that uses parallel rows of straight stitches of different lengths to create a geometric design.
- Net darning, also called filet lace, is a 19th-century technique using stitching on a mesh foundation fabric to imitate lace.
- Needle weaving is a drawn thread work embroidery technique that involves darning patterns into barelaid warp or weft threads.

DARNING CLOTH

In its simplest form, darning consists of anchoring the thread in the fabric on the edge of the hole and carrying it across the gap. It is then anchored on the other side, usually with a running stitch or two. If enough threads are criss-crossed over the hole, the hole will eventually be covered with a mass of thread.

Fine darning, sometimes known as *Belgian darning*, attempts to make the repair as invisible and neat as possible. Often the hole is cut into a square or darn blends into the fabric.

There are many varieties of fine darning. Simple over-and-under weaving of threads can be replaced by various fancy weaves, such as twills, chevrons, etc., achieved by skipping threads in regular patterns.

Invisible darning is the epitome of this attempt at restoring the fabric to its original integrity. Threads from the original weaving are unraveled from a hem or seam and used to effect the repair. Invisible darning is appropriate for extremely expensive fabrics and items of apparel.

In machine darning, lines of machine running stitch are run back and forth across the hole, then the fabric is rotated and more lines run at right angles. This is a fast way to darn, but it cannot match the effects of fine darning.

DARNING TOOLS

There are special tools for darning socks or stockings:

- A darning egg is an egg-shaped ovoid of stone, porcelain, wood, or similar hard material, which is inserted into the toe or heel of the sock to hold it in the proper shape and provide a firm foundation for repairs. When the repairs are finished, the darning egg is removed. A shell of the tiger cowry *Cypraea tigris*, a popular ornament in Europe and elsewhere, was also sometimes used as a ready-made darning egg.
- A darning mushroom is a mushroom-shaped tool usually made of wood. The sock is stretched over the curved top of the mushroom, and gathered tightly around the stalk to hold it in place for darning.
- A darning gourd is a hollow dried gourd with a pronounced neck. The sock can be stretched over the full end of the gourd and held in place around the neck for darning.
- A used light bulb can be used to hold a sock in place for darning.

PATTERN DARNING

Pattern darning is a simple and ancient embroidery technique in which contrasting thread is woven in-and-out of the ground fabric using rows of running stitches which reverse direction at the end of each row. The length of the stitches may be varied to produce geometric designs. Traditional embroidery using pattern darning is found in Africa, Japan, Northern and Eastern Europe, the Middle East, Mexico and Peru. Pattern darning is also used as a *filling stitch* in blackwork embroidery.

DART (SEWING)

Darts are folds sewn into fabric to help provide a three-dimensional shape to a garment. They are frequently used in women's clothing to tailor the garment to the wearer's shape.

Two kinds of darts are common in blouses for women:

- Vertical darts—These are sewn from the bottom of the blouse to a point generally around the bustline. This type of dart may be found in the front, rarely in the back of a garment and are used by the garment maker to pull in the bottom of the blouse towards the wearer's waist.
- Bust darts—These are short triangle folds that provide space for breasts such that the fabric under the breasts isn't hanging, rather is fitting closer to the wearer. There are several subtypes of bust line dart:
  - Center
  - Waistline

- French
- Side seam
- Armhole
- Neckline
- Shoulder
- T-dart
- Inverted T-dart

In the early 1950s, the New York City firm of Evan-Picone pioneered the use of darts in the pockets of women's clothing. The darts help keep the pocket open and thus more easily accessed, reducing the chance of rips or tears.

EMBELLISHMENT

In sewing and crafts an embellishment is anything that adds design interest to the piece.

APPLIQUÉ

In its broadest sense, an appliqué is a smaller ornament or device applied to another surface. An appliqué is usually one piece. In the context of ceramics, for example, an appliqué is a separate piece of clay added to the primary work, generally for the purpose of decoration.

The term is borrowed from French and, in this context, means "applied" or "thing that has been applied." Appliqué is a surface pattern that is used to decorate an aspect of a garment or product. It is highly used with the Textiles industry, but lately is a key trend for make do mend items.

Cloth

In the context of sewing, an appliqué refers to a needlework technique in which patterns or representational scenes are created by the attachment of smaller pieces of fabric to a larger piece of contrasting colour or texture. It is particularly suitable for work which is to be seen from a distance, such as in banner-making. A famous example of appliqué is the Hastings Embroidery. Appliquéd cloth is an important art form in Benin, West Africa, particularly in the area around Abomey, where it has been a tradition since the 18th century and the kingdom of Danhomè.

Quilting

Appliqué is used extensively in quilting. "Dresden Plate" and "Sunbonnet Sue" are two examples of traditional American quilt blocks that are constructed with both patchwork and appliqué. Baltimore album quilts, Broderie perse, Hawaiian quilts, Amish quilts, Egyptian Khayamiya and the ralli quilts of India and Pakistan also use appliqué. In Pakistan, Quilting Ralli Applique work particularly for women's clothing in the name of Islamabad Applique is very famous.

Types

Applied pieces usually have their edges folded under, and are then attached by any of the following:

- Straight stitch, typically 20-30mm in from the edge.
- Satin stitch, all around, overlapping the edge. The patch may be glued or straight stitched on first to ensure positional stability and a neat edge.
- Reverse appliqué: the attached materials are sewn together, then cut away where another material covers it on top, before being sewn down onto the edges of the original material.

Appliqué is also used for school badges.

Appliqué and electronic sewing machines

Modern consumer embroidery machines quickly stitch appliqué designs by following a programme. The programmes have a minimum complexity of two thread colours, meaning the machine stops during stitching to allow the user to switch threads. First, the fabric that will be the background and the appliqué fabric are affixed into the machine's embroidery hoop. The programme is run and the machine makes a loose basting stitch over both layers of fabric. Next, the machine stops for a thread change, or other pre-programmed break. The user then cuts away the excess appliqué fabric from around the basting stitch. Following this, the machine continues on programme, automatically sewing the satin stitches and any decorative stitching over the appliqué for best results

Modern fashion

In this sense,appliqué refers to using fabric shapes/designs usually on the trim of a garment. This can either be sewn or glued. Many appliques are more often imported from China and used here to keep costs down. Since many designers use appliques that are mass-produced, you can easily find matching accessories and such from competitive stores. Each store may carry various items with the same applique.

EMBROIDERY

Embroidery is the handicraft of decorating fabric or other materials with needle and thread or yarn. Embroidery may also incorporate other materials such as metal strips, pearls, beads, quills, and sequins. Embroidery is most often used on caps, hats, coats, blankets, dress shirts, denim, stockings, and golf shirts. Embroidery is available with a wide variety of thread or yarn colour.An interesting characteristic of embroidery is that the basic techniques or stitches on surviving examples of the earliest embroidery—chain stitch, buttonhole or blanket stitch, running stitch, satin stitch, cross stitch—remain the fundamental techniques of hand embroidery today.

Embroidery has been dated to the Warring States period (5th-3rd century BC). The process used to tailor, patch, mend and reinforce cloth fostered the development of sewing techniques, and the decorative possibilities of sewing led to the art of embroidery. In a garment from Migration period Sweden, roughly 300–700 CE, the edges of bands of trimming are reinforced with running stitch, back stitch, stem stitch, tailor's buttonhole stitch, and whipstitching, but it is uncertain whether this work simply reinforced the seams or should be interpreted as decorative embroidery.

The remarkable stability of basic embroidery stitches has been noted:

- It is a striking fact that in the development of embroidery ... there are no changes of materials or techniques which can be felt or interpreted as advances from a primitive to a later, more refined stage. On the other hand, we often find in early works a technical accomplishment and high standard of craftsmanship rarely attained in later times.

In the 16th century, in the reign of the Mughal Emperor Akbar, his chronicler Abu al-Fazl ibn Mubarak wrote in the famous Ain-i-Akbari: "His majesty (Akbar) pays much attention to various stuffs; hence Irani, Ottoman, and Mongolian articles of wear are in much abundance especially textiles embroidered in the patterns of *Nakshi*, *Saadi*, *Chikhan*, *Ari*, *Zardozi*, *Wastli*, *Gota* and *Kohra*. The imperial workshops in the towns of Lahore, Agra, Fatehpur and Ahmedabad turn out many masterpieces of workmanship in fabrics, and the figures and patterns, knots and variety of fashions which now prevail astonish even the most experienced travelers. Taste for fine material has since become general, and the drapery of embroidered fabrics used at feasts surpasses every description."

Embroidery was a very important art in the Medieval Islamic world. One of the most interesting accounts of embroidery were given by the 17th century Turkish traveler Evliya Çelebi, who called it the "craft of the two hands". Because embroidery was a sign of high social status in Muslim societies, it became a hugely popular art. In cities such as Damascus, Cairo and Istanbul, embroidery was visible on handkerchiefs, uniforms, flags, calligraphy, shoes, robes, tunics, horse trappings, slippers, sheaths, pouches, covers, and even on leather belts. Many craftsmen embroidered with gold and silver thread. A number of embroidery cottage industries, each employing over 800 people, grew to supply these items.

Elaborately embroidered clothing, religious objects, and household items have been a mark of wealth and status in many cultures including ancient Persia, India, China, Japan, Byzantium, and medieval and Baroque Europe. Traditional folk techniques are passed from generation to generation in cultures as diverse as northern Vietnam, Mexico, and eastern Europe. Professional workshops and guilds arose in medieval England. The output of these workshops, called *Opus Anglicanum* or "English work," was famous throughout Europe.

Industrial Revolution

The development of machine embroidery on a mass production scale came about in stages. The earliest machine embroidery used a combination of machine looms and teams of women embroidering the textiles by hand. This was done in France by the mid-1800s. The manufacture of machine-made embroideries in St. Gallen in eastern Switzerland flourished in the latter half of the 19th century.

Classification

Embroidery can be classified according to whether the design is stitched *on top of* or *through* the foundation fabric, and by the relationship of stitch placement to the fabric.

In free embroidery, designs are applied without regard to the weave of the underlying fabric. Examples include crewel and traditional Chinese and Japanese embroidery.

Counted-thread embroidery patterns are created by making stitches over a predetermined number of threads in the foundation fabric. Counted-thread embroidery is more easily worked on an even-weave foundation fabric such as embroidery canvas, aida cloth, or specially woven cotton and linen fabrics although non-evenweave linen is used as well. Examples include needlepoint and some forms of blackwork embroidery.

In canvas work threads are stitched through a fabric mesh to create a dense pattern that completely covers the foundation fabric. Traditional canvas work such as bargello is a counted-thread technique. Since the 19th century, printed and hand painted canvases, on which the printed or painted image serves as a guide to the placement of the various thread or yarn colours, have eliminated the need for counting threads. These are particularly suited to pictorial rather than geometric designs such as those deriving from the Berlin wool work craze of the early 19th century.

In drawn thread work and cutwork, the foundation fabric is deformed or cut away to create holes that are then embellished with embroidery, often with thread in the same colour as the foundation fabric. These techniques are the forerunners of needlelace. When created with white thread on white linen or cotton, this work is collectively referred to as whitework.

Materials

The fabrics and yarns used in traditional embroidery vary from place to place. Wool, linen, and silk have been in use for thousands of years for both fabric and yarn. Today, embroidery thread is manufactured in cotton, rayon, and novelty yarns as well as in traditional wool, linen, and silk. Ribbon embroidery uses narrow ribbon in silk or silk/organza blend ribbon, most commonly to create floral motifs.

Surface embroidery techniques such as chain stitch and couching or laid-work are the most economical of expensive yarns; couching is generally used for goldwork. Canvas work techniques, in which large amounts of yarn are buried on the back of the work, use more materials but provide a sturdier and more substantial finished textile.

In both canvas work and surface embroidery an embroidery hoop or frame can be used to stretch the material and ensure even stitching tension that prevents pattern distortion. Modern canvas work tends to follow symmetrical counted stitching patterns with designs emerging from the repetition of one or just a few similar stitches in a variety of hues. In contrast, many forms of surface embroidery make use of a wide range of stitching patterns in a single piece of work.

Machine

Much contemporary embroidery is stitched with a computerized embroidery machine using patterns "digitized" with embroidery software. In machine embroidery, different types of "fills" add texture and design to the finished work. Machine embroidery is used to add logos and monograms to business shirts or jackets, gifts, and team apparel as well as to decorate household linens, draperies, and decorator fabrics that mimic the elaborate hand embroidery of the past.

Industrial embroidery machines

THERE ARE A NUMBER OF BRANDS AVAILABLE ON THE MARKET; THE TOP TWO ARE TAJIMA AND BARUDAN, FOLLOWED BY TOYOTA AND SWF.PIPING (SEWING)

In sewing, piping is a type of trim or embellishment consisting of a strip of folded fabric inserted into a seam to define the edges or style lines of a garment or other textile object. Usually the fabric strip is cut on the bias. It may be made from either self-fabric (the same fabric as the object to be ornamented) or contrasting fabric, or of leather. Today, piping is common on upholstery and decorative pillows, but it is also used on clothing. Piped pocket openings, garment edges, and seams are characteristic of Western wear.

TRIM (SEWING)

Trim or trimming in clothing and home decorating is applied ornament, such as gimp, passementerie, ribbon, ruffles, or, as a verb, to apply such ornament.

Before the industrial revolution, all trim was made and applied by hand, thus making heavily trimmed furnishings and garments expensive and high-status. Machine-woven trims and sewing machines put these dense trimmings within the reach of even modest dressmakers and home sewers, and an abundance of trimming is a characteristic of mid-Victorian fashion. As a predictable reaction, high fashion came to emphasize exquisiteness of cut and construction over denseness of trimming, and applied trim became a signifier of mass-produced clothing by the 1930s. The iconic braid and gold button trim of the Chanel suit are a notable survival of trim in high fashion.

In home decorating, the 1980s and 1990s saw a fashion for dense, elaborately layered trimmings on upholstered furniture and drapery.

Today, most trimmings are commercially manufactured. Scalamandré is known for elaborate trim for home furnishings, and Wrights is a leading manufacturer of trim for home sewing and crafts. Trims are used generally to enhance the beauty of the garments. It attracts buyers. Appropriate use of it creates more value of the product.

- Bias tape
- Braid
- Buttons
- Cord
- Embroidery by hand or machine
- Gimp
- Lace edgings or insertions
- Passementerie
- Piping
- Ribbon
- Rickrack
- Ruffles or frills
- Tassels

LACE

Lace is an openwork fabric, patterned with open holes in the work, made by machine or by hand. The holes can be formed via removal of threads or cloth from a previously woven fabric, but more often open spaces are created as part of the lace fabric. Lace-making is an ancient craft. True lace was not made until the late 15th and early 16th centuries. A true lace is created when a thread is looped, twisted or braided to other threads independently from a backing fabric.

Originally linen, silk, gold, or silver threads were used. Now lace is often made with cotton thread, although linen and silk threads are still available. Manufactured lace may be made of synthetic fibre. A few modern artists make lace with a fine copper or silver wire instead of thread.

Types

There are many types of lace, classified by how they are made. These include:

- Needle lace, such as Venetian Gros Point, is made using a needle and thread. This is the most flexible of the lace-making arts. While some types can be made more quickly than the finest of bobbin laces, others are very time-consuming. Some purists regard needle lace as the height of lace-making. The finest antique needle laces were made from a very fine thread that is not manufactured today.
- Cutwork, or whitework, is lace constructed by removing threads from a woven background, and the remaining threads wrapped or filled with embroidery.
- Bobbin lace, as the name suggests, is made with bobbins and a pillow. The bobbins, turned from wood, bone, or plastic, hold threads which are woven together and held in place with pins stuck in the pattern on the pillow. The pillow contains straw, preferably oat straw or other materials such as sawdust, insulation styrofoam, or ethafoam. Also known as Bone-lace. Chantilly lace is a type of bobbin lace.
- Tape lace makes the tape in the lace as it is worked, or uses a machine- or hand-made textile strip formed into a design, then joined and embellished with needle or bobbin lace.
- Knotted lace includes macramé and tatting. Tatted lace is made with a shuttle or a tatting needle.
- Crocheted lace includes Irish crochet, pineapple crochet, and filet crochet.
- Knitted lace includes Shetland lace, such as the "wedding ring shawl", a lace shawl so fine that it can be pulled through a wedding ring.
- Machine-made lace is any style of lace created or replicated using mechanical means.
- Chemical lace: the stitching area is stitched with embroidery threads that form a continuous motif. Afterwards, the stitching areas are removed and only the embroidery remains. The stitching ground is made of a water-soluble or non-heat-resistant material.

Etymology

The word lace is from Middle English, from Old French *las*, noose, string, from Vulgar Latin **laceum*, from Latin *laqueus*, noose; probably akin to *lacere*, to entice or ensnare.

History

In the late 16th century there was a rapid development in the field of lace. There was an openwork fabric where combinations of open spaces and dense textures formed designs. These forms of lace were dominant in both fashion and home décor during the late 1500s. For enhancing the beauty of collars and cuffs, needle lace was embroidered with loops and picots.

Objects resembling lace bobbins have been found in Roman remains, but there are no records of Roman lace making. Lace was used by clergy of the early Catholic Church as part of vestments in religious ceremonies but did not come into widespread use until the 16th century in the northwestern part of the European continent. The popularity of lace increased rapidly and the cottage industry of lace making spread throughout Europe. Countries like Italy, France, Belgium, Germany (then Holy Roman Empire), Czech Republic (town of Vamberk), Slovenia (town of Idrija), Finland (town of Rauma) England (town of Honiton), Hungary, Ireland, Malta, Russia, Spain, Turkey and others all have an established heritage expressed through lace.

In North America in the 19th century, missionaries spread the knowledge of lace making to the Native American tribes. St. John Francis Regis helped many country girls stay away from the cities by establishing them in the lace making and embroidery trade, which is why he became the Patron Saint of lace making. In 1837, Samuel Ferguson first used jacquard looms with Heathcoat's bobbin net machine, resulting in endless possibilities for lace designs.

Traditionally, lace was used to make tablecloths and doilies and in both men's and women's clothing. The English diarist Samuel Pepys often wrote about the lace used for his, his wife's, and his acquaintances' clothing, and on May 7, 1669, noted that he intended to remove the gold lace from the sleeves of his coat "as it is fit [he] should", possibly in order to avoid charges of ostentatious living.

Industrial revolution

With the passage of time, the way the world produced goods changed. An increase in demand for lace led to the development of machine lace. In 1768, John Heathcoat invented the bobbin net machine, which made it possible to produce complex lace designs more quickly. The Industrial Revolution was the downfall of the handmade lace industry. The teaching of handmade lace making disappeared in schools as emphasis shifted from trades to academics, paving the way for lace making to become a hobby instead of the business it once was.

Military uniforms

The term 'lace' is used by the British to refer to the gold bands sewn onto the sleeves of naval officers' uniforms to indicate rank, and to name the similar decoration elsewhere on other uniforms (such as Italian caps and Polish collars) because of the procedure used to make it. In America, the term is not used for this purpose because the bands are compactly sewn metal thread, and 'lace' seems to imply cloth sewn into patterns with holes in them.

FRINGE (TRIM)

Fringe is an ornamental textile trim applied to an edge of an item, such as drapery, a flag, epaulettes, or decorative tassel.

Fringe originates in the ends of the warp, projecting beyond the woven fabric. In this way, a cut piece of fabric would not require hemming in order to achieve an edge which would not unravel: several strands of weft threads would be removed and the warp threads would remain. More commonly it is made separately and sewn on, consisting sometimes of projecting ends, twisted or plaited together, and sometimes of loose threads of wool, silk, or linen, or narrow strips of leather.

BEAD

A bead is a small, decorative object that is formed in a variety of shapes and sizes of a material such as glass, plastic, or wood, and that is pierced for threading or stringing. Beads range in size from under 1 millimetre (0.039 in) to over 1 centimetre (0.39 in) in diameter. A pair of beads made from *Nassarius* sea snail shells, approximately 100,000 years old, are thought to be the earliest known examples of jewellery. Beadwork is the art or craft of making things with beads. Beads can be woven together with specialized thread, strung onto thread or soft, flexible wire, or adhered to a surface (*e.g.* fabric, clay).

Types of beads

Beads may be divided into several types of overlapping categories based on different criteria such as the materials from which they are made, the process used in their manufacturing, the place or period of origin, the patterns on their surface, or their general shape. In some cases, such as millefiori and cloisonné beads, multiple categories may overlap in an interdependent fashion.

Components

Beads can be made of many different materials. The earliest beads were made of a variety of natural materials which, after they were gathered, could be readily drilled and shaped. As humans became capable of obtaining and working with more difficult materials, those materials were added to the range of available substances. More recently, synthetic materials were added.

In modern manufacturing, the most common bead materials are wood, plastic, glass, metal, and stone.

Natural materials

Beads are still made from many naturally occurring materials, both organic (*i.e.*, of animal- or plant-based origin) and inorganic (purely mineral origin). However, some of these materials now routinely undergo some extra processing beyond mere shaping and drilling such as colour enhancement via dyes or irradiation.

The natural organics include bone, coral, horn, ivory, seeds (such as tagua nuts), animal shell, and wood. For most of human history pearls were the ultimate precious beads of natural origin because of their rarity; the modern pearl-culturing process has made them far more common. Amber and jet are also of natural organic origin although both are the result of partial fossilization.

The natural inorganics include various types of stones, ranging from gemstones to common minerals, and metals. Of the latter, only a few precious metals occur in pure forms, but other purified base metals may as well be placed in this category along with certain naturally occurring alloys such as electrum. There are also paper beads.

Synthetic materials

The oldest-surviving synthetic materials used for beadmaking have generally been ceramics: pottery and glass. Beads were also made from ancient alloys such as bronze and brass, but as those were more vulnerable to oxidation they have generally been less well-preserved at archaeological sites.

Many different subtypes of glass are now used for beadmaking, some of which have their own component-specific names. Lead crystal beads have a high percentage of lead oxide in the glass formula, increasing the refractive index. Most of the other named glass types have their formulations and patterns inseparable from the manufacturing process.

Small, colourful, fusible plastic beads (some brands are Nabbi, Hama, Perler, and Pyssla) can be placed on a solid plastic-backed peg array to form designs and then melted together with a clothes iron; alternatively, they can be strung into necklaces and bracelets or woven into keychains. Fusible beads come in many colours and degrees of transparency/opacity, including varieties that glow in the dark or have internal glitter; peg boards come in various shapes and several geometric patterns. Plastic toy beads, made by chopping plastic tubes into short pieces, were introduced in 1958 by Munkplast AB in Munka-Ljungby, Sweden, under the brand Nabbi. Known as Indian beads, they were originally sewn together to form ribbons. The pegboard for bead designs was invented in the early 1960s (patented 1962, patent granted 1967) by Gunnar Knutsson in Vällingby, Sweden, as a therapy for elderly homes; the pegboard later gained popularity as a toy for children. The bead designs were glued to cardboard or Masonite boards and used as trivets. Later, when the beads were made of polyethylene, it became possible to fuse them with a flat iron. In 2005, Munkplast/Nabbi introduced the Photo Pearls software that converts digital photos to bead designs. Hama come in three sizes: mini (diameter 2.5 mm), midi (5 mm) and maxi (10 mm). Perler beads come in two sizes called classic (5 mm) and biggie (10 mm). Pyssla beads (by IKEA) only come in one size (5 mm).

Manufacturing

Modern mass-produced beads are generally shaped by carving or casting, depending on the material and desired effect. In some cases, more specialized metalworking or glassworking techniques may be employed, or a combination of multiple techniques and materials may be used such as in cloisonné.

Glassworking

Most glass beads are pressed glass, mass-produced by preparing a molten batch of glass of the desired colour and pouring it into molds to form the desired shape. This is also true of most plastic beads.

A smaller and more expensive subset of glass and lead crystal beads are cut into precise faceted shapes on an individual basis. This was once done by hand but has largely been taken over by precision machinery.

"Fire-polished" faceted beads are a less expensive alternative to hand-cut faceted glass or crystal. They derive their name from the second half of a two-part process: first, the glass batch is poured into round bead molds, then they are faceted with a grinding wheel. The faceted beads are then poured onto a tray and briefly reheated just long enough to melt the surface, "polishing" out any minor surface irregularities from the grinding wheel.

Specialized glass techniques and types

There are several specialized glassworking techniques that create a distinctive appearance throughout the body of the resulting beads, which are then primarily referred to by the glass type.

If the glass batch is used to create a large massive block instead of pre-shaping it as it cools, the result may then be carved into smaller items in the same manner as stone. Conversely, glass artisans may make beads by lampworking the glass on an individual basis; once formed, the beads undergo little or no further shaping after the layers have been properly annealed.

Most of these glass subtypes are some form of fused glass, although goldstone is created by controlling the reductive atmosphere and cooling conditions of the glass batch rather than by fusing separate components together.

Dichroic glass beads incorporate a semitransparent microlayer of metal between two or more layers. Fibre optic glass beads have an eyecatching chatoyant effect across the grain.

There are also several ways to fuse many small glass canes together into a multicolored pattern, resulting in millefiori beads or chevron beads (sometimes called "trade beads"). "Furnace glass" beads encase a multicolored core in a transparent exterior layer which is then annealed in a furnace.

More economically, millefiori beads can also be made by limiting the patterning process to long, narrow canes or rods known as murrine. Thin cross-sections, or "decals", can then be cut from the murrine and fused into the surface of a plain glass bead.

Shapes

Hair pipe beads

Elk rib bones were the original material for the long, tubular hair pipe beads.

Today these beads are commonly made of bison and water buffalo bones and are popular for breastplates and chokers among Plains Indians. Black variations of these beads are made from the animals' horns.

Seed beads

Seed beads are uniformly shaped spheroidal or tube shaped beads ranging in size from under a millimetre to several millimetres. "Seed bead" is a generic term for any small bead. Usually rounded in shape, seed beads are most commonly used for loom and off-loom bead weaving.

Place or period of origin

- African trade beads or slave beads may be antique beads that were manufactured in Europe and used for trade during the colonial period, such as chevron beads; or they may have been made in West Africa by and for Africans, such as Mauritanian Kiffa beads, Ghanaian and Nigerian powder glass beads, or African-made brass beads.
- Austrian crystal is a generic term for cut lead-crystal beads, based on the location and prestige of the Swarovski firm.
- Czech glass beads are made in the Czech Republic, centralized around an area called Jablonec nad Nisou. Production of glass beads in the area dates back to the 14th century, though production was depressed under communist rule. Because of this long tradition, their workmanship and quality has an excellent reputation.
- Vintage beads, in the collectibles and antique market, refers to items that are at least 25 or more years old. Vintage beads are available in materials that include lucite, plastic, crystal, metal and glass.

Miscellaneous ethnic beads

Tibetan Dzi beads and Rudraksha beads are used to make Buddhist and Hindu rosaries (malas). Magatama are traditional Japanese beads, and cinnabar was often used for beads in China. Wampum are cylindrical white or purple beads made from quahog or North Atlantic channeled whelk shells by northeastern Native American tribes, such as the Wampanoag and Shinnecock. Job's tears are seed beads popular among southeastern Native American tribes. Heishe are beads made of shells or stones by the Kewa Pueblo people of New Mexico.

Symbolic meaning of beads

In many parts of the world, beads are used for symbolic purposes, for example:

- Use for prayer - *e.g.* rosary beads
- Use for anti-tension devices, *e.g.* worry beads
- Use as currency *e.g.* Aggrey beads from Ghana

- Use for gaming *e.g.* owari beads for mankala
- Greek komboloi beads as in Crete

Surface patterns

After shaping, glass and crystal beads can have their surface appearance enhanced by etching a translucent frosted layer, applying an additional colour layer, or both. *Aurora Borealis*, or AB, is a surface coating that diffuses light into a rainbow. Other surface coatings are vitrail, moonlight, dorado, satin, star shine, and heliotrope. Faux beads are beads that are made to look like a more expensive original material, especially in the case of fake pearls and simulated rocks, minerals and gemstones. Precious metals and ivory are also imitated. Tagua nuts from South America are used as an ivory substitute since the natural ivory trade has been restricted worldwide.

BATIK

Batik is a technique of wax-resist dyeing applied to whole cloth, or cloth made using this technique. Batik is made either by drawing dots and lines of the resist with a spouted tool called a *canting*, or by printing the resist with a copper stamp called a *cap*. The applied wax resists dyes and therefore allows the artisan to colour selectively by soaking the cloth in one colour, removing the wax with boiling water, and repeating if multiple colours are desired.

A tradition of making batik is found in various countries, including Nigeria, China, India, Malaysia, Philippines and Sri Lanka; the batik of Indonesia, however, is the most well-known. Indonesian batik made in the island of Java has a long history of acculturation, with diverse patterns influenced by a variety of cultures, and is the most developed in terms of pattern, technique, and the quality of workmanship. On October 2009, UNESCO designated Indonesian batik as a Masterpiece of Oral and Intangible Heritage of Humanity.

Wax resist dyeing of fabric is an ancient art form. It already existed in Egypt in the 4th century BC, where it was used to wrap mummies; linen was soaked in wax, and scratched using a stylus. In Asia, the technique was practiced in China during the Tang Dynasty (618-907 AD), and in India and Japan during the Nara Period (645-794 AD). In Africa it was originally practiced by the Yoruba tribe in Nigeria, Soninke and Wolof in Senegal. These African version however, uses cassava starch or rice paste, or mud as a resist instead of beeswax.

The art of batik is most highly developed in the island of Java in Indonesia. In Java, all the materials for the process are readily available — cotton and beeswax and plants from which different vegetable dyes are made. Indonesian batik predates written records: G. P. Rouffaer argues that the technique might have been introduced during the 6th or 7th century from India or Sri Lanka. On the other hand, the Dutch archaeologist J.L.A. Brandes and the Indonesian archaeologist F.A. Sutjipto believe Indonesian batik is a native tradition, since regions such as Toraja, Flores, Halmahera, and Papua, which were not directly influenced by Hinduism, have an age-old tradition of batik making.

Rouffaer reported that the *gringsing* pattern was already known by the 12th century in Kediri, East Java. He concluded that this delicate pattern could be created only by using the *canting*, an etching tool that holds a small reservoir of hot wax, and proposed that the *canting* was invented in Java around that time. The carving details of clothes worn by East Javanese Prajnaparamita statues from around the 13th century show intricate floral patterns within rounded margins, similar to today's traditional Javanese *jlamprang* or *ceplok* batik motif. The motif is thought to represent the lotus, a sacred flower in Hindu-Buddhist beliefs. This evidence suggests that intricate batik fabric patterns applied with the *canting* existed in 13th-century Java or even earlier.

In Europe, the technique was described for the first time in the *History of Java*, published in London in 1817 by Stamford Raffles, who had been a British governor for the island. In 1873 the Dutch merchant Van Rijckevorsel gave the pieces he collected during a trip to Indonesia to the ethnographic museum in Rotterdam. Today the Tropenmuseum houses the biggest collection of Indonesian batik in the Netherlands. The Dutch and Chinese colonists were active in developing batik, particularly coastal batik, in the late colonial era. They introduced new patterns as well as the use of the *cap* (copper block stamps) to mass-produce batiks. Displayed at the Exposition Universelle at Paris in 1900, the Indonesian batik impressed the public and artists.

In the 1920s, Javanese batik makers migrating to Malaya (now Malaysia) introduced the use of wax and copper blocks to its east coast.

In Subsaharan Africa, Javanese batik was introduced in the 19th century by Dutch and English traders. The local people there adapted the Javanese batik, making larger motifs with thicker lines and more colours. In the 1970s, batik was introduced to Australia, where aboriginal artists at Erna Bella have developed it as their own craft.

Technique

Firstly, a cloth is washed, soaked and beaten with a large mallet. Patterns are drawn with pencil and later redrawn using hot wax, usually made from a mixture of paraffin or bees wax, sometimes mixed with plant resins, which functions as a dye-resist. The wax can be applied with a variety of tools. A pen-like instrument called a *canting* is the most common. A *canting* is made from a small copper reservoir with a spout on a wooden handle. The reservoir holds the resist which flows through the spout, creating dots and lines as it moves. For larger patterns, a stiff brush may be used. Alternatively, a copper block stamp called a *cap* is used to cover large areas more efficiently.

After the cloth is dry, the resist is removed by scraping or boiling the cloth. The areas treated with resist keep their original colour; when the resist is removed the contrast between the dyed and undyed areas forms the pattern. This process is repeated as many times as the number of colours desired.

The most traditional type of batik, called batik *tulis* (written batik), is drawn using only the *canting*. The cloth need to be drawn on both sides and dipped in a dye bath three to four times. The whole process may take up to a year; it yields considerably finer patterns than stamped batik.

BUTTON

In modern clothing and fashion design, a button is a small fastener, most commonly made of plastic, but also frequently of seashell, which secures two pieces of fabric together. In archaeology, a button can be a significant artifact. In the applied arts and in craft, a button can be an example of folk art, studio craft, or even a miniature work of art.

Buttons are most often attached to articles of clothing but can also be used on containers such as wallets and bags. However, buttons may be sewn onto garments and similar items exclusively for purposes of ornamentation. Buttons serving as fasteners work by slipping through a fabric or thread loop, or by sliding through a buttonhole. Other types of fastenings include zippers, Velcro and magnets.

Buttons in museums and galleries

Some museums and art galleries hold culturally, historically, politically, and/or artistically significant buttons in their collections. The Victoria and Albert Museum has many buttons, particularly in its jewellery collection, as does the Smithsonian Institution.

Hammond Turner and Sons, a button-making company in Birmingham, hosts an online museum with an image gallery and historical button-related articles, including an 1852 article on button-making by Charles Dickens. In the USA, large button collections are on public display at The Waterbury Button Museum of Waterbury, Connecticut, the Keep Homestead Museum of Monson, Massachusetts, which also hosts an extensive online button archive and in Gurnee, Illinois at The Button Room.

Early button history

Buttons and button-like objects used as ornaments or seals rather than fasteners have been discovered in the Indus Valley Civilization during its Kot Yaman phase (circa 2800–2600 BCE) as well as Bronze Age sites in China (circa 2000–1500 BCE), and Ancient Rome.

Buttons made from seashell were used in the Indus Valley Civilization for ornamental purposes by 2000 BCE. Some buttons were carved into geometric shapes and had holes pierced into them so that they could be attached to clothing with thread. Ian McNeil (1990) holds that: "The button, in fact, was originally used more as an ornament than as a fastening, the earliest known being found at Mohenjo-daro in the Indus Valley. It is made of a curved shell and about 5000 years old."

The earliest functional buttons were found in the tombs of conquering Hungarian tribes from the late 9th century. Functional buttons with buttonholes for fastening or closing clothes appeared first in Germany in the 13th century. They soon became widespread with the rise of snug-fitting garments in 13th- and 14th-century Europe.

Materials and manufacture

Because buttons have been manufactured from almost every possible material, both natural and synthetic, and combinations of both, the history of the material composition of buttons reflects the timeline of materials technology.

Buttons can be individually crafted by artisans, craftspeople or artists from raw materials or found objects (for example fossils), or a combination of both. Alternatively, they can be the product of low-tech cottage industry or can be mass-produced in high-tech factories. Buttons made by artists are art objects, known to button collectors as "studio buttons" (or simply "studios", from studio craft).

Nowadays, hard plastic, seashell, metals, and wood are the most common materials used in button-making; the others tending to be used only in premium or antique apparel, or found in collections.

Decoration and coating techniques

Historically, fashions in buttons have also reflected trends in applied aesthetics and the applied visual arts, with buttonmakers using techniques from jewellery making, ceramics, sculpture, painting, printmaking, metalworking, weaving and others. The following are just a few of the construction and decoration techniques that have been used in button-making:

- Arita porcelain
- Cloisonné
- Daguerreotype
- Electroplating
- Embroidery
- Filigree
- Intaglio
- Lacquerware
- Lithography
- Metallizing
- Metal openwork
- Opus interassile
- Passementerie
- Portrait miniatures

- Satsuma ware
- Vitreous enamel

Styles of attachment

- Shank buttons have a hollow protrusion on the back through which thread is sewn to attach the button. Button shanks may be made from a separate piece of the same or a different substance as the button itself, and added to the back of the button, or be carved or moulded directly onto the back of the button, in which latter case the button is referred to by collectors as having a 'self-shank'.
- Flat or sew-through buttons have holes through which thread is sewn to attach the button. Flat buttons may be attached by sewing machine rather than by hand, and may be used with heavy fabrics by working a thread shank to extend the height of the button above the fabric.
- Stud buttons (also pressure buttons, press studs or snap fasteners) are metal (usually brass) round discs pinched through the fabric. They are often found on clothing, in particular on denim pieces such as pants and jackets. They are more securely fastened to the material. As they rely on a metal rivet attached securely to the fabric, stud buttons are difficult to remove without compromising the fabric's integrity. They are made of two couples: the male stud couple and the female stud couple. Each couple has one front (or top) and rear (or bottom) side (the fabric goes in the middle).

Types of fabric buttons

- Covered buttons are fabric-covered forms with a separate back piece that secures the fabric over the knob.
- Mandarin buttons or Frogs are knobs made of intricately knotted strings. Mandarin buttons are a key element in Mandarin dress (*Qi Pao* and *cheongsam* in Chinese), where they are closed with loops. Pairs of mandarin buttons worn as cuff links are called silk knots.
- Worked or cloth buttons are created by embroidering or crocheting tight stitches (usually with linen thread) over a knob or ring called a form. Dorset buttons, handmade from the 17th century to 1750, are of this type.

Button sizes

The size of the button depends on its use. Shirt buttons are generally small, and spaced close together, whereas coat buttons are larger and spaced further apart. Buttons are commonly measured in lignes (also called *lines* and abbreviated *L*), with 40 lignes equal to 1 inch. For example, some standard sizes of buttons are 16 lignes (10.16 mm, standard button of men's shirts) and 32 lignes (20.32 mm, typical button on suit jackets).

The American National Button Society (NBS) has its own button sizing system which divides button sizes into 'small', 'medium' and 'large'.

Buttons as containers

Since at least the seventeenth century, when box-like metal buttons were constructed especially for the purpose, buttons have been one of the items in which drug smugglers have attempted to hide and transport illegal substances. At least one modern smuggler has tried to use this method.

Also making use of the storage possibilities of metal buttons, during the World Wars, British and U.S. military locket buttons were made, containing miniature working compasses.

Buttons in politics

Historically, buttons are a very important part of Western and Near-Eastern culture. They were valued by many European groups for practical and lucrative reasons. Buttons can range from crude buttons made at home out of wood to modern, cheaply made plastic buttons to highly decorative and ornate buttons of precious materials. They are so revered in certain parts of the world that there are some countries where it is illegal to destroy a button.

The mainly American tradition of politically significant clothing buttons appears to have begun with the first presidential inauguration of George Washington in 1789. Known to collectors as "Washington Inaugurals", they were made of copper, brass or Sheffield plate, in large sizes for coats and smaller sizes for breeches. Made in twenty-two patterns and hand-stamped, they are now extremely valuable cultural artifacts.

Between about 1840 and 1916, clothing buttons were used in American political campaigns, and still exist in collections today. Initially, these buttons were predominantly made of brass (though horn and rubber buttons with stamped or moulded designs also exist) and had loop shanks. Around 1860 the badge or pin-back style of construction, which replaced the shanks with long pins, probably for use on lapels and ties, began to appear.

One common practice that survived until recent times on campaign buttons and badges was to include the image of George Washington with that of the candidate in question.

Some of the most famous campaign buttons are those made for Abraham Lincoln. Memorial buttons commemorating Lincoln's inaugurations and other life events, including his birth and death, were also made, and are also considered highly collectible.

ZIPPER

he bulk of a zipper/zip consists of two rows of protruding teeth which may be made to interdigitate, linking the rows, carrying from tens to hundreds of specially shaped metal or plastic teeth. These teeth can be either individual or shaped from a continuous coil, and are also referred to as *elements*. The slider, operated by hand, moves along the rows of teeth. Inside the slider is a Y-shaped channel that meshes together or separates the opposing rows of teeth, depending on the direction of the slider's movement. The word *Zipper* is onomatopoetic, because it was named for the sound the device makes when used, a high-pitched *zip*.In many jackets and similar garments, the opening is closed completely when the slider is at one of the ends of the tape. The mechanism allows for partial fastening where only some of the tape is fastened together, but various movements and pressures may move the slider around the tape. In many kinds of luggage, there are two sliders on the tape, mounted in opposite directions head to head: the part of the zipper between them is unfastened. When the sliders are located at opposite ends of the tape, the zipper is fully unfastened; when the two sliders are located next to each other, which can be at any point along the tape, the zipper is fully closed. Some jackets have double-separating zippers. There are two sliders, mounted in opposite directions tail to tail: the part of the zipper between them is fastened. When the sliders are on opposite ends of the tape then the jacket is closed. If the lower slider is raised then the lower two sides of the jacket may be opened to allow more comfortable sitting or bicycling or whatever. When both sliders are lowered then the zipper may be totally separated. Although potentially convenient, there are often problems getting this type of zipper to start or to separate.

Zippers may

- increase or decrease the size of an opening to allow or restrict the passage of objects, as in the fly of trousers or in a pocket.
- join or separate two ends or sides of a single garment, as in the front of a jacket, or on the front, back or side of a dress or skirt to facilitate dressing.
- attach or detach a separable part of the garment to or from another, as in the conversion between trousers and shorts or the connection or disconnection of a hood and a coat.
- be used to decorate an item.

These variations are achieved by sewing one end of the zipper together, sewing both ends together, or allowing both ends of the zipper to fall completely apart.

A zipper costs relatively little, but if it fails, the garment may be unusable until the zipper is repaired or replaced—which can be quite difficult and expensive. Problems often lie with the zipper slider; when it becomes worn it does not properly align and join the alternating teeth. If a zipper fails, it can either jam (*i.e.* get stuck) or partially break off.

Sometimes zippers are just on garments for decoration and are not meant to be used at all.

In 1851, Elias Howe received a patent for an "Automatic, Continuous Clothing Closure". He did not try seriously to market it, missing recognition he might otherwise have received. Howe's device was more like an elaborate draw-string than a true slide fastener.

Forty-two years later, Whitcomb Judson, who invented a pneumatic street railway, marketed a "Clasp Locker". The device served as a (more complicated) hook-and-eye shoe fastener. With the support of businessman Colonel Lewis Walker, Judson launched the Universal Fastener Company to manufacture the new device. The clasp locker had its public debut at the 1893 Chicago World's Fair and met with little commercial success. Judson is sometimes given credit as the inventor of the zipper, but he never made a practical device.

The company, reorganized as the "Fastener Manufacturing and Machine Company," moved to Hoboken, N.J. in 1901. Gideon Sundback, a Swedish-American electrical engineer, was hired to work for the company in 1906. Good technical skills and a marriage to the plant-manager's daughter Elvira Aronson led Sundbäck to the position of head designer. The company moved to Meadville, PA, where it operated for most of the 20th century under the name "Talon, Inc." After his wife's death in 1911, Sundback devoted himself to improving the fastener, and by December 1913 had designed the modern zipper. The rights to this invention were owned by the Meadville company (operating as the "Hookless Fastener Co."), but Sundback retained non-U.S. rights and used these to set up in subsequent years the Canadian firm 'Lightning Fastner Co.' in St. Catharines, Ont. Sundback's work with this firm has led to the common misperception that he was Canadian and that the zipper originated in that country.

Gideon Sundback increased the number of fastening elements from four per inch (about one every 6.4 mm) to ten or eleven (around every 2.5 mm), introduced two facing rows of teeth that pulled into a single piece by the slider, and increased the opening for the teeth guided by the slider. The patent for the "Separable Fastener" was issued in 1917. Gideon Sundback also created the manufacturing machine for the new device. The "S-L" or "scrapless" machine took a special Y-shaped wire and cut scoops from it, then punched the scoop dimple and nib, and clamped each scoop on a cloth tape to produce a continuous zipper chain. Within the first year of operation, Sundback's machinery was producing a few hundred feet (around 100 meters) of fastener per day. In March of the same year, Mathieu Burri, a Swiss inventor, improved the design by adding a lock-in system attached to the last teeth, but his version never got into production due to conflicting patents.

The popular North American term *zipper*, (UK *zip*, or occasionally *zip-fastener*), came from the B. F. Goodrich Company in 1923. The company opted to use Gideon Sundback's fastener on a new type of rubber boots (or galoshes) and referred to it as the zipper, and the name stuck. The two chief uses of the zipper in its early years were for closing boots and tobacco pouches. Zippers began being used for clothing in 1925 by Schott NYC on leather jackets.

In the 1930s, a sales campaign began for children's clothing featuring zippers. The campaign praised zippers for promoting self-reliance in young children by making it possible for them to dress in self-help clothing. The zipper beat the button in 1937 in the "Battle of the Fly", after French fashion designers raved over zippers in men's trousers. *Esquire* declared the zipper the "Newest Tailoring Idea for Men" and among the zippered fly's many virtues was that it would exclude "The Possibility of Unintentional and Embarrassing Disarray."

The most recent innovation in the zipper's design was the introduction of models that could open on both ends, as on jackets. Today the zipper is by far the most widespread fastener, and is found on clothing, luggage, leather goods, and various other objects.

Types

- Coil zippers now form the bulk of sales of zippers worldwide. The slider runs on two coils on each side; the *teeth* are formed by the windings of the coils. Two basic types of coils are used: one with coils in spiral form, usually with a cord running inside the coils; the other with coils in ladder form, also called the Ruhrmann type. Coil zippers are made of polyester coil and are thus also termed polyester zippers. Nylon was formerly used and though only polyester is used now, the type is still also termed a nylon zipper.
- Invisible zippers have the teeth hidden behind a tape, so that the zipper is *invisible*. It is also called the *Concealed* zipper. The tape's colour matches the garment's, as does the slider's and the puller's. This kind of a zipper is common in skirts and dresses. Invisible zippers are usually coil zippers. They are also seeing increased use by the military and emergency services because the appearance of a button down shirt can be maintained, while providing a quick and easy fastening system. A regular invisible zipper uses a lighter lace-like fabric on the zipper tape, instead of the common heavier woven fabric on other zippers.
- Reverse coil zippers are a variation of the coil zipper. A reverse coil zipper is exactly that - the coil is on the reverse (back) side of the zipper and the slider is engineered to work on the flat side of the zipper (normally the back, now the front). Unlike an invisible zipper where the coil is also on the back, the reverse coil shows stitching on the front side and the slider will accommodate a variety of pulls (the invisible zipper requires a small, tear-drop pull due to the small slider attachment). Water resistant zippers are generally configured as reverse coil so that the pvc coating can cover the stitching. A rubber or PVC coated reverse zipper is called a waterproof zipper.
- Metal zippers are the classic zipper type, found mostly in jeans today.

The teeth are not a coil, but are individual pieces of metal molded into shape and set on the zipper tape at regular intervals. Metal zippers are made in brass, aluminum and nickel, according to the metal used for teeth making. All these zippers are basically made from flat wire. A special type of metal zipper is made from pre-formed wire, usually brass but sometimes other metals too. Only a few companies in the world have the technology. This type of pre-formed metal zippers is mainly used in high grade jeans-wear, work-wear, etc., where high strength is required and zippers need to withstand tough washing.

- Plastic-molded zippers are identical to metallic zippers, except that the teeth are plastic instead of metal. Metal zippers can be painted to match the surrounding fabric; plastic zippers can be made in any colour of plastic. Plastic zippers mostly use polyacetal resin, though other thermoplastic polymers are used as well, such as polyethylene.
- Open-ended zippers use a *box and pin* mechanism to lock the two sides of the zipper into place, often in jackets. Open-ended zippers can be of any of the above described types.
- Two way open-ended zippers Instead of having an insertion pin and pin box at the bottom, a two way open-ended zipper has two pullers on each end of the zipper tape. The user can slide up the bottom puller when a zipped garment is being worn, so that the bottom part of the zipper will be opened for situation where more legs movement is needed, without adding unnecessary tension on the pin and box of a one-way open-ended zipper when it is zipped, which could cause accidental damage on the zipper. It is most commonly used on long coats.
- Two way closed-ended zippers are closed at both ends; they are often used in luggage and can have either one or two pullers on the zipper.
- Magnetic zippers allow for one-handed closure and are used in sportswear.

Air and water tightness

Airtight zippers were first developed by NASA for making high-altitude pressure suits and later space suits, capable of retaining air pressure inside the suit in the vacuum of space.

The airtight zipper is built like a standard toothed zipper, but with a waterproof sheeting (which is made of fabric-reinforced polyethylene and is bonded to the rest of the suit) wrapped around the outside of each row of zipper teeth. When the zipper is closed, the two facing sides of the plastic sheeting are squeezed tightly against one another (between the C-shaped clips) both above and below the zipper teeth, forming a double seal.

This double-mated surface is good at retaining both vacuum and pressure, but the fit must be very tight, to press the surfaces together firmly. Consequently these zippers are typically very stiff when zipped shut and have minimal flex or stretch. They are hard to open and close because the zipper anvil must bend apart teeth that are being held under tension. They can also be derailed (and damage the sealing surfaces) if the teeth are misaligned while straining to pull the zipper shut.

These zippers are very common where airtight or watertight seals are needed, such as on scuba diving dry suits, ocean survival suits, and hazmat suits.

A less common water-resistant zipper is similar in construction to a standard toothed zipper, but includes a molded plastic ridge seal similar to the mating surfaces on a ziploc bag. Such a zipper is easier to open and close than a clipped version, and the slider has a gap above the zipper teeth for separating the ridge seal. This seal is structurally weak against internal pressure, and can be separated by pressure within the sealed container pushing outward on the ridges, which will simply flex and spread apart, potentially allowing air or liquid entry through the spread-open ridges. Ridge-sealed zippers are sometimes seen on lower cost surface dry suits.

Anti-slide zipper locks

Some zippers include a designed ability for the slider to hold in a steady open or closed position, resisting forces that would try to move the slider and open the zipper unexpectedly. There are two commons ways this is accomplished:

The zipper handle can have a short protruding pin stamped into it, which inserts between the zipper teeth through a hole on the slider, when the handle is folded down flat against the zipper teeth. This appears on some brands of pants. The handle of the fly zipper is folded flat against the teeth when it is not in use, and the handle is held down by both slider hinge tension and the fabric flap over the fly.

The slider can also have a two-piece hinge assembly attaching the handle to the slider, with the base of the hinge under spring tension and with protruding pins on the bottom that insert between the zipper teeth. To move the zipper, the handle is pulled outward against spring tension, lifting the pins out from between the teeth as the slider moves. When the handle is released the pins automatically engage between the zipper teeth again.

A three-piece version of the above uses a tiny pivoting arm held under tension inside the hinge. Pulling on the handle from any direction lifts the pivoting arm's pins out of the zipper teeth so that the slider can move.

Components

The components of a zipper are:

1. Top Tape Extension (The fabric part of the zipper, that extends beyond the teeth, at the top of the chain.)
2. Top Stop (Two devices affixed to the top end of a zipper, to prevent the slider from coming off the chain.)
3. Slider (The device that moves up and down the chain to open or close the zipper.)
4. Pull Tab or Puller (The part of the slider that is held to move the slider up or down.)
5. Tape Width (Refers to the width of the fabric on both sides of the zipper chain.)
6. Chain or Zipper Teeth (The continuous piece that is formed when both halves of a zipper are meshed together) and Chain Width (Refers to the specific gauge of the chain - common gauge sizes are #3, #5, #7, #8 and #10, the bigger the number, the wider the teeth/chain width is.)
7. Bottom Stop (A device affixed to the bottom end of a zipper, to prevent further movement of the half of the zipper from separating.)
8. Bottom Tape Extension (The fabric part of the zipper, that extends beyond the teeth, at the bottom of the chain.)
9. Single Tape Width (Refers to the width of the fabric on one side of the zipper chain.)
10. Insertion Pin (A device used on a separating zipper whose function is to allow the joining of the two zipper halves.)
11. Retainer Box or Pin Box (A device used on a separating zipper whose function is to correctly align the pin, to begin the joining of the zipper halves.)
12. Reinforcement Film (A strip of plastic fused to each half of the zipper tape to allow a manufacturer to electronically “weld” the zipper onto the garment or item that is being manufactured, without the need of sewing or stitching.)

Manufacturing

- Forbes reported in 2003 that although the zipper market in the 1960s was dominated by Talon Zipper (USA) and Optilon (Germany), Japanese manufacturer YKK Group grew to become the industry giant by the 1980s. YKK held 45 per cent of the world market share, followed by Optilon (8 per cent) and Talon Zipper (7 per cent).
- Indian Tex Corp has also emerged as a significant supplier to the apparel industry.
- In Europe, Cremalleras Rubi company established in 1926 (Spain), continues to compete with the big multinationals selling over 30 million zippers in 2012.

- In 2005, The Guardian reported that China had 80 per cent of the international market. Most its product is made in Qiaotou, Yongjia County.

BUCKLE

The buckle or clasp is a device used for fastening two loose ends, with one end attached to it and the other held by a catch in a secure but adjustable manner. Often taken for granted, the invention of the buckle has been indispensable in securing two ends before the invention of the zipper. The basic buckle frame comes in a variety of shapes and sizes depending on the intended use and fashion of the era. Buckles are as much in use today as they have been in the past. Used for much more than just securing one's belt, instead it is one of the most dependable devices in securing a range of items.

The word "buckle" enters Middle English via Old French and the Latin *buccula* or "cheek-strap," as for a helmet. Some of the earliest buckles known are those used by Roman soldiers to strap their body armor together and prominently on the balteus and cingulum. Made out of bronze and expensive, these buckles were purely functional for their strength and durability vital to the individual soldier. The baldric was a later belt worn diagonally over the right shoulder down to the waist at the left carrying the sword, and its buckle therefore was as important as that on a Roman soldier's armor.

Bronze Roman buckles came in various types. Not only used for practical purposes, these buckles were also decorated. A Type I Roman buckle was a "buckle-plate" either decorated or plain and consisted of geometric ornaments. Type IA Roman buckles were similar to Type I buckles but differed by being long and narrow, made of double sheet metal, and attached to small D-shaped buckles (primarily had dolphin-heads as decorations). Type IB "buckle-loops" were even more similar to Type IA buckles, only difference being that instead of dolphin-heads, they were adorned with horse-heads. There were also Type II buckles (Type IIA and Type IIB) used by Romans, but all types of Roman buckles could have served purposes for simple clothing as well, and predominantly, as a military purpose. Aside from the practical use found in Roman buckles, Scythian and Sarmatian buckles incorporated animal motifs that were characteristic to their respective decorative arts. These motifs often represented animals engaged in mortal combat. These motifs were imported by many Germanic peoples and the belt buckles were evident in the graves of the Franks and Burgundies. And throughout the Middle Ages, the buckle was used mostly for ornamentation until the second half of the 14th century where the knightly belt and buckle took on its most splendid form.

Buckles remained exclusively for the wealthy until the 15th century where improved manufacturing techniques made it possible to easily produce a cheaper molded item available to the general population.

Components

The buckle essentially consists of four main components: the frame, chape, bar, and prong. The oldest Roman buckles are of a simple "D"-shaped frame, in which the prong or tongue extends from one side to the other. In the 14th century, buckles with a double-loop or "8"-shaped frame emerged. The prongs of these buckles attach to the center post. The appearance of multi-part buckles with chapes and removable pins, which were commonly found on shoes, occurred in the 17th century.

Frame

The frame is the most visible part of the buckle and holds the other parts of the buckle together. Buckle frames come in various shapes, sizes, and decorations. The shape of the frame could be a plain square or rectangle, but may be oval or made into a circular shape. A reverse curve of the frame indicated that the whole buckle was intended to be used for securing a thick material, such as leather. This reverse curve shape made it easier to thread the intended thick material end over the bar. But the shape of the frame is not limited to simply squares and ovals, the decoration of the frame itself defines the shape it will turn out to be. Since the frame is the largest part of the buckle, any and all decorations are placed on it. Decorations range from wedged shapes, picture references to people and animals, and insignia of a desired organization.

The part of the frame that strap goes through prior to putting the tongue/ prong through the hole is often referred to as the 'end bar'. The 'center bar' holds the tongue and the part (if present) that holds the tip of the strap in place is called the 'keeper' or 'keeper bar' these terms are used when additional information is needed to describe a buckle for measurements or design. Note that if an a separate piece of leather or metal is attached to the strap for holding the tip of the belt/strap in place that is sometimes also called a 'keeper'.

Chape

Chapes or "caps" of various designs could be fitted to the bar to enable one strap end to be secured before fastening the other, adjustable end. This made buckles easily removable and interchangeable, leading to a significant advantage since buckles were expensive. Unfortunately, the teeth or spikes on the semi-circular chapes damaged the straps or belts, making frequent repairs of the material necessary. Buckles fitted with "T"-, anchor-, or spade-shaped chapes avoided this problem but needed a slotted end in the belt to accommodate them.

Prong

The prong is typically made out of steel or other types of metal. In conventional belts, the prong fits through the buckle to secure the material at a pre-set length.

The Prong is usually referred to as the tongue of the buckle in America. As in in 'Lock-Tongued buckle'. Prong is only used when the tongue is permanently fixed in position.

Bar

The bar served to hold the chape and prong to the frame. When prongs and chapes are removed from the buckle design, the buckle incorporated a movable bar relying on the tension of the adjusted belt to keep it in place.

Materials

Metal

The first known buckles to be used were made out of bronze for their strength and durability for military usage. For the last few hundred years, buckles have been made from brass (an alloy of copper and zinc). In the 18th century, brass buckles incorporated iron bars, chapes, and prongs due to the parts being made by different manufactures. Silver was also used in buckle manufacturing for its malleability and for being strong and durable with an attractive shine. White metal, any bright metallic compound, was also used in all styles of buckles; however, if iron was present, rust will form if it is allowed to be exposed and remain in damp conditions.

Pearl

Pearl buckles have been made from pearly shells and usually for ladies' dresses. Since a reasonable size flat surface was needed to make a buckle, oyster was commonly used to make these types of buckles. The quality and colour of course vary, ranging from layers of yellow and white to brown or grey.

Wood

When preferred materials were scarce during the Great Depression of the 1930s and the two World Wars, buckles became a low priority and manufactures needed to find ways to continue to produce them cheaply. Makers turned to wood as a cheap alternative since it was easily worked by hand or simple machinery by impressing the designs onto the wood. But there were problems using wood. Any attempt to brighten the wood's dull appearance with painted designs or plasterwork embellishments immediately came off if the buckle were to be washed.

Leather

Buckles were not entirely made out of leather because a frame and bar of leather would not be substantial enough to carry a prong or the full weight of the belt and anything the belt and buckle intend to support. However, leather (or dyed suede, more common to match a lady's garment colour) was used more as a "cover-up" for cheap materials to create a product worthy of buying.

Glass

Buckles were not made out of glass; rather the glass was used as a decorative feature that covered the entire frame of a metal buckle. One method of creating glass buckles was gluing individual discs of glass to the metal frame. Another more intricate method was to set a wire into the back of a glass disc, and then threading the wire through a hole in the fretted frame of the buckle. The glass was further secured by either bending it over the back of the frame or splayed out like a rivet.

Polymers

Compositions refers to polymer materials used for buckles. Celluloid, a type of thermoplastic invented in 1869, was used sparingly and only for decoration until after World War I where it began to be produced on a wider commercial scale. After World War II, the chemical industry saw a great expansion where Celluloid and other plastics such as Casein and Bakelite formed the basis of the buckle-making industry. Many thermoplastic polymers such as nylon are now used in snap-fit buckles for a wide variety of applications.

Types of Buckles

- Clasp (Clearing up some disambiguation): Although any device that serves to secure two loose ends is casually called a buckle, if it consists of two separate pieces with one for a hook and the other for a loop, it should be called a clasp. Clasps became increasingly popular at the turn of the 19th century with one clear disadvantage: since each belt end was fixed to each clasp piece, the size of the belt was typically not adjustable unless an elastic panel was inserted.
- Buckle trim or slide: A buckle without a chape or prongs is called a buckle trim or slide. It may have been designed this particular way or it may have lost its prongs through continuous use. This type was frequently used in home dress-making (belt end being secured with the simple hook-and-eye) and was purely used for decoration for items such as shoe fronts to conceal unattractive elastic fitting.
- Conventional (or Belt buckle): The conventional buckle with a frame, bar and prong gives the most reliable and easy-to-use closure for a belt. It is not meant, by design, to offer much space for decoration, but for its time-tested reliability.
- Side release buckle: A conventional buckle that is formed by a male buckle member (the hook end) and a female buckle member (the catch end). The male buckle member consists of a center guide rod forwardly extending from the front side with two spring arms equally spaced from the center rod. The two spring arms each have a retaining

block that terminates at the front end. The female buckle member has a front open side and two side holes which hold and secure the two spring arms of the male buckle member. This sort of buckle may be found on backpacks, belts, rifle slings, boots, and a host of other common but overlooked items. It is also known as the "parachute buckle".

- Blimp buckle: The bottom part of the blimp, also known as a gondola, is called the buckle.

GROMMET

A grommet is a ring or edge strip inserted into a hole through thin material, typically a sheet of textile fabric, sheet metal and or composite of carbon fibre, wood or honeycomb. Grommets are generally flared or collared on each side to keep them in place, and are often made of metal, plastic, or rubber. They may be used to prevent tearing or abrasion of the pierced material or protection from abrasion of the insulation on the wire, cable, line being routed through the penetration, and to cover sharp edges of the piercing, or all of the above.

A small grommet may also be called an eyelet, used for example on shoes, tarps and sails for lacing purposes. In electrical applications these are referred to as "insulating bushings". Most common are molded rubber that are inserted into small hole diameters up to 2" in diameter. There are many hole configurations from standard round to assorted U-shapes. Larger penetrations that are irregular in shape as well as long straight edges often use extruded or stamped strips of continuous length. These Continuous length materials are referred to as "grommet edging".

These are quite common in applications that range from telecom switches and data center cabinets to complex and dense wire/cable and even hydraulic tubing in aircraft, transportation vehicles and medical equipment.

Grommets as reinforcement or crafting

Grommets are used to reinforce holes in leather, cloth, shoes, canvas and other fabrics. They can be made of metal, rubber, or plastic, and are easily used in common projects, requiring only the grommet itself and a means of setting it with a punch, a metal rod with a convex tip. A simple punch, often sold with the grommets, can be struck with a hammer to set the grommet. There are also dedicated grommet presses with punch and anvil, as shown in the picture, ranging from inexpensive to better-quality tools, which are somewhat faster to use. They are used to strengthen holes; in footwear for boot and shoe laces, in laced clothing such as corsets, and in curtains and other household items that require hanging from hooks, as when they are used in conjunction with tensioner rods for shower curtains. The grommet prevents the cord from tearing through the hole, thereby providing structural integrity. Small grommets are also called eyelets, especially when used in clothing or crafting. Eyelets may be used purely decoratively for crafting. When used in sailing and various other applications they are called cringles. Sometimes field workers may refer to them as grunyons.

Grommets used in electrical equipment

For cable protection

If metal or another hard material has a hole made in it, the hole will probably have sharp edges. Electrical wires, cord, rope, lacings, or other soft vulnerable material passing through the hole can become abraded or cut, or electrical insulation may break due to repeated flexing at the exit point. Rubber, plastic or plastic coated metal grommets are used to avoid this. The grommet could also protect the wiring/cabling from contamination from dirt, air, water, etc. The smooth and sometimes soft inner surface of the grommet shields the wire from damage.

Grommets are generally used whenever wires pass through punched/drilled sheet metal or plastic casings for this reason. Molded and continuous strip grommets, also known as edge grommets, are manufactured in a wide variety of sizes and lengths expressly for this purpose; they are usually a single piece which can be inserted by hand. Two-piece hard plastic devices are available which also grip the wire that passes through. These are called strain relief bushings and are often used to insulate, anchor, and protect power cords where they enter panels. Preventing a tug or twist on the wire from stressing the electrical connections inside the connected equipment. Sleeved grommets have a flexible extension (sleeve), usually tapered or moulded to flex increasingly towards the free end in order to reduce fracturing of electrical insulation.

To minimise vibration

Grommets made of rubber or other elastic material are also used to minimise the transmission of vibration. They were widely used for mounting shock-sensitive computer disk drives, particularly in equipment subject to vibration or jarring, but are not usually used with more robust modern drives. The screws that hold the drive in place pass through grommets that decouple it acoustically from the chassis. Grommets are used in a similar way to acoustically isolate electronic circuit components that are susceptible to microphonism caused by mechanical vibration or jarring.

Surgical grommets

In chronic cases of otitis media with effusions present for months, surgery is sometimes performed to insert a grommet, called a "tympanostomy tube" into the eardrum to allow air to pass through into the middle ear, and thus release any pressure buildup and help clear excess fluid within. This is also a correcting measure for a patulous Eustachian tube (when air moves to and from the middle ear with each breath making the eardrum flap).

SEQUIN

Sequins are disk-shaped beads used for decorative purposes. In earlier centuries, they were made from shiny metals. Today, sequins are most often made from plastic. They are available in a wide variety of colours and geometrical shapes. Sequins are commonly used on clothing, jewelry, bags, shoes and lots of other accessories.

Sequins are sometimes also referred to as spangles, paillettes, or diamantes. Paillettes themselves are commonly very large and flat. Sequins may be stitched flat to the fabric, so that they do not move, and are less likely to fall off; or they may be stitched at only one point, so that they dangle and move easily, to catch more light. Some sequins are made with multiple facets, to increase their reflective ability.

Evidence exists that gold sequins were being used as decoration on clothing or paraphernalia in the Indus Valley as early as 2500BC, during the Kot Diji phase.

The word "sequin" is from a rendition into French of the Italian word *zecchino,* which was a gold coin that was issued in the late medieval and early post-medieval centuries in Republic of Venice and also issued in Ottoman Turkey; see Sequin (coin). By the early 19th century the sequin coins were no longer being issued, and the name sequin was falling out of use in its original sense. It was then that the name was taken up in France to designate what it means today. The 19th century sequins were made of shiny metal. Large sequins, fastened only at the top, have been used on billboards and other signage, particularly prior to the development of lighted and neon signs.

FABRIC TUBE TURNING

Fabrics are often sewn inside out, in a tube form, or in a manner to allow a filler (such as cording) to be added. The process of turning the fabric tube has several variations. Fabric can be turned with a blunt item such as a bobbin or chopstick, or with purchased tools such as the Fasturn, Loop Turners, or Hemostats.

FACING (SEWING)

In sewing and tailoring, facing is fabric applied to a garment edge, on the inside. *Shaped facings* are cut to match the outside shape of the piece to provide a neat finish, and are often cut from the same pattern pieces. Shaped facings are typically made of the same fabric as the garment, but may also be made of lighter-weight fabric or in a contrasting colour as a design element. *Extended facings* are extensions of the garment fabric, folded back and usually stabilized. *Bias facings* are strips of lightweight fabric cut on the true bias (US) or cross-grain (UK), and shaped rather than cut to match the edge to which they are applied.

FLOATING CANVAS

Floating canvas is a term used by tailors to describe the cloth-construction inside a jacket or coat. It consists of two layers of cloth into which horsehair has been woven, which are then sewn together to create a curving of the finished canvas. This is then sewn into the front of a coat.

GATHER (SEWING)

Gathering is a sewing technique for shortening the length of a strip of fabric so that the longer piece can be attached to a shorter piece. It is commonly used in clothing to manage fullness, as when a full sleeve is attached to the armscye or cuff of a shirt, or when a skirt is attached to a bodice.

In simple gathering, parallel rows of running stitches are sewn along one edge of the fabric to be gathered.

The stitching threads are then pulled or "drawn up" so that the fabric forms small folds along the threads.

Gathering seams once tedious hand sewing of basting which was time consuming and inefficient with heavy fabric. Now, a quick and easy way to make a gather is to use a wide zigzag stitch. Both the upper and lower thread are pulled long and placed in front of the sewing machine. Then zigzagging is carefully sewed over top of the two threads without catching the threads as it is sewn. At the end the thread is pulled and is then gathered.

TYPES

- Pleating or plaiting is a type of gathering in which the folds are usually larger, made by hand and pinned in place, rather than drawn up on threads, but very small pleats are often identical to evenly spaced gathers. Pleating is mainly used to make skirts, but can have other uses.(See main article Pleat.)
- Shirring or gauging is a decorative technique in which a panel of fabric is gathered with many rows of stitching across its entire length and then attached to a foundation or lining to hold the gathers in place. It is very commonly used to make larger pieces of clothing with some shape to them.

GODET (SEWING)

A godet is an extra piece of fabric in the shape of a circular sector which is set into a garment, usually a dress or skirt. The addition of a godet causes the article of clothing in question to flare, thus adding width and volume. Adding a godet to a piece of clothing also gives the wearer a wider range of motion.

GORE (SEGMENT)

A gore is a sector of a curved surface or the curved surface that lies between two close lines of longitude on a globe and may be flattened to a plane surface with little distortion. The term has been extended to include similarly shaped pieces such as the panels of a hot-air balloon or parachute, or the triangular insert that allows extra movement in a garment.

Examples:

- Spherical globes of the Earth and Celestial sphere were first mass-produced by Johannes Schöner using a process of printing map details on 12 paper gores that were cut out then pasted to a sphere. This process is still often used. The gores are conveniently made to each have a width of 30 degrees of longitude matching the principal meridians from the South Pole and North Pole to the Equator.
- Parachutes and hot air balloons are made from gores of lightweight material. The gores are cut from flat material, and stitched together to create various shapes.
- Pressure suit joints are often constructed of alternating gores and convolutes of material constrained by cables or straps along the sides of the joint, producing an accordion-like structure that flexes with nearly constant volume to minimize the mechanical work which must be done by the suit occupant.
- Corners in round duct-work can be created by welding or fixing gores of metal sheet to form a bend.
- Some designers use the stretched grid method to design gores that are cut out of weather-resistant fabric and then stitched together to form fabric structures.

GUSSET

In sewing, a gusset is a triangular or rhomboid piece of fabric inserted into a seam to add breadth or reduce stress from tight-fitting clothing. Gussets were used at the shoulders, underarms, and hems of traditional shirts and chemises made of rectangular lengths of linen to shape the garments to the body.

Gussets are used in manufacturing of modern tights and pantyhose to add breadth at the crotch seam. As with other synthetic underwear, these gussets are often made of moisture wicking breathable fabrics such as cotton, to keep the genital area dry and ventilated.

The term "don't bust a gusset" comes from this sewing term; a gusset in this context was usually a piece of fabric sewn between two others to increase mobility or increase the size of the pant waist, the latter being more common in the early 1900s.

Gussets are also used when making three-piece bags. In a Boye Needle Company publication, it is used in a pattern for a bag as a long, wide piece which connects the front piece and back piece. By becoming the sides and bottom of the bag, the gusset opens the bag up beyond what simply attaching the front to the back would do. With reference to the dimension of the gusset, the measurements of a flat bottom bag may be quoted as LxWxG.

HEM

A hem in sewing is a garment finishing method, where the edge of a piece of cloth is folded narrowly and sewn to prevent unravelling of the fabric.

METHODS

There are many different styles of hems of varying complexities. The most common hem folds up a cut edge, folds it up again, and then sews it down. The style of hemming thus completely encloses the cut edge in cloth, so that it cannot unravel. Other hem styles use fewer folds. One of the simplest hems encloses the edge of cloth with a stitch without any folds at all, using a method called an overcast stitch, although an overcast stitch may be used to finish a folded "plain hem" as well.

There are even hems that do not call for sewing, instead using iron-on materials, netting, plastic clips, or other fasteners. These threadless hems are not common, and are often used only on a temporary basis.

The hem may be sewn down with a line of invisible stitches or blind stitch, or sewn down by a sewing machine. The term *hem* is also extended to other cloth treatments that prevent unraveling. Hems can be serged (see serger), hand rolled and then sewn down with tiny stitches (still seen as a high-class finish to handkerchiefs), pinked with pinking shears, piped, covered with binding (this is known as a Hong Kong finish), or made with many other inventive treatments.

Most haute couture hems are sewn by hand. Decorative embroidery embellishment is sometimes referred to as a hem-stitch design.

TYPES OF HEMS AND HEM STITCHES

Hems of different depths (which includes the seam allowance) may have a particular style to achieve, which requires more or less fabric depending upon the style. A handkerchief-style edge requires a hem allowance of 0.6 cm or a quarter inch. A typical skirt or pant hem may be 5-7.6 cm. The hem's depth affects the way the fabric of the finished fabric will drape. Heavier fabric requires a relatively shorter hem. An interface fabric sewn to the fabric in the hem has a useful function in some hem styles. A bias strip is sometimes used as a hem interface. This adds fullness to the finished garment and reduce wrinkling.

The hem stitches that are commonly used for hand-sewn hems include: pick stitch; catch stitch (also called a *herringbone stitch*); slip stitch; and blind stitch.

Sewing machines can make a stitch that appears nearly invisible by using a blind-stitch setting and a blind stitch foot. Blind-stitches are commonly used to finish hems of applique designs on fabric. Modern sewing machines designed for home use can make many decorative or functional stitches, so the number of possible hem treatments is large. These home-use machines can also sew a reasonable facsimile of a hem-stitch, though the stitches will usually be larger and more visible.

Clothing factories and professional tailors use a "blind hemmer", or hemming machine, which sews an invisible stitch quickly and accurately. A blind hemmer sews a chain stitch, using a bent needle, which can be set precisely enough to actually sew through one and a half thicknesses of the hemmed fabric. A rolled hem presser foot on a sewing machine enables quick and easy hemming even by home sewers.

Heavy material with deep hems may be hemmed with what is called a *dressmaker's hem* —an extra line of loose running stitch is added in the middle of the hem, so that all the weight of the cloth does not hang from one line of stitching.

HEM REPAIR

Hem repair tape is available as an alternative solution to sewing a broken hem. To effect a fix, the hem repair tape is laid around the inside of the hem. It is then ironed with a hot iron. The heat causes the tape to bond the two surfaces together.

HEIRLOOM SEWING

Heirloom sewing is a collection of needlework techniques that arose in the last quarter of the 20th century that imitates fine French hand sewing of the period 1890-1920 using a sewing machine and manufactured trims. Heirloom sewing is characterized by fine, often sheer, usually white cotton or linen fabrics trimmed with an assortment of lace, insertions, tucks, narrow ribbon, and smocking, imitating such hand-work techniques as whitework embroidery, Broderie Anglaise, and hemstitching. Typical projects for heirloom sewing include children's garments (especially christening gowns), women's blouses, wedding gowns, and lingerie.

LINING (SEWING)

In sewing and tailoring, a lining is an inner layer of fabric, fur, or other material inserted into clothing, hats, luggage, curtains, handbags and similar items.

Linings provide a neat inside finish and conceal interfacing, padding, the raw edges of seams, and other construction details. A lining reduces the wearing strain on clothing, extending the useful life of the lined garment. A smooth lining allows a coat or jacket to slip on over other clothing easily, and linings add warmth to cold-weather wear.

Linings are typically made of solid colours to coordinate with the garment fabric, but patterned and contrasting-colored linings are also used. Designer Madelaine Vionnet introduced the ensemble in which the coat was lined in the fabric used for the dress worn with it, and this notion remains a characteristic of the Chanel suit, which often features a lining and blouse of the same fabric.

In tailoring, home sewing, and ready-to-wear clothing construction, linings are usually completed as a unit before being fitted into the garment shell. In haute couture, the sleeves and body are usually lined separately before assembly.

- An interlining is an additional layer of fabric between the lining and the outer garment shell. Insulating interlinings for winter garments are usually sewn to the individual lining pieces before the lining is assembled.
- A partial or half lining lines only the upper back and front of the garment, concealing the shoulder pads and interfacings, with or without sleeves.
- A zip-in, zip-out, snap-out or button-in lining (sometimes liner) is a warm removable lining for a jacket, coat, or raincoat that is held in place with a zipper, snap fasteners, or buttons. Garments with removable linings are usually lined with a lightweight fabric as well, to provide a neat finish when the warm lining is not worn.

PLEAT

A pleat (older plait) is a type of fold formed by doubling fabric back upon itself and securing it in place. It is commonly used in clothing and upholstery to gather a wide piece of fabric to a narrower circumference.

Pleats are categorized as *pressed*, that is, ironed or otherwise heat-set into a sharp crease, or *unpressed*, falling in soft rounded folds.

Pleats sewn into place are called tucks.

A vertically hanging piece of fabric such as a skirt or a drape will often be described in terms of its "fullness." Fullness represents the thickness/ depth of the pleats in relation to the original width of the fabric: fabric sewn at "zero fullness" is flat and has no pleats; fabric sewn at "100 per cent fullness" is pleated so that it takes up exactly half as much width as it would if it were not pleated at all (*i.e.*, 24 inches would be pleated down to 12 inches); if sewn at "150 per cent fullness," the unpleated fabric would be two and a half times wider than the final pleated piece (*i.e.*, an unpleated 30 inches would end up as 12 pleated inches of fabric– 12+1.50(12)=30); if fullness were to be "50 per cent", the original fabric would be one and a half times the width of the pleated (*i.e.*, 18 inches of width would end up as 12 pleated inches– 12+0.50(12)=18), etc.

TYPES OF PLEATS

- Accordion pleats or knife pleats are a form of tight pleating which allows the garment to expand its shape when moving. Accordion pleating is also used for some dress sleeves, such as pleating the end of the elbow, with the fullness of the pleat gathered closely at the cuff. This form of pleating inspired the "skirt dancing" of Loie Fuller. Accordion pleats may also be used in hand fans.
- Box pleats are knife pleats back-to-back, and have a tendency to spring out from the waistline. They have the same 3:1 ratio as knife pleats, and may also be stacked to form stacked box pleats. These stacked box pleats create more fullness and have a 5:1 ratio. They also create a bulkier seam. Inverted box pleats have the "box" on the inside rather than the outside.
- Cartridge pleats are used to gather a large amount of fabric into a small waistband or armscye without adding bulk to the seam. This type of pleating also allows the fabric of the skirt or sleeve to spring out from the seam. During the 15th and 16th centuries, this form of pleating was popular in the garments of men and women. Fabric is evenly gathered using two or more lengths of basting stitches, and the top of each pleat is whipstitched onto the waistband or armscye. Cartridge pleating was resurrected in the 1840s to attach the increasingly full bell-shaped skirts to the fashionable narrow waist.
- Fluted pleats or *flutings* are very small, rounded or pressed pleats used as trimmings. The name comes from their resemblance to a pan flute.
- Fortuny pleats are crisp pleats set in silk fabrics by designer Mariano Fortuny in the early 20th century, using a secret pleat-setting process which is still not understood.
- Honeycomb pleats are narrow, rolled pleats used as a foundation for smocking.
- Kick pleats are short pleats leading upwards from the bottom hem of garments such as skirts or coats, usually at the back. They allow the garment to drape straight down when stationary while also allowing freedom of movement.
- Knife pleats are used for basic gathering purposes, and form a smooth line rather than springing away from the seam they have been gathered to. The pleats have a 3:1 ratio–three inches of fabric will create one inch of finished pleat. Knife pleats can be recognized by the way that they overlap in the seam.
- Organ pleats are parallel rows of softly rounded pleats resembling the pipes of a pipe organ. Carl Köhler suggests that these are made by inserting one or more gores into a panel of fabric.

- Plissé pleats are narrow pleats set by gathering fabric with stitches, wetting the fabric, and "setting" the pleats by allowing the wet fabric to dry under weight or tension. Linen chemises or smocks pleated with this technique have been found in the 10th century Viking graves in Birka.
- Rolled pleats create tubular pleats which run the length of the fabric from top to bottom. A piece of the fabric to be pleated is pinched and then rolled until it is flat against the rest of the fabric, forming a tube. A variation on the rolled pleat is the stacked pleat, which is rolled similarly and requires at least five inches of fabric per finished pleat. Both types of pleating create a bulky seam.
- Watteau pleats are one or two box pleats found at the back neckline of 18th century sack-back gowns and some late 19th century tea gowns in imitation of these. The term is not contemporary, but is used by costume historians in reference to these styles as portrayed in the paintings of Antoine Watteau.

MODERN USAGE

Clothing features pleats for practical reasons (to provide freedom of movement to the wearer) as well as for purely stylistic reasons.

Shirts, blouses, jackets

Shirts and blouses typically have pleats on the back to provide freedom of movement and on the arm where the sleeve tapers to meet the cuff. The standard men's shirt has a box pleat in the center of the back just below the shoulder or alternately one simple pleat on each side of the back.

Jackets designed for active outdoor wear frequently have pleats (usually inverted box pleats) to allow for freedom of movement. Norfolk jackets have double-ended inverted box pleats at the chest and back.

Skirts and kilts

Skirts, dresses and kilts can include pleats of various sorts to add fullness from the waist or hips, or at the hem, to allow freedom of movement or achieve design effects.

- One or more kick pleats may be set near the hem of a straight skirt to allow the wearer to walk comfortably while preserving the narrow style line.
- Modern kilts may be made with either box pleats or knife pleats, and can be *pleated to the stripe* or *pleated to the sett* (see main article Kilts: Pleating and stitching).

Trousers

Pleats just below the waistband on the front of the garment are typical of many styles of formal and casual trousers including suit trousers and khakis. There may be one, two, three, or no pleats, which may face either direction. When the pleats open towards the pockets they are called reverse pleats (typical of khakis and corduroy trousers) and when they open towards the zipper, they are known as forward pleats.

Utilitarian or very casual styles such as jeans and cargo pants are flat-front (without pleats at the waistband) but may have bellows pockets.

The pleated business pants reached their height in the business fashion world in the 1980s however their popularity dropped in the 1990s and were unseen starting in 2000.

Pockets

A bellows pocket is patch pocket with an inset box pleat to allow the pocket to expand when filled. Bellows pockets are typical of cargo pants, safari jackets, and other utilitarian garments.

RUFFLE

In sewing and dressmaking, a ruffle, frill, or furbelow is a strip of fabric, lace or ribbon tightly gathered or pleated on one edge and applied to a garment, bedding, or other textile as a form of trimming.

The term flounce is a particular type of fabric manipulation that creates a similar look but with less bulk. The term derives from earlier terms of *frounce* or *fronce*. A wavy effect effected without gathers or pleats is created by cutting a curved strip of fabric and applying the inner or shorter edge to the garment. The depth of the curve as well as the width of the fabric determines the depth of the flounce. A godet is a circle wedge that can be inserted into a flounce to further deepen the outer floating wave without adding additional bulk at the point of attachment to the body of the garment, such as at the hemline, collar or sleeve.

Ruffles appeared at the draw-string necklines of full chemises in the 15th century, evolved into the separately-constructed ruff of the 16th century. Ruffles and flounces remained a fashionable form of trim, off-and-on into modern times.

SHIRRING

In sewing, shirring is two or more rows of gathers used to decorate parts of garments, usually the sleeves, bodice or yoke. The term is also sometimes used to refer to the pleats seen in stage curtains. Shirring is a method of shaping a garment and is done so by controlling fullness. Its technique is similar to gathering. Shirring consists of two or more rows of gathered fabric. Shirring can be a pretty and feminine alternative to darts in small areas of a garment. Shirring can also be done on large areas of a garment like all around the top of a full skirt. Shirring works best on soft fabrics but can also done on stronger fabrics.

STYLE LINE

A style line is a line or curve in a garment that has a visual effect, *e.g.*, the seam between two fabrics of different colours or textures. For comparison, a nearly invisible seam, such as a dart or pleat, would not be considered a style line. A style line is a boundary between two distinguishable areas of fabric, or a visible edge of fabric such as the neckline, waistline or hemline. As the term is generally used in practice, a *style line* is introduced only for fashion or cosmetic purposes. However, a style line may also fulfill a shaping function as a dart, a seam at which a dart may end, or as a way of hiding details of the garment's construction.

2

Stitches

BACKSTITCH

Backstitch or *back stitch* and its variants *stem stitch*, *outline stitch* and *split stitch* are a class of embroidery and sewing stitches in which individual stitches are made backward to the general direction of sewing. In embroidery, these stitches form lines and are most often used to outline shapes and to add fine detail to an embroidered picture. It is also used as a hand-sewing sewing utility stitch to attach definitively and strongly two pieces of fabric together.

APPLICATIONS

Basic backstitch is used to outline shapes in modern cross-stitch, in Assisi embroidery and occasionally in blackwork.

A versatile and easy to work stitch, backstitch is ideal for following both smooth and complicated outlines and as a foundation row for more complex embroidery stitches such as *Herringbone ladder filling stitch*. Although superficially similar to Holbein stitch, commonly used in Blackwork embroidery, backstitch differs in the way it is worked, requiring a single journey only to complete a line of stitching.

Stem stitch is an ancient technique; surviving mantles embroidered with stem stitch by the Paracas people of Peru are dated to the first century BCE. Stem stitch is used in the Bayeux Tapestry, an embroidered cloth probably dating to the later 1070s, for lettering and to outline areas filled with couching or laid-work.

Split stitch in silk is characteristic of Opus Anglicanum, an embroidery style of Medieval England.

DESCRIPTION OF THE TECHNIQUE

Backstitch is most easily worked on an even-weave fabric, where the threads can be counted to ensure regularity, and is generally executed from right to left. The stitches are worked in a 'two steps forward, one step back' fashion, along the line to be filled, as shown in the diagram. Neatly worked in a straight line this stitch resembles machine stitching. The back stitch can also be used as a hand-sewing sewing utility stitch to attach two pieces of fabric together.

VARIANTS

Variants of backstitch include:

- Basic backstitch or *point de sable*.
- Threaded backstitch
- Pekinese stitch, a looped interlaced backstitch
- Stem stitch, in which each stitch overlaps the previous stitch to one side, forming a twisted line of stitching, with the thread passing below the needle. It is generally used for outlining shapes and for stitching flower stems and tendrils.
- Whipped stem stitch
- Outline stitch, sometimes distinguished from stem stitch in that the thread passes above rather than below the needle.
- Split stitch, in which the needle pierces the thread rather than returning to one side.

BAR TACK

Bar tack is a series of hand or machine made stitches used for reinforcing areas of stress on a garment, such as pocket openings, bottom of a fly opening or buttonholes. It consists of a series of close-set zig-zag stitches (machine) or whip stitches (hand), usually 1/16"-1/8" in width and 1/4"-3/8" in length. In denim jeans, it is often in a contrasting colour, such as orange or white.

BLANKET STITCH

The blanket stitch is a stitch used to reinforce the edge of thick materials. Depending on circumstances, it may also be called a *whip stitch* or a *crochet stitch*. It is defined as “A decorative stitch used to finish an unhemmed blanket. The stitch can be seen on both sides of the blanket.”

This stitch has long been both an application by hand and as a machine sewn stitch. When done by hand, it is sometimes considered a crochet stitch, used to join pieces together to make a blanket or other larger item. It is used in sewing leather pieces together, as traditionally done by indigenous American cultures, and even for weaving basket rims. The whipstitch is also a type of surgical suturing stitch.

When done by machine, it may be called a whip stitch or, sometimes, a Merrow Crochet Stitch, after the first sewing machine that was used to sew a blanket stitch. This machine was produced and patented by the Merrow Machine Company in 1877. The defining characteristic of the crochet machine is its ability to sew with yarn and stitch thick goods with a consistent overlock edge. From 1877-1925 the machine evolved dramatically, and consequently so did the capacity of manufacturers to produce goods with the whip stitch.

STYLE

The blanket stitch is commonly used as a decorative stitch on an array of garments. Besides blankets, it is used on sweaters, outerwear, swimsuits, home furnishings, and much more. There are many styles of production blanket stitching, including rolled, narrow, with elastic, and traditional (see photos below). Additionally, the term “blanket stitch” has become a verb, describing the application of the stitch.

BLIND STITCH

A blind stitch in sewing is a method of joining two pieces of fabric so that the stitch thread is invisible, or nearly invisible.

There are several techniques for creating a blind stitch by hand sewing. A common technique used to create a hem, or “blind hem”, hides the stitches on both sides of the garment. The sewer catches only a few threads of the fabric each time the needle is pulled through the fabric. Other techniques hide the stitch within the folds of the fabric, so that the thread is only visible when the folded material is pulled away.

A slip stitch or catch stitch can be used to create the blind stitch, except that they are worked inside the hem, [! to ¼ of an inch away from the edge of the hem fabric. A sewing machine can also create a blind hem. In this case, a specialty presser foot is needed. A zigzag stitch technique may be used with a sewing machine to create a blind stitch.

BUTTONHOLE STITCH

Buttonhole stitch and the related blanket stitch are hand-sewing stitches used in tailoring, embroidery, and needle lace-making.

Buttonhole stitches catch a loop of the thread on the surface of the fabric and needle is returned to the back of the fabric at a right angle to the original start of the thread. The finished stitch in some ways resembles a letter “L” depending on the spacing of the stitches. For buttonholes the stitches are tightly packed together and for blanket edges they are more spaced out. The properties of this stitch make it ideal for preventing raveling of woven fabric.

Buttonhole stitches are structurally similar to featherstitches.

APPLICATIONS

In addition to reinforcing buttonholes and preventing cut fabric from raveling, buttonhole stitches are used to make stems in crewel embroidery, to make sewn eyelets, to attach applique to ground fabric, and as couching stitches. Buttonhole stitch scallops, usually raised or padded by rows of straight or chain stitches, were a popular edging in the 19th century.

Buttonhole stitches are also used in cutwork, including Broderie Anglaise, and form the basis for many forms of needlelace.

VARIANTS

Examples of buttonhole or blanket stitches include:

- Blanket stitch
- Buttonhole stitch
- Closed buttonhole stitch, in which the tops of the stitch touch to form triangles
- Crossed buttonhole stitch, in which the tops of the stitch cross
- Detached buttonhole stitch, in which rows of buttonhole stitches are worked to form a "floating" filling stitch
- Buttonhole shading, in which rows of buttonhole stitch are sewn in related colours to give a naturalistic shaded effect
- Buttonhole stitches combined with knots include:
  - Top knotted buttonhole stitch
  - German knotted buttonhole stitch
  - Tailor's buttonhole stitch
  - Armenian edging stitch

Buttonhole bars are parallel rows of thread laid across an open space in lace or cutwork and then completely covered with closely space buttonhole stitches.

CROSS STITCHES

Cross stitches in embroidery, needlepoint, and other forms of needlework include a number of related stitches in which the thread is sewn in an x or + shape. Cross stitch has been called "probably the most widely used stitch of all" and is part of the needlework traditions of the Balkans, Middle East, Afghanistan, Colonial America and Victorian England.

APPLICATIONS

Cross stitches were typical of 16th century canvas work, falling out of fashion in favour of tent stitch towards the end of the century. Canvas work in cross stitch became popular again in the mid-19th century with the Berlin wool work craze.

Herringbone, fishbone, Van Dyke, and related crossed stitches are used in crewel embroidery, especially to add texture to stems, leaves, and similar objects. Basic cross stitch is used to fill backgrounds in Assisi work.

Cross stitch was widely used to mark household linens in the 18th and 19th centuries, and girls' skills in this essential task were demonstrated with elaborate samplers embroidered with cross-stitched alphabets, numbers, birds and other animals, and the crowns and coronets sewn onto the linens of the nobility.

Much of contemporary cross-stitch embroidery derives from this tradition.

VARIANTS

Common variants of cross stitch include:

- Basic cross stitch
- Long-armed cross stitch
- Double cross stitch
- Italian cross stitch
- Basket stitch
- Leaf stitch
- Herringbone stitch
- Closed herringbone stitch
- Tacked herringbone stitch
- Threaded herringbone stitch
- Tied herringbone stitch
- Montenegrin stitch
- Trellis stitch
- Thorn stitch
- Van Dyke stitch

CHAIN STITCH

Chain stitch is a sewing and embroidery technique in which a series of looped stitches form a chain-like pattern. Chain stitch is an ancient craft - examples of surviving Chinese chain stitch embroidery worked in silk thread have been dated to the Warring States period (5th – 3rd century BC). Handmade chain stitch embroidery does not require that the needle pass through more than one layer of fabric. For this reason the stitch is an effective surface embellishment near seams on finished fabric. Because chain stitches can form flowing, curved lines, they are used in many surface embroidery styles that mimic "drawing" in thread. Chain stitches are also used in making tambour lace, needlelace, macramé and crochet.

The earliest archaeological evidence of chain stitch embroidery dates from 1100 BC in China. Excavated from royal tombs, the embroidery was made using threads of silk. Chain stitch embroidery has also been found dating to the Warring States period. Chain stitch designs spread to Iran through the Silk Road.

APPLICATIONS

Hand embroidery

Chain stitch and its variations are fundamental to embroidery traditions of many cultures, including Kashmiri *numdahs*, Iranian Resht work, Central Asian suzani, Hungarian Kalotaszeg "written embroidery", Jacobean embroidery, and crewelwork.

Machine sewing and embroidery

Chain stitch was the stitch used by early sewing machines; however, as it is easily unravelled from fabric, this was soon replaced with the more secure lockstitch. This ease of unraveling of the single-thread chain stitch, more specifically known as ISO 4915:1991 stitch 101, continues to be exploited for industrial purposes in the closure of bags for bulk products. Machine embroidery in chain stitch, often in traditional hand-worked crewel designs, is found on curtains, bed linens, and upholstery fabrics.

VARIANTS

Hand variants

Variations of the basic chain stitch include:
- Back-stitched chain stitch
- Braided stitching
- Cable chain stitch
- Knotted chain stitch
- Open chain stitch
- Petal chain stitch
- Rosette chain stitch
- Singalese chain stitch
- Twisted chain stitch
- Wheat-ear stitch
- Zig-zag chain stitch

Machine variants[sh]

- The Basic Chain stitch is made by first sending the needle down through the material. Then, as the needle rises upward, the friction of the thread against the fabric is sufficient to form a small loop on the underside of the material. That loop is caught by a circular needle which is beneath the work. The machine then moves the material forward projecting the loop on the underside from the previous stitch. The next drop of the needle goes through the previous loop. The circular needle then releases the first loop and picks up the new loop and the process repeats.
- The Double chain stitch uses two threads. It is rarely used in today's machines except for ornamental purposes because it uses a lot of thread. It is found in bulk material packaging, where it is used to close big bags. In this case it is useful to allow an easy opening of the bag.

CROSS-STITCH

Cross-stitch is a popular form of counted-thread embroidery in which X-shaped stitches in a tiled, raster-like pattern are used to form a picture. Cross-stitch is often executed on easily countable evenweave fabric called aida cloth. The stitcher counts the threads in each direction so that the stitches are of uniform size and appearance. This form of cross-stitch is also called counted cross-stitch in order to distinguish it from other forms of cross-stitch. Sometimes cross-stitch is done on designs printed on the fabric (stamped cross-stitch); the stitcher simply stitches over the printed pattern.

Fabrics used in cross-stitch include aida, linen and mixed-content fabrics called 'evenweave'. All cross stitch fabrics are technically "evenweave," it refers to the fact that the fabric is woven to make sure that there are the same number of threads in an inch both left to right and top to bottom (vertically and horizontally). Fabrics are categorized by threads per inch (referred to as 'count'), which can range from 11 to 40 count. Aida fabric has a lower count because it is made with two threads grouped together for ease of stitching. Cross stitch projects are worked from a gridded pattern and can be used on any count fabric, the count of the fabric determines the size of the finished stitching.

Cross-stitch is the oldest form of embroidery and can be found all over the world. Many folk museums show examples of clothing decorated with cross-stitch, especially from continental Europe and Asia.

Two-dimensional (unshaded) cross-stitch in floral and geometric patterns, usually worked in black and red cotton floss on linen, is characteristic of folk embroidery in Eastern and Central Europe.

The cross stitch sampler is called that because it was generally stitched by a young girl to learn how to stitch and to record alphabet and other patterns to be used in her household sewing. These samples of her stitching could be referred back to over the years. Often, motifs and initials were stitched on household items to identify their owner, or simply to decorate the otherwise-plain cloth. In the United States, the earliest known cross-stitch sampler is currently housed at Pilgrim Hall in Plymouth, Massachusetts. The sampler was created by Loara Standish, daughter of Captain Myles Standish and pioneer of the Leviathan stitch, circa 1653.

Traditionally, cross-stitch was used to embellish items like household linens, tablecloths, dishcloths, and doilies (only a small portion of which would actually be embroidered, such as a border). Although there are many cross-stitchers who still employ it in this fashion, it is now increasingly popular to work the pattern on pieces of fabric and hang them on the wall for decoration. Cross stitch is also often used to make greeting cards, pillowtops, or as inserts for box tops, coasters and trivets.

Multicoloured, shaded, painting-like patterns as we know them today are a fairly modern development, deriving from similar shaded patterns of Berlin wool work of the mid-nineteenth century. Besides designs created expressly for cross stitch, there are software programmes that convert a photograph or a fine art image into a chart suitable for stitching. One stunning example of this is in the cross stitched reproduction of the Sistine Chapel charted and stitched by Joanna Lopianowski-Roberts.

There are many cross-stitching "guilds" and groups across the United States and Europe which offer classes, collaborate on large projects, stitch for charity, and provide other ways for local cross-stitchers to get to know one another. Individually owned local needlework shops (LNS) often have stitching nights at their shops, or host weekend stitching retreats.

Today cotton floss is the most common embroidery thread. It is a thread made of mercerized cotton, composed of six strands that are only loosely twisted together and easily separable. While there are other manufacturers, the two most-commonly used (and oldest) brands are DMC and Anchor, both of which have been manufacturing embroidery floss since the 1800s.

Other materials used are pearl (or perle) cotton, Danish flower thread, silk and Rayon. Different wool threads, metallic threads or other novelty threads are also used, sometimes for the whole work, but often for accents and embellishments. Hand-dyed cross stitch floss is created just as the name implies - it is dyed by hand. Because of this, there are variations in the amount of colour throughout the thread. Some variations can be subtle, while some can be a huge contrast. Some also have more than one colour per thread, which in the right project, creates amazing results. Cross stitch is widely used in traditional Palestinian dressmaking.

RELATED STITCHES AND FORMS OF EMBROIDERY

Other stitches are also often used in cross-stitch, among them ¼, ½, and ¾ stitches and backstitches.

Cross-stitch is often used together with other stitches. A cross stitch can come in a variety of prostational forms. It is sometimes used in crewel embroidery, especially in its more modern derivatives. It is also often used in needlepoint.

A specialized historical form of embroidery using cross-stitch is Assisi embroidery.

There are many stitches which are related to cross-stitch and were used in similar ways in earlier times. The best known are Italian cross-stitch, Celtic Cross Stitch, Irish Cross Stitch, long-armed cross-stitch, Ukrainian cross-stitch and Montenegrin stitch. Italian cross-stitch and Montenegrin stitch are reversible, meaning the work looks the same on both sides. These styles have a slightly different look than ordinary cross-stitch. These more difficult stitches are rarely used in mainstream embroidery, but they are still used to recreate historical pieces of embroidery or by the creative and adventurous stitcher.

The double cross-stitch, also known as a Leviathan stitch or Smyrna cross stitch, combines a cross-stitch with an upright cross-stitch.

Berlin wool work and similar petit point stitchery resembles the heavily shaded, opulent styles of cross-stitch, and sometimes also used charted patterns on paper.

Cross-stitch is often combined with other popular forms of embroidery, such as Hardanger embroidery or blackwork embroidery. Cross-stitch may also be combined with other work, such as canvaswork or drawn thread work. Beadwork and other embellishments such as paillettes, charms, small buttons and speciality threads of various kinds may also be used.

RECENT TRENDS IN THE UK

Cross-stitch has become increasingly popular with the younger generation of the United Kingdom in recent years. The Great Recession has also seen renewal of interest in home crafts. Retailers such as John Lewis experienced a 17 per cent rise in sales of haberdashery products between 2009 and 2010. Hobbycraft, a chain of stores selling craft supplies, also enjoyed an 11 per cent increase in sales over the past year. The chain is said to have benefited from the "make do and mend" mentality of the credit crisis, which has driven people to make their own cards and gifts.

Knitting and cross stitching have become more popular hobbies for a younger market, in contrast to its traditional reputation as a hobby for retirees. Sewing and craft groups such as Stitch and Bitch London have resurrected the idea of the traditional craft club. At Clothes Show Live 2010 there was a new area called "Sknitch" promoting modern sewing, knitting and embroidery.

In a departure from the traditional designs associated with cross stitch, there is a current trend for more postmodern or tongue-in-cheek designs featuring retro images or contemporary sayings. It is linked to a concept known as 'subversive cross stitch', which involves more risque designs, often fusing the traditional sampler style with sayings designed to shock or be incongruous with the old-fashioned image of cross stitch.

Stitching designs on other materials can be accomplished by using a Waste Canvas. This waste canvas is a temporary gridded canvas similar to regular canvas used for embroidery that is held together by a water soluble glue, this is removed after completion of stitch design.

EMBROIDERY STITCH

In everyday language, a stitch in the context of embroidery or hand-sewing is defined as the movement for burgers the embroidery needle from the backside of the fabric to the front side and back to the back side. The thread stroke on the front side produced by this is also called *stitch*. In the context of embroidery, an embroidery stitch means one or more *stitches* that are always executed in the same way, forming a figure of a terrible look. Embroidery stitches are also called *stitches* for short.

Embroidery stitches are the smallest units in embroidery. Embroidery patterns are formed by doing many embroidery stitches, either all the same or different ones, either following a counting chart on paper, following a design painted on the fabric or even working freehand.

COMMON STITCHES

Embroidery uses various combinations of stitches. Each embroidery stitch has a special name to help identify it. These names vary from country to country and region to region. Some embroidery books will include name variations. Taken by themselves the stitches are mostly simple to execute, however when put together the results can be extremely complex.

running stitches===

Straight stitches pass through the fabric ground in a simple up and down motion, and for the most part moving in a single direction. Examples of straight stitches are:

- Running or basting stitch
- Simple satin stitch
- Algerian eye stitch
- Fern stitch

Straight Stitches that have two journeys (generally forwards and backwards over the same path). Examples:

- Holbein stitch, also known as the double running stitch
- Bosnian stitch

Back stitches

Back stitches pass through the fabric ground in an encircling motion. The needle in the simplest backstitch comes up from the back of the fabric, makes a stitch to the right going back to the back of the fabric, then passes behind the first stitch and comes up to the front of the fabric to the left of the first stitch. The needle then goes back to the back of the fabric through the same hole the stitch first came up from. The needle then repeats the movement to the left of the stitches and continues. Some examples of a back stitch are:

- Stem stitch or outline stitch
- Split stitch. The needle pierces the thread as it come back up. that is wrong
- Crewel stitch

Chain stitches

Chain stitches catch a loop of the thread on the surface of the fabric. In the simplest of the looped stitches, the chain stitch, the needle comes up from the back of the fabric and then the needle goes back into the same hole it came out of, pulling the loop of thread almost completely through to the back; but before the loop disappears, the needle come back up (a certain distance from the beginning stitch -the distance deciding the length of the stitch), passes through the loop and prevents it from being pulled completely to the back of the fabric. The needle then passes back to the back of the fabric through the second hole and begins the stitch again. Examples of chain stitches are:

- Chain stitch
- Lazy Daisy stitch, or detached chain. The loop stitch is held to the fabric at the wide end by a tiny tacking stitch.
- Spanish Chain or Zig-zag Chain

Buttonhole stitches

Buttonhole or blanket stitches also catch a loop of the thread on the surface of the fabric but the principal difference is that the needle does not return to the original hole to pass back to the back of the fabric. In the classic buttonhole stitch, the needle is returned to the back of the fabric at a right angle to the original start of the thread. The finished stitch in some ways resembles a letter "L" depending on the spacing of the stitches. For buttonholes the stitches are tightly packed together and for blanket edges they are more spaced out. The properties of this stitch make it ideal for preventing raveling of woven fabric. This stitch is also the basis for many forms of needle lace. Examples of buttonhole or blanket stitches.

- Blanket stitch
- Buttonhole stitch
- Closed buttonhole stitch, the tops of the stitch touch to form triangles
- Crossed buttonhole stitch, the tops of the stitch cross
- Buttonhole stitches combined with knots:
  - Top Knotted Buttonhole stitch
  - German Knotted Buttonhole stitch
  - Tailor's buttonhole stitch

Cross stitches[sh]

Cross stitches or cross-stitch have come to represent an entire industry of pattern production and material supply for the craft person. The stitch is done by creating a line of diagonal stitches going in one direction, usually using the warp and weft of the fabric as a guide, then on the return journey crossing the diagonal in the other direction, creating an "x". Also included in this class of stitches are:

- Herringbone stitches, including the hem stitch
- Breton stitch, here the threads of the "x" are twisted together
- Sprat's Head stitch
- Crow's Foot stitch, these last two stitches are often used in tailoring to strengthen a garment at a point of strain such as a pocket corner or the top of a kick pleat.

Knotted stitches

Knotted stitches are formed by wrapping the thread around the needle, once or several times, before passing it back to the back of the fabric ground. This is a predominate stitch in Brazilian embroidery, used to create flowers. Another form of embroidery that uses knots is Candlewicking, where the knots are created by forming a figure 8 around the needle. Examples of knotted stitches are:

- French knot, or twisted knot stitch
- Chinese knot, which varies from the French knot in that it takes a tiny stitch in the background fabric while creating the knot
- Bullion knots
- Coral stitch
- There are also more complex knotted stitches such as:
  - Knotted Loop stitch
  - Plaited Braid stitch
  - Sorbello stitch
  - Diamond stitch
- Knotted edgings based on buttonhole stitches include:
  - Antwerp edging stitch
  - Armenian edging stitch

Couching and laid work

Couching or laid stitches involve two sets of threads, the set that is being 'laid' onto the surface of the fabric and the set which attach the laid threads. The laid threads may be heavier than the attaching thread, or they may be of a nature that does not allow them to be worked like a regular embroidery thread, such as metal threads. The stitches used to attach the laid thread may be of any nature; cross stitch, buttonhole stitch, straight stitch; but some have specific names:

- Pendant couching,
- Bokhara couching
- Square laid work
- Oriental couching
- Battlement couching
- Klosterstitch
- Roumanian couching

HEMSTITCH

Hemstitch or hem-stitch is a decorative drawn thread work or openwork hand-sewing technique for embellishing the hem of clothing or household linens. Unlike an ordinary hem, hemstitching can employ embroidery thread in a contrasting colour so as to be noticeable.

In hemstitching, one or more threads are drawn out of the fabric parallel and next to the turned hem, and stitches bundle the remaining threads in a variety of decorative patterns while securing the hem in place. Multiple rows of drawn thread work may be used.

Hand hemstitching can be imitated by a hemstitching machine which has a piercer that pierces holes into the fabric and two separate needles that sew the hole open. There are also hemstitcher attachments for home sewing machines, and simple decorative stitches can be used over drawn threads to suggest hand-hemstitching.

LOCKSTITCH

A lockstitch is the most common mechanical stitch made by a sewing machine. The term "single needle stitching", often found on dress shirt labels, refers to lockstitch.

STRUCTURE

The lockstitch uses two threads, an upper and a lower. Lockstitch is so named because the two threads, upper and lower, "lock" (entwine) together in the hole in the fabric which they pass through. The upper thread runs from a spool kept on a spindle on top of or next to the machine, through a tension mechanism, through the take-up arm, and finally through the hole in the needle. Meanwhile the lower thread is wound onto a bobbin, which is inserted into a case in the lower section of the machine below the material.

To make one stitch, the machine lowers the threaded needle through the cloth into the bobbin area, where a rotating hook (or other hooking mechanism) catches the upper thread at the point just after it goes through the needle. The hook mechanism carries the upper thread entirely around the bobbin case, so that it has made one wrap of the bobbin thread. Then the take-up arm pulls the excess upper thread (from the bobbin area) back to the top, forming the lockstitch. Then the feed dogs pull the material along one stitch length, and the cycle repeats.

Ideally, the lockstitch is formed in the center of the thickness of the material—that is to say: ideally the upper thread entwines the lower thread in the middle of the material. The thread tension mechanisms, one for the upper thread and one for the lower thread, prevent either thread from pulling the entwine point from out of the middle of the material.

GEOMETRY

The geometry of the lockstitch is controlled by the presence or absence of:

- Sideways movements of the machine's needle, and
- Backwards movements of the machine's feed dogs.

In older machines, the needle and feed motion is controlled by mechanical cams. Some modern household machines offer a slot for user-replaceable custom stitch cams. In more recent designs, the needle and feed motion are directly motorized.

Straight

Straight stitch geometry is produced when the needle has no sideways movements and when the feed dogs are following only in the normal forward "four motion" movement. Because its two threads run straight and parallel, a straight stitch is not natively stretchable.

Zigzag

Zigzag stitch geometry is produced when the needle moves rhythmically side to side while stitching, while the feed dogs are following only in the normal forward "four motion" movement. Most lockstitch machines are capable of doing this. Zigzag stitches are used when a stretchable stitch is required, such as when sewing stretchy fabrics.

Blind

Blind stitch geometry is a derivative of the zigzag. It is created in the same manner, except that the needle zigs to the side and then zags back only once every fourth or fifth stitch. It is used to reduce the visibility of hems and other seam edges.

Stretch

Stretch stitch geometry is specifically for stretchability. While the needle is moving, as for straight or zigzag stitches, the feed dogs automatically move the fabric forward and backward. As with zigzag stitches, stretch stitching is controlled by mechanical cams, but because of the dual action, stretch stitch machines have *double* cams. As the double cam rotates, the first follower rides along one track to move the needle bar from side to side, while the second follower rides along a different track to move the feed dogs forward and reverse.

Decorative

By adding controlled motion of the material being sewn through an additional set of motors, arbitrary customized patterns of 100 cm or more in each direction can be sewn, opening the door to the very popular category of programmable household embroidery machines.

PREVALENCE

Home

Most home sewing machines are lockstitch machines, although overlockers (aka sergers) have entered the home market in the past ten years or so.

Industrial

Of a typical garment factory's sewing machines, half might be lockstitch machines and the other half divided between overlock machines, chain stitch machines, and various other specialized machines. Industrial lockstitch machines with two needles, each forming an independent lockstitch with their own bobbin, are also very common. There are different types of lockstitch industrial machines. The most commonly used are the drop feed for light and medium duty, and walking foot for medium and heavy duty like the Class 7 with an impressive 3/4" foot lift. This makes the Class 7 able to stitch through heavy materials up to 3/4" with threads as strong as 57 lbs. Originally made by Singer in the US and Europe for supplying the demand of heavy duty clothing for the troops, for many years after the war this class was not available as new because the market was filled. With the outsourcing of many sewing manufacturing jobs, nowadays many Chinese Class 7 machines are available and built by Federal Specifications giving them equal performance to the original ones (FSN:3530-3111-1556, FSN: 3530-3111-3675, FSN: 3530-311-1556, FSN: 3530-3111-3075).

Most industrial lockstitch machines sew only a straight line of stitches. Industrial *zig-zag* machines are available, but uncommon, and there are essentially no fancy-pattern stitching industrial machines other than dedicated embroidery and edge decoration machines. Even something as simple as a bar-tack or a buttonhole stitch is usually done by a dedicated machine incapable of doing anything else. When a variety of decorative stitching is required rather than a single stitch, a "commercial" machine (basically a heavy duty household machine) is usually employed.

OVERLOCK

An overlock is a kind of stitch that sews over the edge of one or two pieces of cloth for edging, Hemming, or seaming. Usually an overlock sewing machine will cut the edges of the cloth as they are fed through (such machines being called "sergers" in North America), though some are made without cutters. The inclusion of automated cutters allows overlock machines to create finished seams easily and quickly. An overlock sewing machine differs from a lockstitch sewing machine in that it uses loopers fed by multiple thread cones rather than a bobbin. Loopers serve to create thread loops that pass from the needle thread to the edges of the fabric so that the edges of the fabric are contained within the seam.

Overlock sewing machines usually run at high speeds, from 1000 to 9000 rpm, and most are used in industry for edging, hemming and seaming a variety of fabrics and products. Overlock stitches are extremely versatile, as they can be used for decoration, reinforcement, or construction. Overlocking is also referred to as "overedging," "merrowing," or "serging." Though "serging" technically refers to overlocking with cutters, in practice the four terms are used interchangeably.

Overlock stitching was invented by the Merrow Machine Company in 1881.

J. Makens Merrow and his son Joseph Merrow, who owned a knitting mill established in Connecticut in 1838, developed a number of technological advancements to be used in the mill's operations. Merrow's first patent was a machine for crochet stitching, and the Merrow Machine Company still produces crochet machines based on this original model. This technology was a starting point for the development of the overlock machine, patented by Joseph Merrow in 1889. Unlike standard lockstitching, which uses a bobbin, overlock sewing machines utilize loopers to create thread loops for the needle to pass through, in a manner similar to crocheting. Merrow's original three-thread overedge sewing machine is the forerunner of contemporary overlocking machines. Over time, the Merrow Machine Company pioneered the design of new machines to create a variety of overlock stitches, such as two- and four-thread machines, the one-thread butted seam, and the cutterless emblem edger.

A landmark lawsuit between Wilcox and Gibbs and the Merrow Machine Company in 1905 established the ownership and rights to the early mechanical development of overlocking to the Merrow Machine Company.

Throughout the early 20th Century, the areas of Connecticut, USA and New York, USA were the centers of textile manufacturing and machine production. Consequently many overlock machine companies established themselves in the Northeastern United States.

In 1964, Juki Corporation was formed; it was a precursor of the modern industrial overlock sewing machine company. Throughout the 1980s, Japanese and Chinese sewing machine production came to dominate the industry.

TERMINOLOGY

In the United States, the term "overlocker" has largely been replaced by "serger." However, in other parts of the world such as Australia and the UK, the term "overlocker" is still in use.

TYPES OF OVERLOCK STITCHES

Overlock stitches are classified in a number of ways. The most basic classification is by the number of threads used in the stitch. Industrial overlock machines are generally made in 1, 2, 3, 4, or 5 thread formations. Each of these formations has unique uses and benefits:

- 1-thread: End-to-end seaming or 'butt-seaming' of piece goods for textile finishing.
- 2-thread: Edging and seaming, especially on knits and wovens, finishing seam edges, stitching flatlock seams, stitching elastic and lace to lingerie, and hemming. This is the most common type of overlock stitch.

- 3-thread: Sewing pintucks, creating narrow rolled hems, finishing fabric edges, decorative edging, and seaming knit or woven fabrics.
- 4-thread: Decorative edging and finishing, seaming high-stress areas, mock safety stitches which create extra strength while retaining flexibility.
- 5-thread: In apparel manufacturing, safety stitches utilizing 2 needles create a very strong seam. For every 1 cm of seam length you would require 20 cm of thread to sew it.

Two- and three-thread formations are also known as “merrowing” after the Merrow Machine Company.

Additional variables in the types of overlock stitches are the stitch eccentric, and the stitch width. The stitch eccentric indicates how many stitches per inch there are, which is adjustable and can vary widely within one machine. Different stitch eccentrics create more or less dense and solid-looking edges. The stitch width indicates how wide the stitch is from the edge of the fabric. Lightweight fabrics often require a wider stitch to prevent pulling.

Adding extra variation in stitch types is the ‘differential feed’ feature, which allows feed to be adjusted; extra-fast feed creates a ruffled or ‘lettuce-leaf’ effect. Finally, some merrowing machines contain parts to roll the fabric edge into the stitch for added durability.

FORMATION OF AN OVERLOCK STITCH

1. When the needle enters the fabric, a loop is formed in the thread at the back of the needle.
2. As the needle continues its downward motion into the fabric, the lower looper begins its movement from left to right. The tip of the lower looper passes behind the needle and through the loop of thread that has formed behind the needle.
3. The lower looper continues along its path moving towards the right of the serger. As it moves, the lower thread is carried through the needle thread.
4. While the lower looper is moving from left to right, the upper looper advances from right to left. The tip of the upper looper passes behind the lower looper and picks up the lower looper thread and needle thread.
5. The lower looper now begins its move back into the far left position. As the upper looper continues to the left, it holds the lower looper thread and needle thread in place.
6. The needle again begins its downward path passing behind the upper looper and securing the upper looper thread. This completes the overlock stitch formation and begins the stitch cycle all over again.

USES OF THE OVERLOCK STITCH

Overlock stitches are traditionally used for edging and light seaming. Other applications include:

- Sewing netting
- Butt-seaming is used for joining the ends of piece goods.
- Flat-locking
- Edging emblems
- Purl stitching
- Rolled hemming
- Decorative edging

PAD STITCH

Pad stitches are a type of running stitch made by placing small stitches perpendicular to the line of stitching. Pad stitches secure two or more layers of fabric together and give the layers more firmness; smaller and denser stitches create more firmness. They may also be used to enforce an overall curvature of the layers.

USAGE

Tailors pad stitch a jacket's lapel and undercollar to give them additional firmness, and maintain their curvature. The line of stitching usually runs parallel to the direction of the most important curve of the layers. For example, pad stitches in a suit's lapel run parallel to the lapel's roll line; pad stitches in the under collar of a tailored jacket run parallel to the collar's back edge.

PICK STITCH

A pick stitch in sewing is a simple running stitch that catches only a few threads of the fabric, showing very little of the thread on the right side (outer side) of the garment. It is also sometimes known as 'stab stitch'. A pick stitch can be made from either the inside of the garment or the outside, depending upon how much thread is meant to show on the outside of the garment. A pick stitch is commonly used for making hems, although it is also used with contrasting thread to create a decorative finish on some garments. It has decorative uses in embroidery. It is exceedingly useful for inserting zips and is surprisingly strong for this purpose. Many home-sewers and new dressmakers find this much easier than inserting zips by sewing machine. A pick stitch along the outside of a lapel is a hallmark of a "high-end, hand-made" men's suit or blazer. A finely made pick stitch is difficult to accomplish but can be achieved with practice.

RUNNING STITCH

The running stitch or straight stitch is the basic stitch in hand-sewing and embroidery, on which all other forms of sewing are based on. The stitch is worked by passing the needle in and out of the fabric. Running stitches may be of varying length, but typically more thread is visible on the top of the sewing than on the underside. So, a running stitch runs through the fabric.

USES

Running stitches are used in hand-sewing and tailoring to sew basic seams, in hand patchwork to assemble pieces, and in quilting to hold the fabric layers and batting or wadding in place. Loosely spaced rows of short running stitches are used to support padded satin stitch. Running stitches are a component of many traditional embroidery styles, including kantha of India and Bangladesh, and Japanese sashiko quilting.

RELATED STITCHES

- Basting stitches, also called *tailor's tack*, are long running stitches used to keep two pieces of fabric or trim aligned during final sewing, or to otherwise temporarily sew two pieces together.
- Darning stitches are closely spaced parallel rows of running stitches used to fill or reinforce worn areas of a textile, or as decoration.
- Double-running or Holbein stitches have a second row of running stitches worked in a reverse direction in between the stitches of the first pass, to make a solid line of stitching.
- Double darning stitches are closely spaced (but not overlapping) rows of Holbein stitches.

SASHIKO STITCHING

Sashiko is a form of decorative reinforcement stitching (or functional embroidery) from Japan. Traditionally used to reinforce points of wear, or to repair worn places or tears with patches, this running stitch technique is often used for purely decorative purposes in quilting and embroidery. The white cotton thread on the traditional indigo blue cloth gives sashiko its distinctive appearance, though decorative items sometimes use red thread. Many Sashiko patterns were derived from Chinese designs, but just as many were developed by the Japanese themselves. The artist Katsushika Hokusai (1760–1849) published the book *New Forms for Design* in 1824, and these designs have inspired many Sashiko patterns.

Patterns:

- Tate-Jima— Vertical stripes
- Yoko-Jima— Horizontal stripes
- Kôshi —Checks
- Nakamura Kôshi —Plaid of Nakamura family
- Hishi-moyô —Diamonds
- Yarai — Bamboo Fence
- Hishi-Igeta /Tasuki — Parallel diamonds / crossed cords

- Kagome — Woven Bamboo
- Uroko — Fish Scales
- Tate-Waku — Rising steam
- Fundô — Counterweights
- Shippô — Seven Treasures of Buddha
- Amime — Fishing nets
- Toridasuki — Interlaced circle of two birds
- Chidori — Plover
- Kasumi— Haze
- Asa no Ha — Hemp leaf
- Mitsuba — Trefoil
- Hirayama-Michi — Passes in the mountains
- Kaki no Hana — Persimmon flower
- Kaminari — Thunderbolts
- Inazuma— Flash of Lightning
- Sayagata— Key pattern
- Matsukawa-Bishi — Pine Bark
- Yabane — Fletching, and many more.

TOPSTITCH

Topstitching is a sewing technique. It is used most often on garment edges such as necklines and hems, where it helps facings to stay in place and gives a crisp edge. Decorative topstitching is designed to show, and may be done in a fancy thread or with a special type of stitch. Otherwise, topstitching is generally done using a straight stitch with a thread that matches the fashion fabric.

ZIGZAG STITCH

A zigzag stitch is variant geometry of the lockstitch. It is a back-and-forth stitch used where a straight stitch will not suffice, such as in reinforcing buttonholes, in stitching stretchable fabrics, and in temporarily joining two work pieces edge-to-edge.

When creating a zigzag stitch, the back-and-forth motion of the sewing machine's needle is controlled by a cam. As the cam rotates, a fingerlike follower, connected to the needle bar, rides along the cam and tracks its indentations. As the follower moves in and out, the needle bar is moved from side to side.

Very old sewing machines lack this hardware and so cannot natively produce a zigzag stitch, but there are often shank-driven attachments available which enable them to do so.

ZIGZAGGER ATTACHMENTS

Older sewing machines which only sew a straight stitch can be adapted to sew a zigzag by means of an attachment. The attachment replaces the machine's presser foot with its own, and draws mechanical power from the machine's needle clamp (which requires the needle clamp to have a side-facing thumbscrew). It creates a zigzag by mechanically moving the fabric side to side as the machine runs. The zigzagger's foot has longitudinal grooves on its underside, facing the material, which confer traction only sideways. This allows the zigzagger to move the material side to side while the machine's feed dogs are simultaneously moving the material forward or backward in the usual manner.

TACK (SEWING)

In sewing, to tack or baste is to make quick, temporary stitching intended to be removed. Tacking is used in a variety of ways:

- To temporarily hold a seam or trim in place until it can be permanently sewn, usually with a long running stitch made by hand or machine called a tacking stitch or basting stitch.
- X-shaped tacking stitches are also very common on vents (slits) on the back of men's suit jackets, or at the bottom of kick pleats on a woman's skirt. They are meant to hold the flaps in place during shipping and when on display in the store. They should be removed before being worn; however many buyers do not realize it. Brand labels loosely basted on the outer edges of the sleeves of suits as well as women's winter coat should also be removed after purchase. They are meant to help customers to easily identify the brands in the store without reaching into the collar.
- To temporarily attach a lace collar, ruffles, or other trim to clothing so that the attached article may be removed easily for cleaning or to be worn with a different garment. For this purpose, tacking stitches are sewn by hand in such a way that they are almost invisible from the outside of the garment.
- To transfer pattern markings to fabric, or to otherwise mark the point where two pieces of fabric are to be joined. A special loose looped stitch used for this purpose is called a tack or tailor's tack. This is often done through two opposing layers of the same fabric so that when the threads are snipped between the layers the stitches will be in exactly the same places for both layers thus saving time having to chalk and tack the other layer.
- A basting stitch is essentially a straight stitch, sewn with long stitches and unfinished ends. The basting stitch is used for temporarily holding sandwiched pieces of fabric in place. The stitch is removed after the piece is finished. Often used in quilting or embroidery.

3

Seams

In sewing, a seam is the join where two or more layers of fabric, leather, or other materials are held together with stitches. Prior to the invention of the sewing machine, all sewing was done by hand. Seams in modern mass-produced household textiles, sporting goods, and ready-to-wear clothing are sewn by computerized machines, while home shoemaking, dressmaking, quilting, crafts, haute couture and tailoring may use a combination of hand and machine sewing.

In clothing construction, seams are classified by their *type* (plain, lapped, abutted, or French seams) and *position* in the finished garment (center back seam, inseam, side seam). Seams are *finished* with a variety of techniques to prevent raveling of raw fabric edges and to neaten the inside of garments.

TYPES

All basics seams used in clothing construction are variants on four basic types of seams:

- Plain seams
- French seams
- Flat or abutted seams
- Lapped seams

A plain seam is the most common type of machine-sewn seam. It joins two pieces of fabric together face-to-face by sewing through both pieces, leaving a seam allowance with raw edges inside the work. The seam allowance usually requires some sort of seam finish to prevent raveling.

Either piping or cording may be inserted into a plain seam.

In a French seam, the raw edges of the fabric are fully enclosed for a neat finish. The seam is first sewn with wrong sides together, then the seam allowances are trimmed and pressed. A second seam is sewn with right sides together, enclosing the raw edges of the original seam.

In a flat or abutted seam, two pieces of fabric are joined edge-to edge with no overlap and sewn with hand or machine stitching that encloses the raw edges. Antique or old German seam is the 19th century name for a hand-sewn flat seam that joins two pieces of at their selvages. This type of construction is found in traditional linen garments such as shirts and chemises, and in hand-made sheets pieced from narrow loom widths of linen.

In a lapped seam, the two layers overlap with the wrong side of the top layer laid against the right side of the lower layer. Lapped seams are typically used for bulky materials that do not ravel, such as leather and felt.

FINISHES

A seam finish is a treatment that secures and neatens the raw edges of a plain seam to prevent raveling, by sewing over the raw edges or enclosing them in some sort of binding.

On mass-produced clothing, the seam allowances of plain seams are usually trimmed and stitched together with an overlock stitch using a serger. Plain seams may also be pressed open, with each seam allowance separately secured with an overlock stitch. Traditional home sewing techniques for finishing plain seams include trimming with pinking shears, oversewing with a zig-zag stitch, and hand or machine overcasting.

A bound seam has each of the raw edges of its seam allowances enclosed in a strip of fabric, lace or net 'binding' that has been folded in half lengthwise. An example of binding is double-fold bias tape. The binding's fold is wrapped around the raw edge of the seam allowance and is stitched, through all thicknesses, catching underside of binding in stitching. Bound seams are often used on lightweight fabrics including silk and chiffon and on unlined garments to produce a neat finish.

A Hong Kong seam or Hong Kong finish is a home sewing term for a type of bound seam in which each raw edge of the seam allowance is separately encased in a fabric binding. In couture sewing or tailoring, the binding is usually a bias-cut strip of lightweight lining fabric; in home sewing, commercial bias tape is often used.

In a Hong Kong finish, a bias strip of fabric is cut to the width of the seam allowance plus 1/4". The bias strip is placed on top of the seam allowance, right sides together, and stitched 1/8" from raw edges. The bias strip is then folded over the raw edge and around to the underside and stitched in place.

POSITION

In clothing construction, seams are identified by their position in the finished garment.

A center front seam runs vertically down the front of a garment.

A center back seam or back seam runs vertically down the center-back of a garment. It can be used to create anatomical shaping to the back portion of a garment particularly through the waist area and hips. It can also be used for styling and functional purposes involving pleats, vents, flare towards the hem or for back closures such as buttoned plackets or zippers.

A side seam runs vertically down the side of a garment.

A side-back seam runs from the armscye to the waist, and fits the garment to the curve below the shoulder blades. Side-back seams may be used instead of, or in combination with, side and center back seams.

A shoulder seam runs from the neckline to the armscye, usually at the highest point of the shoulder.

Princess seams in the front or back runs from the shoulder or armscye to the hem at the side-back or side-front. Princess seams shape the garment to the body's curves and eliminate the need for darting at the bust, waist, and shoulder.

An inseam is the seam that binds the length of the inner trouser leg. The distance from the bottom crotch to the lower ankle is also known as the inseam. The inseam length determines the length of the inner pant leg to appropriately fit the wearer. In the UK this is usually known as the inside-leg measurement (for trousers fit).

NOTCHING OR CLIPPING A CURVED SEAM

When making an outward-curved seam, the material will have reduced bulk and lie flat if notches are cut into the seam allowance. Alternatively, when making an inward-curved seam, clips are cut into the seam allowance to help the seam lie flat with reduced bulk in the fabric. Once seam allowances are pinned or basted, they are often pressed flat with an iron before sewing the final seam. Pressing the seam allowances makes it easier to sew a consistent finished seam.

ARMSCYE

In sewing, the armscye is the armhole, the fabric edge to which the sleeve is sewn. The length of the armscye is the total length of this edge; the width is the distance across the hole at the widest point. Multiple theories for the etymology of "armscye" have been proposed. The scholarly etymology has the origin as "arm" + "scye." The first documented use of "scye" in print is by Jamieson (1825) Suppl.: "sey," a Scots and Ulster dialect word (written also scy, sci, si, sie, sy in glossaries) meaning 'the opening of a gown, etc., into which the sleeve is inserted; the part of the dress between the armpit and the chest (of obscure etymology, and sometimes confused with "scythe" due to similarly curved shapes). A more fanciful folk etymology is as follows. Because the expression "arm's eye" was used in some older sewing texts (*e.g.* "Gynametry," published in 1887) it is conjectured that in poor prints the apostrophe and the crossbar of the lower case "e" were indistinct, and the neologism "armscye" was created by readers who concatenated the orphaned fragments "arm" and "s" with the corrupt "cye". According to this undocumented theory, until the beginning of the 20th century writers favoured the original term or at least a more logical variation (*e.g.* "armeye" in *The Perfect Dressmaking System," published in 1914), but as self-proclaimed experts copied each other, the term "armscye" eventually became widely enough used by home sewers to gain general acceptance.* The latter theory clearly contradicts evidence that the term "scye" was already in use at least as early as 1825. Therefore the erroneous folk analysis was not in the direction from "arm's eye" to "armcye", but rather from the original "armcye" to "arm's eye" (which made more sense to modern English speakers, with a later adjustment more recently back to the correct "armscye").

FELLED SEAM

Felled seam, or flat-fell seam, is a seam made by placing one edge inside a folded edge of fabric, then stitching the fold down. It includes a topstitched finish. It is useful for keeping seam allowances flat and covering raw edges. The flat-felled seam is the type of seam used in making denim jeans, although it appears inside-out to reduce stitching. It is also used in traditional tipi construction. There are flat-felled seams and lap-felled seams.

SEAM ALLOWANCE

Seam allowance (sometimes called inlays) is the area between the edge and the stitching line on two (or more) pieces of material being stitched together. Seam allowances can range from 1/4 inch wide (6.35 mm) to as much as several inches. Commercial patterns for home sewers have seam allowances ranging from 1/4 inch to 5/8 inch. Sewing industry seam allowances range from 1/4 inch for curved areas (*e.g.* neck line, armscye) or hidden seams (*e.g.* facing seams), to one inch or more for areas that require extra fabric for final fitting to the wearer (*e.g.* center back).

4

Notions and Trims

NOTIONS

In sewing and haberdashery, notions is the collective term for a variety of small objects or accessories. Notions can include items that are sewn or otherwise attached to a finished article, such as buttons, snaps, and collar stays, but the term also includes small tools used in sewing, such as thread, pins, marking pens, and seam rippers. The noun is almost always used in the plural. The term is chiefly found in the United States, and was formerly used in the construction Yankee notions.

TRIM (SEWING)

Trim or trimming in clothing and home decorating is applied ornament, such as gimp, passementerie, ribbon, ruffles, or, as a verb, to apply such ornament.

Before the industrial revolution, all trim was made and applied by hand, thus making heavily trimmed furnishings and garments expensive and high-status. Machine-woven trims and sewing machines put these dense trimmings within the reach of even modest dressmakers and home sewers, and an abundance of trimming is a characteristic of mid-Victorian fashion. As a predictable reaction, high fashion came to emphasize exquisiteness of cut and construction over denseness of trimming, and applied trim became a signifier of mass-produced clothing by the 1930s. The iconic braid and gold button trim of the Chanel suit are a notable survival of trim in high fashion.

In home decorating, the 1980s and 1990s saw a fashion for dense, elaborately layered trimmings on upholstered furniture and drapery.

Today, most trimmings are commercially manufactured. Scalamandré is known for elaborate trim for home furnishings, and Wrights is a leading manufacturer of trim for home sewing and crafts. Trims are used generally to enhance the beauty of the garments. It attracts buyers. Appropriate use of it creates more value of the product.

- Bias tape

- Braid
- Buttons
- Cord
- Embroidery by hand or machine
- Gimp
- Lace edgings or insertions
- Passementerie
- Piping
- Ribbon
- Rickrack
- Ruffles or frills
- Tassels

BIAS TAPE

Bias tape or bias binding is a narrow strip of fabric, cut on the bias (UK *cross-grain*). The strip's fibres, being at 45 degrees to the length of the strip, makes it stretchier as well as more fluid and more drapeable compared to a strip that is cut on the grain. Many strips can be pieced together into a long "tape." The tape's width varies from about 1/2" to about 3" depending on applications. Bias tape is used in making piping, binding seams, finishing raw edges, etc. It is often used on the edges of quilts, placemats, and bibs, around armhole and neckline edges instead of a facing, and as a simple strap or tie for casual bags or clothing. Commercially available bias tape is available as a simple bias tape, single-fold bias tape, and double-fold bias tape. Single-fold bias tape is bias tape with each raw edge folded in towards the center, wrong sides together, and pressed.Double-fold bias tape is single-fold bias tape which has been folded in half and pressed, with the single folds to the inside. (Another way to think of it is to fold a Single-fold bias tape in half along its center-line.)Devices are available commercially to aid the home sewer in making folded bias tape. The fabric strip is fed through the device, which folds the fabric. The folds are then pressed into place. The resulting folded tape will be 1/4 the width of the original fabric strip.

COLLAR STAYS

Collar stays (sometimes known as collar sticks, bones, knuckles, tabs, in the UK, collar stiffeners, and in Eastern Canada collar stiffs) are shirt accessories.

Collar stays are smooth, rigid strips of metal (such as brass, stainless steel, or sterling silver), horn, baleen, mother of pearl, or plastic, rounded at one end and pointed at the other, inserted into specially made pockets on the underside of a shirt collar to stabilize the collar's points. The stays ensure that the collar lies flat against the collarbone, looking crisp and remaining in the correct place.

Often shirts come with plastic stays which may eventually need to be replaced if they bend; metal replacements do not have this problem.

Collar stays can be found in haberdashers, fabric- and sewing-supply stores and men's clothing stores. They are manufactured in multiple lengths to fit varying collar designs, or may be designed with a means to adjust the length of the collar stay.

Collar stays are removed from shirts before dry cleaning or pressing, as they could damage the shirt in the process, and then replaced prior to wearing. Shirts that are press ironed with the collar stays are vulnerable to damage, as this results in a telltale impression of the collar stay in the fabric of the collar. Some shirts have stays which are sewn into the collar and are not removable.

ELASTOMER

An elastomer is a polymer with viscoelasticity (having both viscosity and elasticity) and very weak inter-molecular forces, generally having low Young's modulus and high failure strain compared with other materials. The term, which is derived from *elastic polymer*, is often used interchangeably with the term rubber, although the latter is preferred when referring to vulcanisates. Each of the monomers which link to form the polymer is usually made of carbon, hydrogen, oxygen and/or silicon. Elastomers are amorphous polymers existing above their glass transition temperature, so that considerable segmental motion is possible. At ambient temperatures, rubbers are thus relatively soft (E~3MPa) and deformable. Their primary uses are for seals, adhesives and molded flexible parts. Application areas for different types of rubber are manifold and cover segments as diverse as tires, shoe soles as well as dampening and insulating elements. The importance rubbers have can be judged from the fact that global revenues are forecast to rise to US$56 billion in 2020.

Elastomers are usually thermosets (requiring vulcanization) but may also be thermoplastic. The long polymer chains cross-link during curing, *i.e.*, vulcanizing. The molecular structure of elastomers can be imagined as a 'spaghetti and meatball' structure, with the meatballs signifying cross-links. The elasticity is derived from the ability of the long chains to reconfigure themselves to distribute an applied stress. The covalent cross-linkages ensure that the elastomer will return to its original configuration when the stress is removed. As a result of this extreme flexibility, elastomers can reversibly extend from 5-700 per cent, depending on the specific material. Without the cross-linkages or with short, uneasily reconfigured chains, the applied stress would result in a permanent deformation.

Temperature effects are also present in the demonstrated elasticity of a polymer. Elastomers that have cooled to a glassy or crystalline phase will have less mobile chains, and consequentially less elasticity, than those manipulated at temperatures higher than the glass transition temperature of the polymer.

It is also possible for a polymer to exhibit elasticity that is not due to covalent cross-links, but instead for thermodynamic reasons.

EXAMPLES OF ELASTOMERS

Unsaturated rubbers that can be cured by sulfur vulcanization:

- Natural polyisoprene: cis-1,4-polyisoprene natural rubber (NR) and trans-1,4-polyisoprene gutta-percha
- Synthetic polyisoprene (IR for Isoprene Rubber)
- Polybutadiene (BR for Butadiene Rubber)
- Chloropene rubber (CR), polychloroprene, Neo-prene, Baypren etc.
- Butyl rubber (copolymer of isobutylene and isoprene, IIR)
  - Halogenated butyl rubbers (chloro butyl rubber: CIIR; bromo butyl rubber: BIIR)
- Styrene-butadiene Rubber (copolymer of styrene and butadiene, SBR)
- Nitrile rubber (copolymer of butadiene and acrylonitrile, NBR), also called Buna N rubbers
  - Hydrogenated Nitrile Rubbers (HNBR) Therban and Zetpol

Saturated rubbers that cannot be cured by sulfur vulcanization:

- EPM (ethylene propylene rubber, a copolymer of ethylene and propylene) and EPDM rubber (ethylene propylene diene rubber, a terpolymer of ethylene, propylene and a diene-component)
- Epichlorohydrin rubber (ECO)
- Polyacrylic rubber (ACM, ABR)
- Silicone rubber (SI, Q, VMQ)
- Fluorosilicone Rubber (FVMQ)
- Fluoroelastomers (FKM, and FEPM) Viton, Tecnoflon, Fluorel, Aflas and Dai-El
- Perfluoroelastomers (FFKM) Tecnoflon PFR, Kalrez, Chemraz, Perlast
- Polyether block amides (PEBA)
- Chlorosulfonated polyethylene (CSM), (Hypalon)
- Ethylene-vinyl acetate (EVA)

"The definitions are not authentic as the Rubber which is classified in World Customs Organisation Books in Chapter 40, where as the above definitions stating all rubber and different polymers in same chapter which is classified in Chapter 39 of the World Custom Organisation's Harmonised Commodity for Description and coding system. One should go through all differentiation while editing between Plastics and articles thereof and Rubber and articles thereof."

Various other types of elastomers:

- Thermoplastic elastomers (TPE)
- The proteins resilin and elastin
- Polysulfide rubber
- Elastolefin, elastic fibre used in fabric production

GROMMET

A grommet is a ring or edge strip inserted into a hole through thin material, typically a sheet of textile fabric, sheet metal and or composite of carbon fibre, wood or honeycomb. Grommets are generally flared or collared on each side to keep them in place, and are often made of metal, plastic, or rubber. They may be used to prevent tearing or abrasion of the pierced material or protection from abrasion of the insulation on the wire, cable, line being routed through the penetration, and to cover sharp edges of the piercing, or all of the above.

A small grommet may also be called an eyelet, used for example on shoes, tarps and sails for lacing purposes. In electrical applications these are referred to as "insulating bushings". Most common are molded rubber that are inserted into small hole diameters up to 2" in diameter. There are many hole configurations from standard round to assorted U-shapes. Larger penetrations that are irregular in shape as well as long straight edges often use extruded or stamped strips of continuous length. These Continuous length materials are referred to as "grommet edging".

These are quite common in applications that range from telecom switches and data center cabinets to complex and dense wire/cable and even hydraulic tubing in aircraft, transportation vehicles and medical equipment.

GROMMETS AS REINFORCEMENT OR CRAFTING

Grommets are used to reinforce holes in leather, cloth, shoes, canvas and other fabrics. They can be made of metal, rubber, or plastic, and are easily used in common projects, requiring only the grommet itself and a means of setting it with a punch, a metal rod with a convex tip. A simple punch, often sold with the grommets, can be struck with a hammer to set the grommet. There are also dedicated grommet presses with punch and anvil, as shown in the picture, ranging from inexpensive to better-quality tools, which are somewhat faster to use. They are used to strengthen holes; in footwear for boot and shoe laces, in laced clothing such as corsets, and in curtains and other household items that require hanging from hooks, as when they are used in conjunction with tensioner rods for shower curtains. The grommet prevents the cord from tearing through the hole, thereby providing structural integrity. Small grommets are also called eyelets, especially when used in clothing or crafting. Eyelets may be used purely decoratively for crafting. When used in sailing and various other applications they are called cringles. Sometimes field workers may refer to them as grunyons.

GROMMETS USED IN ELECTRICAL EQUIPMENT

For cable protection

If metal or another hard material has a hole made in it, the hole will probably have sharp edges. Electrical wires, cord, rope, lacings, or other soft vulnerable material passing through the hole can become abraded or cut, or electrical insulation may break due to repeated flexing at the exit point. Rubber, plastic or plastic coated metal grommets are used to avoid this. The grommet could also protect the wiring/cabling from contamination from dirt, air, water, etc. The smooth and sometimes soft inner surface of the grommet shields the wire from damage.

Grommets are generally used whenever wires pass through punched/drilled sheet metal or plastic casings for this reason. Molded and continuous strip grommets, also known as edge grommets, are manufactured in a wide variety of sizes and lengths expressly for this purpose; they are usually a single piece which can be inserted by hand. Two-piece hard plastic devices are available which also grip the wire that passes through. These are called strain relief bushings and are often used to insulate, anchor, and protect power cords where they enter panels. Preventing a tug or twist on the wire from stressing the electrical connections inside the connected equipment. Sleeved grommets have a flexible extension (sleeve), usually tapered or moulded to flex increasingly towards the free end in order to reduce fracturing of electrical insulation.

To minimise vibration

Grommets made of rubber or other elastic material are also used to minimise the transmission of vibration. They were widely used for mounting shock-sensitive computer disk drives, particularly in equipment subject to vibration or jarring, but are not usually used with more robust modern drives. The screws that hold the drive in place pass through grommets that decouple it acoustically from the chassis. Grommets are used in a similar way to acoustically isolate electronic circuit components that are susceptible to microphonism caused by mechanical vibration or jarring.

Surgical grommets

In chronic cases of otitis media with effusions present for months, surgery is sometimes performed to insert a grommet, called a "tympanostomy tube" into the eardrum to allow air to pass through into the middle ear, and thus release any pressure buildup and help clear excess fluid within. This is also a correcting measure for a patulous Eustachian tube (when air moves to and from the middle ear with each breath making the eardrum flap).

INTERFACING

Interfacing is a textile used on the unseen or "wrong" side of fabrics to make an area of a garment more rigid.

Interfacings can be used to:

- Stiffen or add body to fabric, such as the interfacing used in shirt collars

- Strengthen a certain area of the fabric, for instance where buttonholes will be sewn
- Keep fabrics from stretching out of shape, particularly knit fabrics

Interfacings come in a variety of weights and stiffnesses to suit different purposes. They are also available in different colours, although typically interfacing is white. Generally, the heavier weight a fabric is, the heavier weight an interfacing it will use. Interfacing is sold at fabric stores by the yard or metre from bolts, similar to cutting fabric. Sewing patterns specify if interfacing is needed, the weight of interfacing that is required, and the amount. Some patterns use the same fabric as the garment to create an interfacing, as with sheer fabrics.

FUSIBLE INTERFACING

Most modern interfacings have heat-activated adhesive on one side. They are affixed to a garment piece using heat and moderate pressure, from a hand iron for example. This type of interfacing is known as "fusible" interfacing. Non-fusible interfacings do not have adhesive and must be sewn by hand or machine.

PASSEMENTERIE

Passementerie or passementarie is the art of making elaborate trimmings or edgings (in French, passements) of applied braid, gold or silver cord, embroidery, colored silk, or beads for clothing or furnishings.

Styles of passementerie include the tassel, fringes (applied, as opposed to integral), ornamental cords, galloons, pompons, rosettes, and gimps as other forms. Tassels, pompons, and rosettes are *point* ornaments, and the others are linear ornaments.

Passementerie worked in white linen thread is the origin of bobbin lace, and *passement* is an early French word for lace.

Today, passementerie is used with clothing, such as the gold braid on military dress uniforms, and for decorating couture clothing and wedding gowns. They are also used in furniture trimming, such as the Centripetal Spring Armchair of 1849 and some lampshades, draperies, fringes and tassels.

In the West, tassels were originally a series of windings of thread or string around a suspending string until the desired curvature was attained. Decades later, turned wooden moulds, which were either covered in simple wrappings or much more elaborate coverings called "satinings", were used. This involved an intricate binding of bands of filament silk vertically around the mould by means of an internal "lacing" in the bore of the mould. A tassel is primarily an ornament, and was at first the casual termination of a cord to prevent unraveling with a knot. As time went on, various peoples developed variations on this.

In the 16th century, the Guild of Passementiers was created in France. In France practitioners of the art were called "passementiers", and an apprenticeship of seven years was required to become a master in one of the subdivisions of the guild.

The Guild documented the art of passementerie. The tassel was its primary expression, but it also included fringes (applied, as opposed to integral), ornamental cords, galloons, pompons, rosettes, and gimps as other forms. Tassels, pompons, and rosettes are *point* ornaments; the others are linear ornaments. These constructions were varied and augmented with extensive ornamentations. These constructions were each assigned an idiosyncratic term by their French practitioners.

The French widely exported their very artistic work, and at such low prices that no other nation developed a mature "trimmings" industry. Tassels and their associated forms changed style throughout the years, from the small and casual of Renaissance designs, through the medium sizes and more staid designs of the Empire period, and to the Victorian Era with the largest and most elaborate.

Passementerie with clothing was for a long time reserved for the elites as a sign of social distinction among royalty, aristocracy, religious, and military. Since the 18th century, the use became obsolete with the simplification of clothing. Some of the historic designs are returning today from European and American artisans.

PIPING (SEWING)

In sewing, piping is a type of trim or embellishment consisting of a strip of folded fabric inserted into a seam to define the edges or style lines of a garment or other textile object. Usually the fabric strip is cut on the bias. It may be made from either self-fabric (the same fabric as the object to be ornamented) or contrasting fabric, or of leather. Today, piping is common on upholstery and decorative pillows, but it is also used on clothing. Piped pocket openings, garment edges, and seams are characteristic of Western wear.

RICKRACK

Rickrack is a "flat narrow braid woven in zigzag form, used as a trimming for clothing or curtains." Made of cotton or polyester rickrack is stitched or glued to the edges of an item. Its zig-zag configuration repeats every third of an inch (about one centimeter) and is sold in multiple colours and textures. Rickrack's popularity peaked in the 1970s and is associated with the *Little House on the Prairie* and the pioneer sentiment brought about by the 1976 American bicentennial.

SELF-FABRIC

Self-fabric is a term used in sewing. It refers to a fabric piece or embellishment made from the same fabric as the main fabric, as opposed to contrast fabric. Self-fabric is usually used in certain pattern pieces such as facings and linings to produce clean garment lines and make the fabric piece blend in with the rest of the garment. A special type of button is often covered in self-fabric to minimize its visibility. Self-fabric can also be used to make design details stand out. For example, a patch pocket on a coat could be made of contrasting fabric, but have an appliqué made of self-fabric on the pocket. A very common use of self-fabric as an embellishment is to make two garments that are to be worn together out of different fabrics and use self-fabric from one garment as a trim on the other (such as piping).

SOUTACHE

A soutache is narrow flat decorative braid, a type of galloon, used in the trimming of drapery or clothing. In clothing soutache is used to conceal a seam. Often woven of metallic bullion thread, silk, or a blend of silk and wool, soutache began to be made of rayon and other synthetic fibres in the 20th century. In military uniforms a particular width or colour of soutache is used to indicate rank, particularly in a hat. In athletic uniforms a contrasting soutache is used to outline numbers or players' names. The term is also used in bookbinding, where a narrow soutache is applied at the top and bottom of a book back to reinforce the spine and provide a barrier to keep dust out of the binding.

TWILL TAPE

Twill tape is a flat twill-woven ribbon of cotton, linen, polyester, or wool. Twill tape is available in various widths, generally up to 1 inch (2.5 cm), and a wide range of colours. Twill tape is used in sewing and tailoring to reinforce seams, make casings, bind edges, and make sturdy ties for closing garments (for example, on hospital gowns). Twill tape is also used in theatre to tie curtains, cable and scenery to various objects, or to tie cable coils so that they do not unroll. A form of twill tape is sometimes used to wrap the handlebars of road bicycles.

WRIGHTS (TEXTILE MANUFACTURERS)

Wrights is a brand of trim and other textiles for home sewing, owned since 2001 by the Simplicity Creative Group.

The Wrights brand was introduced in 1897 when William E. Wright and Sons was founded in Massachusetts.

Wright and Sons remained independent until 1985, when a group of shareholders—including a grandson of the founder—enabled the Newell Company to acquire a minority share in the company; within month that share grew until it acquire control over the company. In 1989 Boye Needle Company was merged into Wrights. In 2001 Conso International Corporation, a South Carolina manufacturer of trims to the wholesale trade, and owners of the Simplicity Pattern brand, bought the company.

5

Closures

BUTTONHOLE

Buttonholes are holes in fabric which allow buttons to pass through, securing one piece of the fabric to another. The raw edges of a buttonhole are usually finished with stitching. This may be done either by hand or by a sewing machine. Some forms of button, such as a frog, use a loop of cloth or rope instead of a buttonhole. Buttonholes can also refer to flowers worn in the lapel buttonhole of a coat or jacket, which are referred to simply as "buttonholes" or *boutonnières*.

Buttonholes for fastening or closing clothing with buttons appeared first in Germany in the 13th century. However it is believed that ancient Persians used it first. They soon became widespread with the rise of snug-fitting garments in 13th- and 14th-century Europe.

ASPECTS OF BUTTONHOLES

Buttonholes often have a bar of stitches at either side of them. This is a row of perpendicular hand or machine stitching to reinforce the raw edges of the fabric, and to prevent it from fraying.

Traditionally, men's clothing buttonholes are on the left side, and women's clothing buttonholes are on the right. The lore of this 'opposite' sides buttoning is that the practice came into being as 'women of means' had chamber maids who dressed them. So as not to confuse the poor chamber maids, the wealthy began having women's garments made with the buttons and holes 'switched'; the birth of the modern ladies' blouse. It is interesting to note that the chamber maids themselves, as did most all the common class, both male and female, actually wore 'shirts' with buttons and holes placed as on men's clothing. There appears to be no concrete reference to prove or disprove this story, but its plausibility bears noting.

There is also the theory that if a man is driving his ox cart or carriage or car, he can see inside her blouse and she can see inside of his. (Of course this assumes the driver is on the left hand side.)

TYPES OF BUTTONHOLES

Hand stitching

- A plain buttonhole, by far the most common type. In plain buttonholes, the raw (cut) edges of the textile are finished with thread in very closely spaced stitches (if made by hand, often the buttonhole stitch) with a gimp cord at the edges to act as a reinforcement. When stitched by hand, a slit is made in the fabric first and the result is called a worked buttonhole.

Machined stitching

Sewing machines offer various levels of automation to creating plain buttonholes. When made by machine, the slit between the sides of the buttonhole is opened after the stitching is completed.

- A machine-made buttonhole is usually sewn with two parallel rows of machine sewing in a narrow zig-zag stitch, with the ends finished in a broader zig-zag stitch. (One of the first automatic buttonhole machines was invented by Henry Alonzo House in 1862.)
- A bound buttonhole is one which has its raw edges encased by pieces of fabric or trim instead of stitches.
- A keyhole buttonhole is a special case of a thread-finished buttonhole that is normally machine-made due to the difficulty of achieving it by hand working. It is characterized by a round hole at the end of the slit. Because a button-closed gap in a garment is normally under some stress, the button will tend to move towards the end of the buttonhole closest to the gap in the garment. A keyhole at the end of the buttonhole closest to the gap will accommodate the button's shank without distorting the fabric.

Keyhole buttonholes are most often found on tailored coats and jackets.

FLY (CLOTHING)

A fly on clothing is a covering over an opening concealing the mechanism, such as a zip, velcro, or buttons, used to close the opening. The term is most frequently applied to a short opening over the groin in trousers, shorts, and other garments, which makes them easier to put on or take off, allows men and boys to urinate without lowering the garment, and prevents exhibitionism. The term is also used of overcoats, where a design of the same shape is used to hide a row of buttons. This style is common on a wide range of coats, from single-breasted Chesterfields to covert coats. An open fly is a fly that has been left unzipped or unbuttoned. Trousers have varied historically in whether or not they have flies. Originally, trousers did not have flies or other openings, being pulled down for sanitary functions. The use of a codpiece, a separate covering attached to the trousers, became popular in 16th-century Europe, eventually evolving into an attached fall-front (or broad fall). The fly-front (split fall) emerged later. The panelled front returned as a sporting option, such as in riding breeches, but is now hardly used, flies being by far the most common fastening. Most flies now use a zip, though button flies continue in use.

FROG (FASTENING)

A frog (sometimes referred to as a Chinese frog) is an ornamental braiding for fastening the front of a garment that consists of a button and a loop through which it passes. The usual purpose of frogs is to provide a closure for a garment while decorating it at the same time. These frogs are usually used on garments that appear oriental in design. Tops with a mandarin collar often use frogs at the shoulder and down the front to keep the two sections of the front closed. Frogs are usually meant to be a design detail that "stands out". Where many frogs are repeated beyond any reasonable needs for fastening, this purely decorative form is termed 'frogging'. Frogs and frogging became an important decorative feature on military uniforms from the 17th–19th centuries. This was particular evident for prestigious regiments, especially cavalry or hussars, and gave rise to the German term for frogging in general, 'Husarentressen'. These dolman jackets were tight-fitting and dominated by extensive frogging, often in luxurious materials such as metallic cording or brocades. The frogging was usually far more than was necessary simply for fastening. In some cases the frogging even became non-functional, with a concealed opening beneath it and the original jacket opening becoming a false detail. By the later 19th century, for lower grade uniforms down to postmen, telegraph boys and hotel pages, the frogging cordage would be retained as a decoration but there would be no corresponding toggle or opening with it. Many sewers make their own frogs because supplies are inexpensive and the results are customizable. Using larger or smaller size cording or fabric tubes will result in larger and smaller frogs. Also, self-fabric can be used to create frogs that are the same colour as the garment, though frogs are usually chosen to be a contrasting colour to that of the garment. Frogs are made by looping and interlocking the cording or fabric tube into the desired design, then securing the places where the cords touch by hand-sewing. The frog is then stitched onto a garment, usually by hand. When a fabric tube is used, the fabric is cut on bias. This allows the fabric tube to remain smooth and flex easily when bent into curves.

HOOK-AND-EYE CLOSURE

A hook-and-eye closure is a very simple and secure method of fastening garments together. It consists of a metal hook, commonly made of flattened wire bent to the required shape, and an eye (or "eyelet") of the same material into which the hook fits.

The hook and eye closure has a long history and is still used today, primarily on brassieres.

This form of fastening first appears under the name of "crochet and loop" in 14th century England.

The first reference to the modern term appears in *Aubrey's Brief Lives* in 1697, which describes a doublet and breeches being attached with "hook and eies". Hooks and eyes were made by hand from wire, until the city of Redditch, England, already famous for needle manufacture, was the first to machine-manufacture them. In 1643 a woman in the American colony of Maryland is recorded to have paid £10 worth of tobacco for hooks and eyes.

The hook and eye played an important role in women's corsetry; used in rows, they distribute the stress involved in restrictive garments.

It was not until the first part of the 19th century that the industry was furthered in the United States. One of the greatest improvements in the attachment was the "Delong hump", patented in 1889 by the Richardson and Delong Hook and Eye Company of Philadelphia, Pennsylvania which was a raised elevation or "hump" in the wire hook that prevented the eye from slipping out of the hook, "except at the will of the wearer". In 1893, Marie Tucek patented the "Breast Supporter" – the first garment similar to the modern-day bra, which used separate pockets for the breasts and straps that went over the shoulder and fastened by hook-and-eye closures to the center front of the garment. E.C. Beecher patented his hook-and-eye in June 1900 with the U.S. Patent Office; in 1902, an updated version was submitted that consisted of an attachable hook-and-eye, without any stitching required. A similar hook and eye for brassieres was patented in 1902 by the M.E. Company. The fasteners were eventually manufactured in the form of hook-and-eye tape, consisting of two tapes, one equipped with hooks and the other equipped with eyelets so that the two tapes could be "zipped" together side-by-side. To construct the garment, sections of hook-and-eye tape were sewn into either side of the garment closure. Today this labour-saving method comes on either silk or cotton tape, depending on the firmness and strength needed. In addition to their application on brassieres, bustiers, corsets and other fine lingerie, a single hook-and-eye closure is often sewn above the top of the zipper to "finish" it and take stress off the fastening on a skirt, dress or pants. They are generally provided at one gross to a box and range in size from No. 1 small, to No. 10 large.

HOOK AND LOOP FASTENER

Hook-and-loop fasteners, hook-and-pile fasteners, or touch fasteners (colloquially known as Velcro) consist of two components: typically, two lineal fabric strips (or, alternatively, round "dots" or squares) which are attached (*e.g.*, sewn, adhered, etc.) to the opposing surfaces to be fastened. The first component features tiny hooks; the second features even smaller and "hairier" loops. When the two components are pressed together, the hooks catch in the loops and the two pieces fasten or bind temporarily during the time that they are pressed together. When separated, by pulling or peeling the two surfaces apart, the strips make a distinctive "ripping" sound.

Touch fasteners are available in several strengths and constructions by several manufacturers.

The hook-and-loop fastener was conceived in 1941 by Swiss engineer, George de Mestral

The idea came to him one day after returning from a hunting trip with his dog in the Alps. He took a close look at the burrs (seeds) of burdock that kept sticking to his clothes and his dog's fur. He examined them under a microscope, and noted their hundreds of "hooks" that caught on anything with a loop, such as clothing, animal fur, or hair. He saw the possibility of binding two materials reversibly in a simple fashion if he could figure out how to duplicate the hooks and loops. Velcro is viewed by some like Steven Vogel or Werner Nachtigall as a key example of inspiration from nature or the copying of nature's mechanisms (called bionics or biomimesis).

Originally people refused to take de Mestral seriously when he took his idea to Lyon, which was then a center of weaving. He did manage to gain the help of one weaver, who made two cotton strips that worked. However, the cotton wore out quickly, so de Mestral turned to synthetic fibres. He settled on nylon as being the best synthetic, which had several advantages. Nylon doesn't break down, rot, or attract mold, and it could be produced in threads of various thickness. Nylon had only recently been invented, and through trial and error he eventually discovered that, when sewn under hot infrared light, nylon forms hooks that were perfect for the hook side of the fastener. Though he had figured out how to make the hooks, he had yet to figure out a way to mechanize the process, and to make the looped side. Next he found that nylon thread, when woven in loops and heat-treated, retains its shape and is resilient; however, the loops had to be cut in just the right spot so that they could be fastened and unfastened many times. On the verge of giving up, a new idea came to him. He bought a pair of shears and trimmed the tops off the loops, thus creating hooks that would match up perfectly with the loops in the pile.

Mechanizing the process of weaving the hooks took eight years, and it took another year to create the loom that trimmed the loops after weaving them. In all, it took ten years to create a mechanized process that worked. He submitted his idea for patent in Switzerland in 1951 and the patent was granted in 1955. Within a few years he obtained patents and began to open shops in Germany, Switzerland, Great Britain, Sweden, Italy, the Netherlands, Belgium, and Canada. In 1957 he branched out to the textile center of Manchester, New Hampshire in the United States. Columnist Sylvia Porter made the first mention of the product in her column *Your Money's Worth* of August 25, 1958, writing

"It is with understandable enthusiasm that I give you today an exclusive report on this news: A 'zipperless zipper' has been invented — finally. The new fastening device is in many ways potentially more revolutionary than was the zipper a quarter century ago." A Montreal firm, Velek, Ltd., acquired the exclusive right to market the product in North and South America, as well as in Japan, with American Velcro, Inc. of New Hampshire, and Velcro Sales of New York, marketing the "zipperless zipper" in the United States.

De Mestral obtained patents in many countries right after inventing Velcro as he expected a high demand immediately. Partly due to its cosmetic appearance, though, Velcro's integration into the textile industry took time. At the time, Velcro looked like it had been made from leftover bits of cheap fabric, and thus was not sewn into clothing or used widely when it debuted in the early 1960s. It was also viewed as impractical.

A number of Velcro products were displayed at a fashion show at the Waldorf-Astoria hotel in New York in 1959, and the fabric got its first break when it was used in the aerospace industry to help astronauts maneuver in and out of bulky space suits. However, this reinforced the view among the populace that Velcro was something with very limited utilitarian uses. The next major use Velcro saw was with skiers, who saw the similarities between their costume and the astronauts, and thus saw the advantages of a suit that was easier to don and remove. Scuba and marine gear followed soon after. Having seen astronauts storing food pouches on walls, children's clothing makers came on board. As Velcro only became widely used after NASA's adoption of it, NASA is popularly — and incorrectly — credited with its invention. By the mid-1960s Velcro was used in the futuristic creations of fashion designers such as Pierre Cardin, André Courrèges and Paco Rabanne.

Later improvements included strengthening the filament by adding polyester.

In 1978 de Mestral's patent expired, prompting a flood of low-cost imitations from Taiwan, China and South Korea onto the market. Today, the trademark is the subject of more than 300 trademark registrations in over 159 countries. George de Mestral was inducted into the national inventors hall of fame for his invention.

The big breakthrough George de Mestral made was to think about hook-and-eye closures on a greatly reduced scale. Hook and eye fasteners have been common for centuries, but what was new about Velcro was the miniaturisation of the hooks and eyes. Shrinking the hooks led to the two other important differences. Firstly, instead of a single-file line of hooks, Velcro has a two-dimensional surface. This was needed, because in decreasing the size of the hooks, the strength was also unavoidably lessened, thus requiring more hooks for the same strength. The other difference is that Velcro has indeterminate match-up between the hooks and eyes. With larger hook and eye fasteners, each hook has its own eye. On a scale as small as that of Velcro, matching up each of these hooks with the corresponding eye is impractical, thus leading to the indeterminate matching.

STRENGTH

Some touch fasteners are strong enough that a two inch square piece is enough to support a load of 175-pound (79 kg). Fasteners made of Teflon loops, polyester hooks, and glass backing are used in aerospace applications, *e.g.* on space shuttles. The strength of the bond depends on how well the hooks are embedded in the loops, how much surface area is in contact with the hooks, and the nature of the force pulling it apart. If Velcro is used to bond two rigid surfaces, *e.g.* auto body panels and frame, the bond is particularly strong because any force pulling the pieces apart is spread evenly across all hooks. Also, any force pushing the pieces together is disproportionally applied to engaging more hooks and loops. Vibration can cause rigid pieces to improve their bond. Full-body Velcro suits have been made that can hold a person to a suitably covered wall.

When one or both of the pieces is flexible, *e.g.* a pocket flap, the pieces can be pulled apart with a peeling action that applies the force to relatively few hooks at a time. If a flexible piece is pulled in a direction parallel to the plane of the surface, then the force is spread evenly as it is with rigid pieces.

Three ways to maximize the strength of a bond between the two flexible pieces are:

- Increase the area of the bond, *e.g.* using larger pieces
- Ensure that the force is applied parallel to the plane of the fastener surface, *e.g.* bending around a corner or pulley.
- Increase the number of hooks and loops per area unit.

Shoe closures can resist a large force with only a small amount of hook and loop fasteners. This is because the strap is wrapped through a slot, halving the force on the bond by acting as a pulley system (thus gaining a mechanical advantage), and further absorbing some of the force in friction around the tight bend. This layout also ensures that the force is parallel to the strips.

ADVANTAGES AND DISADVANTAGES

Touch fasteners are easy to use, safe, and maintenance free. There is only a minimal decline in effectiveness even after many fastening and unfastenings. The tearing noise it makes can also be useful against pickpockets.

There are also some deficiencies: it tends to accumulate hair, dust, and fur in its hooks after a few months of regular use. The loops can become elongated or broken after extended use. Velcro often becomes attached to articles of clothing, especially loosely woven items like sweaters. Additionally, the clothing may be damaged when one attempts to remove the Velcro, even if they are separated slowly. The tearing noise made by unfastening hook and loop fasteners make it inappropriate for some applications. For example, a soldier hiding from the enemy would not want to alert the enemy to his position by opening a Velcro pocket. It also absorbs moisture and perspiration when worn next to the skin, which means it will smell if not washed.

Textiles can contain chemicals or compounds, *e.g.* dyes, that may be allergenic to sensitive populations. Some products have been tested according to the Oeko-tex certification standard which imposes limits on the chemical content of textiles to address the issue of human ecological safety.

APPLICATIONS

Because its ease of use, hook-and-loop fasteners have been used for a wide variety of applications where a temporary bond is required. It is especially popular in clothing where it replaces buttons or zippers, and as a shoe fastener for children who have not yet learned to tie shoelaces and for those who choose Velcro over laces. Touch fasteners are used in adaptive clothing, which is clothing designed for people with physical disabilities, the elderly, and the infirm who may experience difficulty dressing themselves due to an inability to manipulate closures, such as buttons and zippers.

Touch fasteners held together a human heart during the first artificial heart surgery, and it is used in nuclear power plants and army tanks to hold flashlights to walls. Cars use it to bond headliners, floor mats and speaker covers. It is used in the home when pleating draperies, holding carpets in place and attaching upholstery, among many other things. It closes backpacks, briefcases and notebooks, secures pockets, and holds disposable diapers, and diaper covers for cloth diapers, on babies. It is an integral part of the game tag rugby, is used in surfboard leashes and orthopaedic braces.

NASA makes significant use of touch fasteners. Each space shuttle flew equipped with ten thousand inches of a special fastener made of Teflon loops, polyester hooks, and glass backing. Touch fasteners are widely used, from the astronauts' suits, to anchoring equipment. In the near weightless conditions in orbit, hook and loop fasteners is used to temporarily hold objects and keep them from floating away. A patch is used inside astronauts' helmets where it serves as a nose scratcher. During mealtimes astronauts use trays that attach to their thighs using spring and fasteners.

The US Army is another user. It uses hook and loop fasteners on combat uniforms to attach name tapes, rank insignia, shoulder pockets for unit patches, skill tabs, and recognition devices, such as the infrared (IR) feedback American flag. There are a silent version of touch fasteners sometimes called Quiet Closures.

VARIATIONS ON TOUCH FASTENERS

Slidingly Engaging Fastener was developed to address several problems with common hook and loop fasteners. Heavy duty variants (*e.g.*, "Dual Lock" or “Duotec”) feature mushroom shaped stems on each face of the fastener, providing an audible snap when the two faces mate. A strong pressure sensitive adhesive bonds each component to its substrate.

PLACKET

A placket is an opening in the upper part of trousers or skirts, or at the neck or sleeve of a garment. Plackets are almost always used to allow clothing to be put on or removed easily, but are sometimes used purely as a design element. Modern plackets often contain fabric facings or attached bands to surround and reinforce fasteners such as buttons, snaps, or zippers.

In modern usage, the term *placket* often refers to the double layers of fabric that hold the buttons and buttonholes in a shirt. Plackets can also be found at the neckline of a shirt, the cuff of a sleeve, or at the waist of a skirt or pair of trousers.

Plackets are almost always made of more than one layer of fabric, and often have interfacing in between the fabric layers. This is done to give support and strength to the placket fabric because the placket and the fasteners on it are often subjected to stress when the garment is worn. The two sides of the placket often overlap. This is done to protect the wearer from fasteners rubbing against their skin and to hide underlying clothing or undergarments.

A button front shirt without a separate pieced placket is called a “French placket.” The fabric is simply folded over and the buttonhole stitching secures the two layers (or three layers if there is an interlining). This method affords a very clean finish, especially if heavily patterned fabrics are being used. This method is normally only used in stiff-fronted formal evening (“white-tie”) shirts. However, the normal, separate placket on a shirt gives a more symmetrical appearance.

If the buttons are concealed by a separate flange or flap of the shirting fabric running the length of the placket, it is called a “fly front.” The inner placket of a fly front shirt can be made as a less constructed French placket or as a fully constructed regular placket.

Historically, a *placket* may also be:

- A decorative front-panel used to fill in the opening of a doublet or gown (later called a stomacher). Also spelled *placard.*
- A decorative panel or “forepart” (see 1500–1550 in Fashion) attached to a woman’s petticoat.
- An opening or slit in a skirt or petticoat to access a separate hanging pocket.
- A petticoat or skirt pocket.

SHANK (SEWING)

A shank is a device for providing a small amount of space in between a garment and a button. Shanks are necessary to provide space for fabric to sit in between the button and the garment when the garment is buttoned. Shanks also allow a garment to hang and drape nicely.

TYPES OF SHANKS

Button shank

Shank ons have a hollow protrusion on the back through which thread is sewn to attach the button. Button shanks may be a separate piece added to the back of a button, or be carved or moulded directly onto the back of the button, in which case the button is referred to by collectors as having a 'self-shank'. This is a common construction for older shell and glass buttons, for example. Buttons with shanks have no holes in the button blank (the main part of the button) itself because they are not needed for sewing. Buttons with shanks are more expensive to produce than shankless buttons.

Thread shank

A *thread shank* is made of thread and is intended to be used with a shankless button (a button with typically two or four holes). It is created while a button is stitched onto a garment.

Creating a thread shank

A *thread shank* is created by loosely stitching a shankless button onto fabric. This is usually done by keeping a toothpick or other small object in between the button and fabric while the button is stitched on. Once the button has been sewn through a few times, the toothpick is removed and the needle is moved down through one of the buttonholes, placing the needle and its thread in between the button and fabric. The sewer takes care to not tighten the thread too much.

While holding the button away from the garment, the thread is then firmly wrapped around the button (in between the button and fabric) a few times to form a sturdy wrapper for the other threads. The needle is then pushed through the fabric to the underside of the garment, where it can then be securely fastened off. A thread shank's length depends on the thickness of the fabric that will be buttoned. The ideal thread shank is long enough to button the fabric and still have the garment draping nicely, but short enough that the button does not flop around when buttoned.

SPECIALTY SHANKS

- Regular thread is used on lightweight garments such as shirts which are subject to little stress, but a specialty thread called buttonhole twist is used on items that experience more wear and tear, such as coats and pants. If a sewer uses regular thread, they usually use a double thread to strengthen the quality of the stitching.

- Long coats that are good-quality or better often have a special shank at the bottom button. Because the bottom button of a coat is subject to a lot of stress, shanks are typically longer than usual and are sometimes made with elastic-thread. In addition to the special shank, an additional (shankless) button with no thread shank is often used on the underside of the fabric when sewing the button on. As a result, the stress to the button, shank, and thread pulls on this second button instead of the fabric. This prevents the thread from damaging and ripping the fabric, and is one of two ways to prevent "popping a button" (the other method is to use good strong thread such as buttonhole twist).
- In addition to the typical shank and shankless buttons, *covered buttons* are shank buttons that have fabric completely covering the back. They are stitched on like regular shank buttons, but the sewer has to "feel" for the shank while they are sewing the button on. Covered buttons are handmade (to some degree), expensive, and are used in high-end and specialty garments.

SNAP FASTENER

A snap fastener (also called press stud, popper, snap or tich) is a pair of interlocking discs, made out of a metal or plastic, commonly used in place of buttons to fasten clothing and for similar purposes. A circular lip under one disc fits into a groove on the top of the other, holding them fast until a certain amount of force is applied. Different types of snaps can be attached to fabric or leather by riveting with a punch and die set specific to the type of rivet snaps used (striking the punch with a hammer to splay the tail), sewing, or plying with special snap pliers.

Snap fasteners are a noted detail in American Western wear and are also often chosen for children's clothing, as they are relatively easy for children to use.

Modern snap fasteners were first patented by German inventor Heribert Bauer in 1885 as the "Federknopf-Verschluss", a novelty fastener for men's trousers. Some attribute the invention to Bertel Sanders, of Denmark. These first versions had an S-shaped spring in the "male" disc instead of a groove. Australian inventor Myra Juliet Farrell is also credited with inventing a 'stitchless press stud' and the 'stitchless hook and eye.' In America, Jack Weil (1901–2008) put snaps on his iconic Western shirts, which spread the fashion for them.In the famous Chinese Terracotta Army, dating from 210 BC, the horse halters of wagons, made of a gold tube and a silver tube, were joined with a form of snap fasteners.

Press studs were worn by rodeo cowboys from the 1930s onwards, because these could be quickly removed if, in the event of a fall, the shirt became snagged in the saddle. Pearl snaps entered American mainstream Western fashion during the 1950s, when singing cowboys like Gene Autry and Roy Rogers incorporated them into their embroidered and fringed stage shirts. The most desirable shirts were unique creations tailored by Nudie Cohn or Rodeo Ben, but commercially produced Western clothing could be purchased from companies like Wrangler, Levi Strauss, Panhandle Slim, Rockmount Ranch Wear, H Bar C, or Roper.

Due to the popularity of Spaghetti Westerns, cowboy shirts with oversized collars were widely worn by teenagers and young adults from the mid 1960s until the early 1980s. By the 1990s, however, press studs had become associated with the adaptable clothing worn by pensioners and the disabled. During the late 2000s and 2010s, however, shirts with Western detailing made a comeback in Europe and the southern US due to the popularity of indie rock and a resurgence of interest in vintage Americana.

VELCRO

Velcro is a fabric hook and loop fastener, invented in 1948 by the Swiss electrical engineer George de Mestral. De Mestral patented Velcro in 1955, subsequently refining and developing its practical manufacture until its commercial introduction in the late 1950s. The name "Velcro" is also that of the company that first made and commercially marketed this type of fastener, and continues to do so.

The word *Velcro* is a portmanteau of the two French words *velours* ("velvet"), and *crochet* ("hook"). Hook-and-loop fasteners consist of two components: typically, two lineal fabric strips (or, alternatively, round "dots" or squares) which are attached (*e.g.*, sewn, adhered, etc.) to the opposing surfaces to be fastened. The first component features tiny hooks; the second features even smaller and "hairier" loops. When the two components are pressed together, the hooks catch in the loops and the two pieces fasten or bind temporarily during the time that they are pressed together. When separated, by pulling or peeling the two surfaces apart, the velcro strips make a distinctive "ripping" sound. The first Velcro sample was made of cotton, which proved impractical and was replaced by nylon and polyester. Velcro fasteners made of Teflon loops, polyester hooks, and glass backing are used in aerospace applications, *e.g.* on space shuttles. The term *Velcro* is commonly used to mean any type of hook-and-loop fastener, but remains a registered trademark in many countries used by the Velcro company to distinguish their brand of fasteners from their competitors. The Velcro company headquarters is in Amsterdam, the Netherlands.

On August 26, 2013, YKK Corporation, a subsidiary of YKK Group, filed a patent infringement complaint in the Middle District of Georgia against Velcro USA Inc. relating to a fastener strip used in foam molded products like a cushion body used for an automobile seat. In the fictional universe of *Star Trek*, Velcro is, albeit indirectly, invented by the Vulcans. In the *Star Trek: Enterprise* episode "Carbon Creek", a sample of Velcro is taken from a crashed Vulcan starship and given to a patent clerk to raise money for a teenage boy's college career. Also, one of the Vulcan crewmembers in that episode is named Mestral. In 1988 Velcro was mentioned in the comic-strip *Peanuts*. In the March 21 cartoon Sally brings a Praying Doll into school for "show and tell". The dolls hands are held together by Velcro, which leads to the question if Velcro is mentioned in the New Testament. In 1992 Velcro was mentioned in the sitcom *Seinfeld*, episode "The Wallet", where Jerry's father, Morty Seinfeld, states that he hates Velcro, due to the distinctive sound it makes when the two sides are being separated: "The Velcro! I can't stand Velcro! That tearing sound".

6

Materials

TEXTILE

A textile is a flexible material consisting of a network of natural or artificial fibres often referred to as thread or yarn. Yarn is produced by spinning raw wool fibres, linen, cotton, or other material on a spinning wheel to produce long strands. Textiles are formed by weaving, knitting, crocheting, knotting, or pressing fibres together (felt).

The words fabric and cloth are used in textile assembly trades (such as tailoring and dressmaking) as synonyms for *textile*. However, there are subtle differences in these terms in specialized usage. *Textile* refers to any material made of interlacing fibres. *Fabric* refers to any material made through weaving, knitting, crocheting, or bonding. *Cloth* refers to a finished piece of fabric that can be used for a purpose such as covering a bed.

USES

Textiles have an assortment of uses, the most common of which are for clothing and containers such as bags and baskets. In the household, they are used in carpeting, upholstered furnishings, window shades, towels, covering for tables, beds, and other flat surfaces, and in art. In the workplace, they are used in industrial and scientific processes such as filtering. Miscellaneous uses include flags, backpacks, tents, nets, cleaning devices such as handkerchiefs and rags, transportation devices such as balloons, kites, sails, and parachutes, in addition to strengthening in composite materials such as fibreglass and industrial geotextiles. Children can learn using textiles to make collages, sew, quilt, and make toys.

Textiles used for industrial purposes, and chosen for characteristics other than their appearance, are commonly referred to as *technical textiles.* Technical textiles include textile structures for automotive applications, medical textiles (*e.g.* implants), geotextiles (reinforcement of embankments), agrotextiles (textiles for crop protection), protective clothing (*e.g.* against heat and radiation for fire fighter clothing, against molten metals for welders, stab protection, and bullet proof vests). In all these applications stringent performance requirements must be met. Woven of threads coated with zinc oxide nanowires, laboratory fabric has been shown capable of "self-powering nanosystems" using vibrations created by everyday actions like wind or body movements.

FASHION AND TEXTILE DESIGNERS

Fashion designers commonly rely on textile designs to set their fashion collections apart from others. Armani, Marisol Deluna, Nicole Miller, Lilly Pulitzer, the late Gianni Versace and Emilio Pucci can be easily recognized by their signature print driven designs.

SOURCES AND TYPES

Textiles can be made from many materials. These materials come from four main sources: animal (Wool, Silk), plant (Cotton, Flax, Jute), mineral (Asbestose), and synthetic (Nylon, Polyester, Acrylic). In the past, all textiles were made from natural fibres, including plant, animal, and mineral sources. In the 20th century, these were supplemented by artificial fibres made from petroleum.

Textiles are made in various strengths and degrees of durability, from the finest gossamer to the sturdiest canvas. The relative thickness of fibres in cloth is measured in deniers. Microfibre refers to fibres made of strands thinner than one denier.

Animal textiles

Animal textiles are commonly made from hair or fur.

Wool refers to the hair of the domestic goat or sheep, which is distinguished from other types of animal hair in that the individual strands are coated with scales and tightly crimped, and the wool as a whole is coated with a wax mixture known as lanolin (aka wool grease), which is waterproof and dirtproof. Woollen refers to a bulkier yarn produced from carded, non-parallel fibre, while worsted refers to a finer yarn which is spun from longer fibres which have been combed to be parallel. Wool is commonly used for warm clothing. Cashmere, the hair of the Indian cashmere goat, and mohair, the hair of the North African angora goat, are types of wool known for their softness.

Other animal textiles which are made from hair or fur are alpaca wool, vicuña wool, llama wool, and camel hair, generally used in the production of coats, jackets, ponchos, blankets, and other warm coverings. Angora refers to the long, thick, soft hair of the angora rabbit.

Wadmal is a coarse cloth made of wool, produced in Scandinavia, mostly 1000~1500CE.

Silk is an animal textile made from the fibres of the cocoon of the Chinese silkworm. This is spun into a smooth, shiny fabric prized for its sleek texture.

Plant textiles

Grass, rush, hemp, and sisal are all used in making rope. In the first two, the entire plant is used for this purpose, while in the last two, only fibres from the plant are utilized. Coir (coconut fibre) is used in making twine, and also in floormats, doormats, brushes, mattresses, floor tiles, and sacking.

Straw and bamboo are both used to make hats. Straw, a dried form of grass, is also used for stuffing, as is kapok.

Fibres from pulpwood trees, cotton, rice, hemp, and nettle are used in making paper.

Cotton, flax, jute, hemp, modal and even bamboo fibre are all used in clothing. Piña (pineapple fibre) and ramie are also fibres used in clothing, generally with a blend of other fibres such as cotton.

Acetate is used to increase the shininess of certain fabrics such as silks, velvets, and taffetas.

Seaweed is used in the production of textiles. A water-soluble fibre known as alginate is produced and is used as a holding fibre; when the cloth is finished, the alginate is dissolved, leaving an open area

Lyocell is a man-made fabric derived from wood pulp. It is often described as a man-made silk equivalent and is a tough fabric which is often blended with other fabrics - cotton for example.

Mineral textiles

Asbestos and basalt fibre are used for vinyl tiles, sheeting, and adhesives, “transite” panels and siding, acoustical ceilings, stage curtains, and fire blankets.

Glass Fibre is used in the production of spacesuits, ironing board and mattress covers, ropes and cables, reinforcement fibre for composite materials, insect netting, flame-retardant and protective fabric, soundproof, fireproof, and insulating fibres.

Metal fibre, metal foil, and metal wire have a variety of uses, including the production of cloth-of-gold and jewelry. Hardware cloth is a coarse weave of steel wire, used in construction.

Synthetic textiles

All synthetic textiles are used primarily in the production of clothing.

Polyester fibre is used in all types of clothing, either alone or blended with fibres such as cotton.

Aramid fibre (*e.g.* Twaron) is used for flame-retardant clothing, cut-protection, and armor.

Acrylic is a fibre used to imitate wools, including cashmere, and is often used in replacement of them.

Nylon is a fibre used to imitate silk; it is used in the production of pantyhose. Thicker nylon fibres are used in rope and outdoor clothing.

Spandex (trade name *Lycra*) is a polyurethane fibre that stretches easily and can be made tight-fitting without impeding movement. It is used to make activewear, bras, and swimsuits.

Olefin fibre is a fibre used in activewear, linings, and warm clothing. Olefins are hydrophobic, allowing them to dry quickly. A sintered felt of olefin fibres is sold under the trade name Tyvek. Ingeo is a polylactide fibre blended with other fibres such as cotton and used in clothing. It is more hydrophilic than most other synthetics, allowing it to wick away perspiration.

Lurex is a metallic fibre used in clothing embellishment.

Milk proteins can also be used to create synthetic fabric. Milk or casein fibre cloth was developed during World War I in Germany, and further developed in Italy and America during the 1930s. Milk fibre fabric is not very durable and wrinkles easily, but has a pH similar to human skin and possesses anti-bacterial properties. It is marketed as a biodegradable, renewable synthetic fibre.

PRODUCTION METHODS

Weaving is a textile production method which involves interlacing a set of longer threads (called the warp) with a set of crossing threads (called the weft). This is done on a frame or machine known as a loom, of which there are a number of types. Some weaving is still done by hand, but the vast majority is mechanised. Knitting and crocheting involve interlacing loops of yarn, which are formed either on a knitting needle or on a crochet hook, together in a line. The two processes are different in that knitting has several active loops at one time, on the knitting needle waiting to interlock with another loop, while crocheting never has more than one active loop on the needle.

Braiding or plaiting involves twisting threads together into cloth. Knotting involves tying threads together and is used in making macrame.

Lace is made by interlocking threads together independently, using a backing and any of the methods described above, to create a fine fabric with open holes in the work. Lace can be made by either hand or machine.

Carpets, rugs, velvet, velour, and velveteen, are made by interlacing a secondary yarn through woven cloth, creating a tufted layer known as a nap or pile.

Felting involves pressing a mat of fibres together, and working them together until they become tangled. A liquid, such as soapy water, is usually added to lubricate the fibres, and to open up the microscopic scales on strands of wool.

Non-woven textiles are manufactured by the bonding of fibres to make fabric. Bonding may be thermal, mechnical or adhessives can be used.

TREATMENTS

Textiles are often dyed, with fabrics available in almost every colour. The dying process often requires several dozen gallons of water for each pound of clothing. Coloured designs in textiles can be created by weaving together fibres of different colours (tartan or Uzbek Ikat), adding coloured stitches to finished fabric (embroidery), creating patterns by resist dyeing methods, tying off areas of cloth and dyeing the rest (tie-dyeing), or drawing wax designs on cloth and dyeing in between them (batik), or using various printing processes on finished fabric. Woodblock printing, still used in India and elsewhere today, is the oldest of these dating back to at least 220CE in China. Textiles are also sometimes bleached, making the textile pale or white.

Textiles are sometimes finished by chemical processes to change their characteristics. In the 19th century and early 20th century starching was commonly used to make clothing more resistant to stains and wrinkles. Since the 1990s, with advances in technologies such as permanent press process, finishing agents have been used to strengthen fabrics and make them wrinkle free. More recently, nanomaterials research has led to additional advancements, with companies such as Nano-Tex and NanoHorizons developing permanent treatments based on metallic nanoparticles for making textiles more resistant to things such as water, stains, wrinkles, and pathogens such as bacteria and fungi.

More so today than ever before, textiles receive a range of treatments before they reach the end-user. From formaldehyde finishes (to improve crease-resistance) to biocidic finishes and from flame retardants to dyeing of many types of fabric, the possibilities are almost endless. However, many of these finishes may also have detrimental effects on the end user. A number of disperse, acid and reactive dyes (for example) have been shown to be allergenic to sensitive individuals . Further to this, specific dyes within this group have also been shown to induce purpuric contact dermatitis . Although formaldehyde levels in clothing are unlikely to be at levels high enough to cause an allergic reaction , due to the presence of such a chemical, quality control and testing are of utmost importance. Flame retardants (mainly in the brominated form) are also of concern where the environment, and their potential toxicity, are concerned . Testing for these additives is possible at a number of commercial laboratories, it is also possible to have textiles tested for according to the Oeko-tex Certification Standard which contains limits levels for the use of certain chemicals in textiles products.

HISTORY OF CLOTHING AND TEXTILES

Clothing and textiles have been enormously important throughout human history—so have their materials, production tools and techniques, cultural influences, and social significance .

Textiles, defined as felt or spun fibres made into yarn and subsequently netted, looped, knit or woven to make fabrics, appeared in the Middle East during the late stone age. From ancient times to the present day, methods of textile production have continually evolved, and the choices of textiles available have influenced how people carried their possessions, clothed themselves, and decorated their surroundings.

Sources available for the study of the history of clothing and textiles include material remains discovered via archaeology; representation of textiles and their manufacture in art; and documents concerning the manufacture, acquisition, use, and trade of fabrics, tools, and finished garments. Scholarship of textile history, especially its earlier stages, is part of material culture studies.

Prehistoric development

Interest in prehistoric developments of textile and clothing manufacture has resulted in a number of scholarly studies since the late twentieth century, including *Prehistoric Textiles: The Development of Cloth in the Neo-lithic and Bronze Ages with Special Reference to the Aegean*, as well as *Women's Work: The First 20,000 Years: Women, Cloth, and Society in Early Times*. These sources have helped to provide a coherent history of these prehistoric developments. Evidence suggests that human beings may have begun wearing clothing as far back as 100,000 to 500,000 years ago.

Genetic analysis suggests that the human body louse, which lives in clothing, may have diverged from the head louse some 107,000 years ago, evidence that humans began wearing clothing at around this time.

Possible sewing needles have been dated to around 40,000 years ago. The earliest definite examples of needles originate from the Solutrean culture, which existed in France from 19,000 BC to 15,000 BC. The earliest dyed flax fibres have been found in a cave the Republic of Georgia and date back to 36,000 BP.

The earliest evidence of weaving comes from impressions of textiles and basketry and nets on little pieces of hard clay, dating from 27,000 years ago and found in Dolni Vestonice in the Czech Republic.

At a slightly later date (25,000 years) the Venus figurines were depicted with clothing. Those from western Europe were adorned with basket hats or caps, belts worn at the waist, and a strap of cloth that wrapped around the body right above the breast. Eastern European figurines wore belts, hung low on the hips and sometimes string skirts.

Archaeologists have discovered artifacts from the same period that appear to have been used in the textile arts: net gauges, spindle needles and weaving sticks.

Ancient textiles and clothing

The first actual textile, as opposed to skins sewn together, was probably felt. Surviving examples of Nålebinding, another early textile method, date from 6500 BC. Our knowledge of ancient textiles and clothing has expanded in the recent past thanks to modern technological developments. Our knowledge of cultures varies greatly with the climatic conditions to which archeological deposits are exposed; the Middle East and the arid fringes of China have provided many very early samples in good condition, but the early development of textiles in the Indian subcontinent, sub-Saharan Africa and other moist parts of the world remains unclear. In northern Eurasia peat bogs can also preserve textiles very well. Early woven clothing was often made of full loom widths draped, tied, or pinned in place.

Ancient Near East

The earliest known woven textiles of the Near East may be fabrics used to wrap the dead excavated at a Neo-lithic site at Çatalhöyük in Anatolia, carbonized in a fire and radiocarbon dated to c. 6000 BC. Flax cultivation is evidenced from c. 8000 BC in the Near East, but the breeding of sheep with a wooly fleece rather than hair occurs much later, c. 3000 BC

But Çayönü in Turkey has also been claimed as the site of the oldest known cloth, a piece of woven linen wrapped around an antler and reported to be from around 7000 BC.

Ancient India

The inhabitants of the Indus Valley Civilization used cotton for clothing as early as the 5th millennium BC – 4th millennium BC.

According to The Columbia Encyclopaedia, Sixth Edition:

"Cotton has been spun, woven, and dyed since prehistoric times. It clothed the people of ancient India, Egypt, and China. Hundreds of years before the Christian era cotton textiles were woven in India with matchless skill, and their use spread to the Mediterranean countries. In the 1st cent. Arab traders brought fine Muslin and Calico to Italy and Spain. The Moors introduced the cultivation of cotton into Spain in the 9th cent. Fustians and dimities were woven there and in the 14th cent. in Venice and Milan, at first with a linen warp. Little cotton cloth was imported to England before the 15th cent., although small amounts were obtained chiefly for candlewicks. By the 17th cent. the East India Company was bringing rare fabrics from India. Native Americans skillfully spun and wove cotton into fine garments and dyed tapestries. Cotton fabrics found in Peruvian tombs are said to belong to a pre-Inca culture. In colour and texture the ancient Peruvian and Mexican textiles resemble those found in Egyptian tombs."

Ancient Egypt

Evidence exists for production of linen cloth in Ancient Egypt in the Neo-lithic period, c. 5500 BC. Cultivation of domesticated wild flax, probably an import from the Levant, is documented as early as c. 6000 BC Other bast fibres including rush, reed, palm, and papyrus were used alone or with linen to make rope and other textiles. Evidence for wool production in Egypt is scanty at this period..

Spinning techniques included the drop spindle, hand-to-hand spinning, and rolling on the thigh; yarn was also spliced.. A horizontal ground loom was used prior to the New Kingdom, when a vertical two-beam loom was introduced, probably from Asia.

Linen bandages were used in the burial custom of mummification, and art depicts Egyptian men wearing linen kilts and women in narrow dresses with various forms of shirts and jackets, often of sheer pleated fabric.

Ancient China

The earliest evidence of silk production in China was found at the sites of Yangshao culture in Xia, Shanxi, where a cocoon of bombyx mori, the domesticated silkworm, cut in half by a sharp knife is dated to between 5000 and 3000 BC. Fragments of primitive looms are also seen from the sites of Hemudu culture in Yuyao, Zhejiang, dated to about 4000 BC. Scraps of silk were found in a Liangzhu culture site at Qianshanyang in Huzhou, Zhejiang, dating back to 2700 BC. Other fragments have been recovered from royal tombs in the Shang Dynasty (ca. 1600 BC – c. 1046 BC).

Under the Shang Dynasty, Han Chinese clothing or Hanfu consisted of a *yi*, a narrow-cuffed, knee-length tunic tied with a sash, and a narrow, ankle-length skirt, called *shang*, worn with a *bixi*, a length of fabric that reached the knees. Clothing of the elite was made of silk in vivid primary colours.

Ancient Japan

The earliest evidence of weaving in Japan is associated with the Yayoi period (per cent_uBfN, *Yayoi-jidai*) , from about 300 BC to 250.

The textile trade in the ancient world

The exchange of luxury textiles was predominant on the Silk Road, a series of ancient trade and cultural transmission routes that were central to cultural interaction through regions of the Asian continent connecting East and West by linking traders, merchants, pilgrims, monks, soldiers, nomads and urban dwellers from China to the Mediterranean Sea during various periods of time. The trade route was initiated around 114 BC by the Han Dynasty, although earlier trade across the continents had already existed. Geographically, the Silk Road or Silk Route is an interconnected series of ancient trade routes between Chang'an (today's Xi'an) in China, with Asia Minor and the Mediterranean extending over 8,000 km (5,000 miles) on land and sea. Trade on the Silk Road was a significant factor in the development of the great civilizations of China, Egypt, Mesopotamia, Persia, the Indian subcontinent, and Rome, and helped to lay the foundations for the modern world.

Classical antiquity

Dress in classical antiquity favoured wide, unsewn lengths of fabric, pinned and draped to the body in various ways.

Ancient Greek clothing consisted of lengths of wool or linen, generally rectangular and secured at the shoulders with ornamented pins called fibulae and belted with a sash. Typical garments were the peplos, a loose robe worn by women; the chlamys, a cloak worn by men; and the chiton, a tunic worn by both men and women. Men's chitons hung to the knees, whereas women's chitons fell to their ankles. A long cloak called a himation was worn over the peplos or chlamys.

The toga of ancient Rome was also an unsewn length of wool cloth, worn by male citizens draped around the body in various fashions, over a simple tunic. Early tunics were two simple rectangles joined at the shoulders and sides; later tunics had sewn sleeves. Women wore the draped stola or an ankle-length tunic, with a shawl-like palla as an outer garment. Wool was the preferred fabic, although linen, hemp, and small amounts of expensive imported silk and cotton were also worn.

Medieval clothing and textiles

The history of Medieval European clothing and textiles has inspired a good deal of scholarly interest in the twenty-first century. Elisabeth Crowfoot, Frances Pritchard, and Kay Staniland authored *Textiles and Clothing: Medieval Finds from Excavations in London, c.1150-c.1450* (Boydell Press, 2001). The topic is also the subject of an annual series *Medieval Clothing and Textiles* (Boydell Press) edited by Robin Netherton and Professor Gale R. Owen-Crocker of Anglo-Saxon Culture at the University of Manchester.

Byzantium

The Byzantines made and exported very richly patterned cloth, woven and embroidered for the upper classes, and resist-dyed and printed for the lower. By Justinian's time the Roman toga had been replaced by the tunica, or long *chiton*, for both sexes, over which the upper classes wore various other garments, like a *dalmatica* (dalmatic), a heavier and shorter type of tunica; short and long cloaks were fastened on the right shoulder.

Leggings and hose were often worn, but are not prominent in depictions of the wealthy; they were associated with barbarians, whether European or Persian.

Early medieval Europe

European dress changed gradually in the years 400 to 1100. People in many countries dressed differently depending on whether they identified with the old Romanised population, or the new invading populations such as Franks, Anglo-Saxons, and Visigoths. Men of the invading peoples generally wore short tunics, with belts, and visible trousers, hose or leggings. The Romanised populations, and the Church, remained faithful to the longer tunics of Roman formal costume.

The elite imported silk cloth from the Byzantine, and later Moslem, worlds, and also probably cotton. They also could afford bleached linen and dyed and simply patterned wool woven in Europe itself. But embroidered decoration was probably very widespread, though not usually detectable in art. Lower classes wore local or homespun wool, often undyed, trimmed with bands of decoration, variously embroidery, tablet-woven bands, or colorful borders woven into the fabric in the loom..

High middle ages and the rise of fashion

Clothing in 12th and 13th century Europe remained very simple for both men and women, and quite uniform across the subcontinent. The traditional combination of short tunic with hose for working-class men and long tunic with overgown for women and upper class men remained the norm. Most clothing, especially outside the wealthier classes, remained little changed from three or four centuries earlier. The 13th century saw great progress in the dyeing and working of wool, which was by far the most important material for outer wear. Linen was increasingly used for clothing that was directly in contact with the skin. Unlike wool, linen could be laundered and bleached in the sun. Cotton, imported raw from Egypt and elsewhere, was used for padding and quilting, and cloths such as buckram and fustian.

Crusaders returning from the Levant brought knowledge of its fine textiles, including light silks, to Western Europe. In Northern Europe, silk was an imported and very expensive luxury. The well-off could afford woven brocades from Italy or even further afield. Fashionable Italian silks of this period featured repeating patterns of roundels and animals, deriving from Ottoman silk-weaving centres in Bursa, and ultimately from Yuan Dynasty China via the Silk Road.

Cultural and costume historians agree that the mid-14th century marks the emergence of recognizable "fashion" in Europe. From this century onwards Western fashion changes at a pace quite unknown to other civilizations, whether ancient or contemporary. In most other cultures only major political changes, such as the Muslim conquest of India, produced radical changes in clothing, and in China, Japan, and the Ottoman Empire fashion changed only slightly over periods of several centuries. In this period the draped garments and straight seams of previous centuries were replaced by curved seams and the beginnings of tailoring, which allowed clothing to more closely fit the human form, as did the use of lacing and buttons. A fashion for *mi-parti* or *parti-coloured* garments made of two contrasting fabrics, one on each side, arose for men in mid-century, and was especially popular at the English court. Sometimes just the hose would be different colours on each leg.

Renaissance and early modern period

Renaissance Europe

Wool remained the most popular fabric for all classes, followed by linen and hemp. Wool fabrics were available in a wide range of qualities, from rough undyed cloth to fine, dense broadcloth with a velvety nap; high-value broadcloth was a backbone of the English economy and was exported throughout Europe. Wool fabrics were dyed in rich colours, notably reds, greens, golds, and blues.

Silk-weaving was well-established around the Mediterranean by the beginning of the 15th century, and figured silks, often silk velvets with silver-gilt wefts, are increasingly seen in Italian dress and in the dress of the wealthy throughout Europe. Stately floral designs featuring a pomegranate or artichoke motif had reached Europe from China in the previous century and became a dominant design in the Ottoman silk-producing cities of Istanbul and Bursa, and spread to silk weavers in Florence, Genoa, Venice, Valencia and Seville in this period.

As prosperity grew in the 15th century, the urban middle classes, including skilled workers, began to wear more complex clothes that followed, at a distance, the fashions set by the elites. National variations in clothing increased over the century.

Early Modern Europe

By the first half of the 16th century, the clothing of the Low Countries, German states, and Scandinavia had developed in a different direction than that of England, France, and Italy, although all absorbed the sobering and formal influence of Spanish dress after the mid-1520s.. Elaborate slashing was popular, especially in Germany. Black was increasingly worn for the most formal occasions. Bobbin lace arose from passementerie in the mid-16th century, probably in Flanders. This century also saw the rise of the ruff, which grew from a mere ruffle at the neckline of the shirt or chemise to immense cartwheel shapes. At their most extravagant, ruffs required wire supports and were made of fine Italian reticella, a cutwork linen lace.

By the turn of the 17th century, a sharp distinction could be seen between the sober fashions favoured by Protestants in England and the Netherlands, which still showed heavy Spanish influence, and the light, revealing fashions of the French and Italian courts. The great flowering of needlelace occurred in this period. Geometric reticella deriving from cutwork was elaborated into true needlelace or *punto in aria* (called in England "point lace"), which reflected the scrolling floral designs popular for embroidery. Lacemaking centers were established in France to reduce the outflow of cash to Italy.

According to Dr. Wolf D. Fuhrig, "By the second half of the 17th century, Silesia had become an important economic pillar of the Habsburg monarchy, largely on the strength of its textile industry."

Industrial revolution and modern times

During the industrial revolution, production was mechanised with machines powered by waterwheels and steam-engines.

Sewing machines emerged in the nineteenth century.

Synthetic fibres such as nylon were invented during the twentieth century.

Clothing and textile manufacture expanded as an industry so that such unions as the Amalgamated Clothing Workers of America and the Textile Workers Union of America formed early in the twentieth century. Later in the twentieth century, the industry had expanded to such a degree that such educational institutions as UC Davis established a Division of Textiles and Clothing, The University of Nebraska-Lincoln also created a Department of Textiles, Clothing and Design that offers a Masters of Arts in Textile History, and Iowa State University established a Department of Textiles and Clothing that featurs a History of costume collection, 1865–1948. Even high school libraries have collections on the history of clothing and textiles.

Alongside these developments were changes in the types and style of clothing worn by humans. During the 1960s, had a major influence on subsequent developments in the industry.

Textiles were not only made in factories. Before this that they were made in local and national markets. Dramatic change in transportation throughout the nation is one source that encouraged the use of factories. New advances such as steamboats, canals, and railroads lowered shipping costs which caused people to buy cheap goods that were produced in other places instead of more expensive goods that were produced locally. Between 1810 and 1840 the development of a national market prompted manufacturing which tripled the output's worth. This increase in production created a change in industrial methods, such as the use of factories instead of hand made woven materials that families usually made.

The vast majority of the people that worked in the factories were women. Women went to go work in textile factories because of some of the following reasons. Crowding at home was indeed a cause for them to leave and be on their own. The need to save for future marriage portions also motivated these women to decide to work in the millhouses. The work enabled them to see more of the world, to earn something in anticipation of marriage, and to ease the crowding within the home. They also did it to make money for family back home. The money they sent home was to help out with the trouble some of the farmers were having. They also worked in the millhouses because they could gain a sense of independence and growth as a personal goal.

TEXTILE CARE: STRUCTURE, STORAGE, AND DISPLAY

INTRODUCTION

Textiles have been used in various human endeavors for thousands of years and have the potential to be highly symbolic and culturally important. This is especially true in the United States where even mundane textiles such as handkerchiefs and bandannas have held political and cultural significance . Due to this intimate link with historical events, items such as flags, campaign banners and bandannas, pennants, and other flat textiles stand a reasonable chance of being included in library, archive, or museum collections.

Ideally, a textile conservator should be consulted in the care and repair of a historic textile; however, this is not always immediately possible because of budgetary concerns or a lack of local or in-house specialists. In some cases, the cost of a conservator's services may greatly exceed the monetary value of the piece . When professional repair services are unavailable or impractical, preservation should be the focus as "the first and safest line of defence against all the causes and some of the effects of deterioration" . To this end, this paper offers a brief overview of the structure, storage, and display of flat textiles for libraries, archives, museums, and private collectors who may not have much experience in textile care and who lack immediate access to professional textile conservation services.

WHAT IS A TEXTILE?

The term textile can be applied to several types of materials under a couple of related definitions. The most basic definition of a textile is a material that has been fabricated by some type of weaving process. This definition is derived from the Latin root of the work "textile," *textere*, which means "to weave." The term textile can also be applied to materials manufactured by the interlacing of yarn-like materials, such as objects made by braiding, knitting, and lacing, as well as some non-yarn based materials, such as felts, in which the fibres have gained coherence by mechanical treatments or chemical processes. In rare cases, pelts, hides, and plastics may also be considered textiles, especially when they are used in the manufacture of clothing items .

TEXTILE FIBRES

All textiles are made of *fibres*, that are technically defined as "a unit of matter with a length at least 100 times its diameter, a structure of long chain molecules having a definite preferred orientation, a diameter of 10-200 microns, and flexibility" . Variations in fibres on both the microscopic and the visible levels can have a great impact on the behaviour and deterioration of a textile object, and learning the basic properties of textiles can greatly aid in caring for them. There are three major factors that determine the final characteristics of any textile- the fibre form, the source of the fibre, and the method of constructing the final product.

Fibre Sources and Forms

Fibres come in one of two forms based on the length of the fibre. A *filament* is a fibre of continuous length. Both natural and man made filaments can be extremely long. Silk worm cocoons, for example, can contain about two miles of continuous twin filaments, and man made filaments from spinning machines can be even longer. Filament yarns are typically thin, smooth, and lustrous. A *staple*, on the other hand, is a fibre of limited length ranging from about one-quarter of an inch to many inches in length. Staple fibre yarns tend to be thicker, fibrous, and non-lustrous.

There are three catagories of fibres based on source- natural fibres, mineral fibres, and man made fibres. Mineral fibres include glass and asbestos and are normally not directly involved in textile production so only the natural and man-made fibres will be discussed here. All natural and man-made fibres on a microscopic level are built of organic polymers, large carbon based molecules composed of a single unit repeated many times. Different types of polymers result in different fibre, and eventually different textile characteristics.

Natural Fibres

Among the natural fibres, silk and wool come from animal sources while the common vegetable sources are cotton and flax . The silkworm, *Bombyx mori*, produces silk fibres when it spins a cocoon to protect itself in the pupa stage . The fibres are constructed from amino acids that are cross-linked and generally oriented parallel to the fibre axis. This is referred to as a crystalline chain structure, and this structure is responsible for the strength of silk fibres.

Wool fibres are also constructed of amino acids except they are arranged into long helical molecules making wool much more extensible than silk . The fibres, because of this structure, also tend to shrink and mat together when washed in hot, soapy water. This is referred to as *felting* . Wool fibres, like human hairs, are difficult to press into sharp folds, and permanent folds can only be achieved through chemical processes. The natural function of wool is to keep the animal on which is grows dry. Even when incorporated into a textile object, wool fibres retain the ability to absorb up to one-third of their own weight in water before feeling damp to the touch.

Vegetable fibres are constructed of cellulose polymers which join together to form long, flexible, and very strong long-chain molecules . The function of flax is to hold the flax plant upright and carry moisture through the plant, thus linen (fabric that is made from flax fibres) will have a tendency to draw moisture to itself. Cotton fibres come from the seed heads of the cotton plant and surround the seed before it drops. Both cotton and flax are stronger when wet and humidity is a requirement for weaving cotton fibres.

Man-Made and Metal Fibres

Man-made fibres were first developed in an attempt to make artificial silk, and typically have a high degree of crystallinity like silk. While no true substitutes for silk were ever developed, the research did lead to the development of several types of manmade fibres that can be produced via various chemical processes. These fibres can be divided into two categories-regenerated fibres and synthetic fibres.

Regenerated fibres are made from natural materials that have been dissolved and then extruded as filaments. Regenerated fibres made from cellulose, commonly termed rayon, have become the most commercially important. Synthetic fibres include polyamides , polyesters, and polyvinyls .

Metal can also be fashioned into a filament like form and used in textiles. Consequently, metal threads are sometimes classified as a type of fibre. Gold and silver alloyed with baser metals such as copper are the most common materials used for metal thread production. The metal is beaten or drawn into very thin laminates and usually wound around a central fibre core that can either be silk, linen, or, in rare cases, cotton. Sometimes the laminate is attached to paper or an animal membrane before it is used. Metal fibres are typically more resistant to deterioration than organic fibres and are often the only intact parts of very ancient textiles.

FROM FIBRE TO FABRIC

Yarn Based Structures

In all fabrics except bonded fabrics and felt, fibres are twisted into thicker structures called *yarns* or *threads* before being used. The process of creating yarns is called *spinning*. Yarns can be spun in either the clockwise or counter-clockwise direction. One direction is termed the Z direction and one is termed the S direction. After the initial yarn is spun, several yarns can then be twisted together to form ply yarns. These types of yarns are typically thicker and stronger than single ply yarns.

Yarns can be woven, knitted, braided, and laced or netted to create fabric. Each type of structure has an effect on the elasticity and durability of the final product. Woven fabrics consist of two series of threads that are interlaced at right angles to one another. The two thread series are termed the *warp* and the *weft*, with the warp threads running the length of the fabric and the weft threads running the width of the fabric.

The edge on the long sides of a piece of woven fabric is termed the *selvedge*. The selvedge provides a neat edge to the fabric as well as a secure grip for finishing machinery in machine made fabric. It is often different in appearance and structure to the rest of the fabric. Depending on the method of weaving, the density and type of interlacing can vary, both of which affect the final appearance and handle of the fabric. In general, no matter what method of weaving is used, the fabric will show little capacity for stretching beyond the natural elasticity of the materials in either the warp or weft direction. Instead, a woven piece of fabric will stretch more easily in the bias direction, the diagonal of the fabric that is normally at a forty-five degree angle between the warp and the weft.

Knitted structures are formed by interlocking loops of yarn, and, like weaving, there are several methods of knitting fabrics. Vertical rows of interlocked loops are termed *wales* and horizontal rows are termed *courses*. Knitted fabrics are much more susceptible to stretching and distortion than woven fabrics because any tension exerted on the fabric will distort the individual loops that form the fabric. Knitted fabrics are also easily unraveled, and significant damage can be caused by simply breaking one loop that, in turn, causes other loops to be released.

Lacing and netting were formerly hand techniques in which yarns are twined or knotted around each other to form various open structures. Items made by lacing and netting are even more dimensionally unstable than knitted fabrics are and the uses of such fabrics are limited; however, several types of banners were constructed with net bases in the nineteenth century.

Braiding involves the interlacing of yarns diagonally to form a narrow flat or tubular structure. It is difficult to form large braided pieces either by hand or by machine due to the fact that all the constituent yarns must be kept in motion simultaneously and separately. Shoe laces and other kinds of cording as well as decorative braiding are common braided products. The diagonal direction of the yarns allows braids to be somewhat extensible in length and width.

Fibre Based Structures

Under the influence of heat, moisture, and mechanical pressure some types of fibre can be made to mat together to form fabric without the need for yarn. Fabrics made in this way are called felts. Wool and a few other animal fibres are most suited to this type of fabric construction. Felt fabrics have no grain because the fibres do not lie in any particular direction, and because of this, felt can be cut in any direction without fraying or unraveling. Dense felts can be very strong and durable, but are generally stiff and do not drape well. Softer and suppler felts result from less dense fibre structures but there is also a loss of strength and a vulnerability to distortion associated with thinner felts that makes them unsuitable for most purposes . Pennants are a common type of historical felt textile in the United States .

Fibres other than wool can also be bonded together through chemical rather than mechanical processes. These types of fabrics are referred to as bonded fibre fabrics. Bonded fibre fabrics are similar in structure to felt, although some types can be made with the majority of the fibres lying in one direction creating a fabric with a noticeable grain. Bonded fibre research has not, however, been able to overcome the suppleness and durability problems shared with felt .

TEXTILE FINISHES

Any given textile will probably undergo one or more finishing processes before it is used and many processes have been in use for hundreds of years. These processes are too numerous to list here, but they all serve at least one of the following purposes-

- To enhance the appearance of the fabric
- To improve the texture or weight
- To increase flexibility, durability, or ease of care

Finishing processes can be carried out either before or after the textile construction process. Mercerizing, sizing, and weighing are some examples of finishing processes that have been widely used for several centuries . Mercerizing is a finishing technique used on cotton yarn and cloth. Various concentrations of sodium hydroxide, an alkali substance, are applied to make the finished textile piece more lustrous, stronger, more absorbent, and easier to dye. Sizing is also a finishing technique for cotton. Gelatin sizing can be used to give the cotton a coated a papery look, and animal glue sizing made from fish skins can be used to give a greater luster.

Weighting is "the process of loading either yarns or fabric with minerals, sugar, or other foreign matters mixed with the dyes, to make the goods look thick or feel heavy" . In silks this finish compensates for the loss of the natural compound sericin which is lost during the manufacturing process . Other finishing methods were also used on silks to make lower quality silks appear more costly. These finishes employed gum, starch, oil, and wax based materials, most of which will not withstand washing.

Dyeing is another of the more common finishes with certain dyes becoming more popular during certain periods. Early bandannas are often referred to as "turkey red" bandannas because they were dyed a solid red colour before a pattern was applied via bleaching or printing.. Early dyes were obtained from natural sources and varied greatly in quality and ease of use. In 1856, W.H. Perkin, a British scientist discovered the fist synthetic dye by accident and this lead to the development of a wide range of synthetic dyes that eventually replaced natural dyes. Even with synthetic dyes, however, dying is a difficult and complicated process due to the fact that many dyes on their own are not inclined to be colorfast and most fabrics on their own are not capable of absorbing dyes, especially in the case of man made fibres . Natural dyes, for example, almost always required a metal salt to be applied to the cloth before dying to increase the affinity of the dye for the cloth and, in some cases, to increase colorfastness or change the colour of the dye.

TEXTILE DECORATIONS

After being manufactured, textiles can be decorated in a variety of ways including printing, painting, and embroidery. Printing is a decorative method used to apply colour to one side of a textile. There are two classes of printing- the block method and the stencil method. The block method entails tracing a design onto a block of hard material and then either cutting away the portions that are not part of the design for relief printing or cutting away the portions that are part of the design for intaglio printing.

In stenciling a design is cut out of a piece of thin, hard, and non-absorbent material. This stencil is then used to paint or spray a design onto the cloth. Screenprinting is a variation of stencil printing in which a design is drawn on to a fine mesh screen. The parts of the screen that are not going to be printed are then blocked with a lacquer compound so that only the design is printed when ink is applied. In both block and stencil printing, a thick dye paste is required to prevent colour migration during printing. After the dye is fixed by finishing processes, unwanted residues can be removed .

When a banner or flag is painted, one of two painting substances may have been used. One type of paint employs thickening agents of animal or vegetable origin that are at least partly washed out after the paint dries. This type of painting leaves the structure of the textile visible and is usually employed on wool or linen. The other type of painting is similar to ordinary canvas painting. A paint paste with binding agents of animal or vegetable origin entirely covers the fabric and is sometimes covered with varnish as well. These types of flags are often painted on both sides because the paint does not penetrate the cloth.

Embroidery is another decorative method that also comes in two forms. In the first, images are sewn directly into the flag or banner cloth usually with silk or metal threads, but wool and linen yarns are not unheard of. In this type of embroidery, two pieces of cloth are always used and embroidered separately. One or two linings may also be used to strengthen the piece. In the other embroidery method, separately manufactured embroidered pieces are sewn onto the cloth. Attaching the embroidered pieces to a heavy inner lining is almost always necessary because thin cloth is not strong enough to support them .

DETERIORATION MECHANISMS AND AGENTS

All fabrics and dyes and many other materials previously mentioned that are used in textile production are organic compounds that are subject to the same deterioration from external sources that other organic substances are subject to. Specifically, the polymer chains that form textile fibres are broken down, weakening the fibres on the microscopic level and eventually leading to brittleness of the entire textile.

While some breakdown is unavoidable, the rate of this process can be greatly slowed by controlling the agents in the textile environment that accelerate this process. Deterioration factors include finishing processes, light, humidity, heat, and exposure to acids and alkalis. Some of these factors work in combination to accelerate deterioration. For example humidity can cause mold growth that leads to acid production. All of these factors when alone or when combined with mechanical and physical stress can lead to irreversible damage.

Finishing Processes

The process of deterioration may begin before the textile even reaches the consumer. Certain dyeing processes and weighting are the two most damaging finishing processes . As mentioned above, natural dyes often had to be used in conjunction with metal salts to make the textile accept the dye . In the case of very dark browns and blacks, iron was used as a mordant. Oxidation reactions involving the iron in the dye will eventually rot the fibres to such an extent that loss of the fibre is inevitable.

Weighting processes, most often used on silk, also have the potential to be extremely damaging. Weighting of silk was carried out extensively starting in 1870. In 1909 one company noted that "coloured silks could be weighted to between fifty and one hundred per cent and black silk up to four to five hundred per cent". While this may have a desirable effect on the product for the consumer, weighting makes silk more susceptible to light and mechanical damage.

Light

Light can have a dramatic detrimental affect on all textiles as well as some dyes. Both visible light and UV light from sunlight and fluorescent tube lighting have the energy needed to activate chemical reactions that lead to deterioration, though humidity and oxygen are also needed . The deteriorating effect of light depends on the strength of the light and how long the item is exposed. It takes a surprisingly small amount of light to initiate deterioration. Experiments at the Doerner Institute in Munich demonstrated that fading of colored textiles can be observed after being exposed to an amount of light equivalent to 50 days of illumination at 500 lux for only 8 hours a day . 500 lux is about the amount of light present in a typical office or home kitchen. Indirect sunlight averages 10,000 to 20,000 lux while direct sunlight averages 100,000 to 130,000 lux .

Humidity

As well as participating in light activated reactions, high humidity causes fibres to swell and distort changing the shape of the textile object and causing some types of dyes to run or fade. Humid conditions also promote mold, fungus, and insect growth. Low humidity can also lead to damage by causing fibres to shrink and become brittle. Wool is especially susceptible to damage in dry conditions because its natural function is to absorb water. At the other extreme, in the same amount of humidity ideal for wool, silk will rot.

Heat

Heat damage manifests itself in brittleness as well as a brown discoloration caused by the products of polymer breakdown. Heat can come from lighting mechanisms that are placed too close to the object as well as from heating systems. Storage in attics or storerooms without climate control can also lead to heat damage. Infrared light has also been known to have a detrimental warming effect.

Acids and Alkalis

In general protein polymers fibres will tolerate small amounts of acids while cellulose polymer fibres will tolerate small amounts of alkali, but it is best to limit exposure to both. There are several internal sources of acids and alkalis including dark brown or black dyes that have iron in them, and mold or insect waste . External sources include unsealed wood, non-archival paper goods, and urban pollution . In sufficient amounts, both acids and alkalis will cause irreparable damage in any type of textile by provoking hydrolysis reactions that result in shorter, weaker polymer chains. Alkalis produce a particularly dramatic effect on wool, and a five- percent caustic soda, at the boil, will destroy wool fibres completely. Silks, because they are constructed of filaments, suffer the most from acid attacks that can break the filament polymer chains at any point.

PRESERVING TEXTILE OBJECTS

There are several techniques that can be used to avoid exposing textiles to the above deterioration agents. Many ways of repairing and cleaning textiles also exist; however, aside from careful vacuuming, these techniques are not recommended for those who are not trained in textile conservation. Many conservation methods have the potential to damage the textile further, and even professional conservators consider many factors before making the decisions on how to treat and repair a textile object. Wet cleaning is a widely used cleaning method, but it is extremely hazardous when done improperly. Loss of dyes or paints, the removal of water-soluble sizes, dimensional changes, and the production of a kind of "textile mud" when highly degraded textiles are exposed to water are among the possible results of wet cleaning.

Another situation that requires caution and expert advice is when folded, aged textiles are discovered. If a textile object is found folded, it is best to contact a professional conservator for advice before attempting to unfold it. Even though the object may appear to be in stable condition, deep creases can cause damage over time causing the fabric to split when it is unfolded. Professional treatments such as humidifying may be necessary before the textile can be flattened.

Non-professionals should focus on preservation methods that carry as little risk as possible for causing further damage. Fortunately, adequate preservation has become increasingly easy for the general public and entities with smaller budgets due to the availability of relatively affordable archival products over the Internet and elsewhere.

Vacuuming

In some cases, a textile may have surface dirt and dust that can be removed by vacuuming if the textile is not excessively brittle. The advice of a professional is recommended before this is performed to ensure that the textile is stable enough to endure this process. Dust removal is an important first step because layers of dust can carry aggressive corrosive substances such as sulphur dioxide, nitorgen monoxide, or various oily substances . Regular surface cleaning by vacuuming is necessary for any textile that is on display as well as for textiles with large amounts of surface soiling. A monofilament screening attached to an open frame should be placed screen side down on top of the area to be vacuumed to prevent fibre loss. The screen should be heavy enough to lie firmly on the textile during vacuuming, and the edges of the screen should be taped to prevent snagging of the textile. The hand held nozzle attachment of a vacuum with a low suction setting or a low powered hand vacuum can then be used directly on top of the screen to remove dust and surface dirt . Any textile object that is at a point where small pieces are breaking off should not be vacuumed at all even under a screen because textile loss may occur.

It is also possible to surface clean a textile using a hand held nozzle or vacuum with the opening covered with fine net. This method is not usually recommended, however, because the vacuuming apparatus must be held slightly above the textile and cannot touch the textile directly without risking damage to the object. Much more caution is required for this method than for the screen method. Brushing is not recommended because it only moves dust around instead of removing it and has the potential to cause damage even when done lightly.

STORAGE

The overall goal of any storage method should be to protect the object from the agents of deterioration mentioned above. The particular method chosen for storing any textile should be based on the textile's condition and size, and all materials that come in direct contact with the object should be of archival quality. For example, sealed woods or metal, acid-free boards and tissues, unbleached muslin, and Mylar are all acceptable archival materials when used properly. Adhesives of any kind should never come in contact with the textiles. Light is usually the greatest threat to textiles and dyes and wherever the textile is stored it should be protected from both natural and artificial light sources. Ventilation is also necessary to avoid creating an atmosphere of concentrated corrosive substances that may originate from agents in the textiles or from environmental pollutants.

Relative humidity should be kept constant at a level between 50 and 60 percent to prevent fibres from expanding and shrinking repeatedly causing dimensional distortion . Silks should be stored at a slightly narrower range of relative humidity levels ranging from 50 to55 percent . Rapid changes in temperature should be avoided because they usually cause changes in humidity and because high temperatures can cause heat damage and organism growth. Temperatures of 60 to 68 degrees Fahrenheit are considered best for textiles.

People are also a source of both heat and humidity, and storage areas that are not accessed frequently and not by more than a few people at a time are preferable.

Excess handling and mechanical stresses are another source of damage. This damage can be minimized by limiting the amount of contact between items in storage by keeping detailed records of where each item is stored to reduce the amount of rummaging necessary and the number of items that must be moved when a particular object is requested. In some cases, it may be possible to provide high quality pictures to aid in this endeavor and to offer the user in order as a substituted for the item . Whenever items are handled, they should be placed on a support such as a rigid piece of archival quality cardboard.

Items that have become detached from the textile such as fringe, tassels, ribbons, or original poles should ideally be stored with the original object; however, care should be taken to assure that any original hardware attached to the item will not cause further damage to the textile while it is in storage .

Rolled Storage

Rolling and flat storage are the two best storage methods. Rolling is the most space efficient for large textiles that are in fairly good condition. It is important to note, however, that painted textiles should not be stored rolled because the painted areas will or are already stiffened by age. If these types of textiles are rolled, cracking, splitting, and paint loss can occur .

Archival quality cardboard tubes are available in a variety of diameters and lengths for rolled storage. If archival quality tubes are unavailable other types may be used after first wrapping them in Mylar, buffered paper, and unbleached muslin in that order with the Mylar closest to the tube . Tubes with wider diameters should be used for thicker items. Textiles should always be gently rolled with the decorated side outwards because the side closest to the tube is prone to wrinkling if too much material is present. Acid free tissue paper is rolled between the layers and can also be used to wrap the outside of the rolled item . Unbleached muslin cloth can also be used as a cover in cases where light exposure may be a problem. Wide fabric strips can be used to secure items that may come unrolled easily. These strips are best secured with Velcro, pins, or buckles instead of knots or tying.

After rolling, items should be stored in areas where they are protected from sunlight and dust but still well ventilated as previously noted. Tubes should never be stored in the vertical position, and the textile should never support its own weight . The easiest storage method for smaller rolled textiles is inside a box with the tube ends supported. A variety of archival boxes are available for purchase from archival suppliers that could be used in this way. In cases where archival boxes cannot be purchased, other types of boxes or drawers lined with several layers of acid free tissue paper or unbleached muslin can be used as a substitute. The method shown uses shaped Ethafoam for the tube supports. Non-archival types of foam should not be used because they could emit corrosive substances as they decompose. Sealed wood brackets could also be used but will add more weight to the storage system. Larger rolled textiles can be stored on shelves in support brackets or horizontally mounted on a wall in brackets of sealed wood or metal. Dowels can be placed inside the tubes to provide extra support if needed. These types of systems can be protected from light by muslin covers over each object or by cabinets that cover several rolls.

Flat Storage

Small textiles (or even fairly large ones, space permitting) can be stored flat in properly lined and sealed metal or wooden drawers . This is also the best storage method for painted textiles . Narrow drawers such as those in map cases or engineer drawing cases are the most space efficient for larger collections. Otherwise, boxes or other kinds of drawers can be used for storage. Heavy textile objects should not be stacked. Lighter objects of similar sizes may be stored in a few layers with acid free tissue paper in between items if no mounts are involved. Acid free tissue paper or unbleached fabric should also be used on top of the objects to prevent dust accumulation.

In some cases, there may be no other option but to fold a very large or heavy textile in order to store it. Sharp folds should never be used on historic textiles. Instead, rounded folds with crumpled acid free tissue paper inside the fold can be used. Acid free tissue paper should also be used between fabric layers. Folded textiles should be occasionally refolded in different areas to prevent weakening of any one area.

Mounting

Items that are fragile, frequently handled, or destined for display might benefit from lining or mounting before being stored. Due to the fact that this involves sewing the object onto a fabric support, this should only be done with the professional assistance of a textile specialist. Improper mounting can cause significant damage over time due to mechanical stress and weakening of yarns in places where the stitches are made. Fairly detailed instructions are given here to aid in identifying any improper mounts that may be already be present. These should be professionally removed before they cause distortion or damage.

The material chosen for the support fabric should be identical or very similar to the textile material so that it will shrink or expand at the same rate as the piece being supported. The support fabric should also be pre-washed and shrunk before use. Threads may be silk, nylon, mercerized cotton, or a yarn pulled from the same kind of fabric being used for lining. Finer threads are usually preferable. Very thin curved needles are normally used for sewing and have less potential for damaging any yarns of the textile object while sewing. The stitching can be done in two ways. One method is to sew one-half to threefourth inch long stitches in rows starting at the center of piece and continuing through the entire piece. This allows for even weight distribution across several yarns for each stitch . The other method is to only sew only around the edge of the piece using stitches that are perpendicular to the object. If this method is used, the board should be acid free and non-rubber based glue should be used to attach the cloth on the backside of the board only. Before the cloth is attached to the board, the corners can be mitered to assure a better fit. The cloth should be kept square to the board and attached before the object is sewn on to prevent the stitches from pulling holes in the textile object . In both mounting systems, the stitches should be loose enough so that the older textile is not strained, but tight enough so that the mounted textile will not move or abrade on the mounting.

If a textile is so fragile that mounting with a needle and thread could potentially cause damage, the sandwich method of mounting may be used. In this method a backing material is stretched over a frame with the fragile textile placed on top. Then, crepelinesilk or polyester netting is placed over the textile and either sewn to the backing material or attached to the frame . Note that a textile should never be directly mounted in an open frame. Objects framed like this are subject to a great deal of mechanical stress. The will also age unevenly because the parts of the textile that are protected by the frame will not age as quickly as the parts that are open to the air.

DISPLAY

There are many perils involved in putting textiles on display including distortion from physical stress, exposure to deterioration agents, and vandalism. No historic textile should be on display for extended periods of time . Display of textiles carries even more risk than mounting and, again, professional advice from a conservator or textile display specialist should be sought before making display decisions. Many textiles should not be displayed at all. The following descriptions of display methods are offered as a starting point for discussing options with a specialist. Additionally, as with mounting, it is advisable to be able to identify improper display techniques so those textile objects found in such situations can be immediately removed from danger.

As in storage, exposure to light is the chief concern when display is considered. Ideally, textiles should be displayed in an area where lighting can be removed or turned off when the objects are not being viewed, and textiles should never be displayed in an area where they will be exposed to sunlight of any kind. Fluorescent lights, because they emit UV light, should be fitted with UV filters or not used at all, and light sources should not be mounted close to the textile to prevent heat damage. In general, 50 lux is considered to be a safe amount of light for textile display . While it is hard to determine the lux of light sources without the proper equipment, this is light that is slightly brighter than twilight.

Non-Hanging Displays

Small mounted textiles can be easily displayed on their mounts in protective cases, preferably in a horizontal or angled position. If suitable cases for small textiles are not available and the items are in good condition, framing might be an option. The items should be mounted in the way described above in the "storage" section for a fabric covered paperboard mount. For framing, hardboard can be used in place of paperboard as an extra barrier to humidity. When the piece is inserted into the frame, it will be necessary to insert very thin pieces of wood or an archival mat around the edges between the glass and the mounted textile object so that the glass will never come in contact with the textile. As usual, all the wood used in framing should be sealed to prevent damaging chemicals from seeping into the textile. Taping the glass into the frame with framing tape or other tape of archival quality, and taping any gaps that are present after the backing is put into the frame will prevent dust from reaching the textile while it is framed.

If the items will only be on display for a short period of time and are not already mounted, pin couching may be ideal. In pin couching, a block of foam is used as a base for directly pinning a textile with stainless steel needles. The foam should be of archival quality to prevent hazardous gases from contaminating the object and it should be covered in pre-washed and shrunk fabric before use. The pins can be pushed almost entirely into the foam and should be inserted carefully through the weave of the textile object to prevent damaging any yarns. The positions of the pins should be varied across the textile so that the same warps and wefts do not come in contact with the pins multiple times. Spacing of 1 to 2 and one-half inches is ideal.

Hanging Displays

Larger textiles that are difficult to display horizontally can be vertically mounted, but this type of mounting is the most hazardous of all techniques and should be approached with caution and with advice from a professional conservator. Woven materials are the best candidates for vertical hanging because they are the most resistant to distortion. Almost all textiles that will be hung should first be lined using the all-over stitching method described above. Only very heavy textiles can be hung unlined.

It is especially important to consider carefully what kind of lining will be used when textile objects are displayed. When large textiles like flags and banners were displayed in the past, coarse cotton netting was often sewn directly onto the object before being displayed. While this method of display did provide some support and allowed the objects to be easily viewed from both sides, it also caused a great deal of damage through abrasion. Silks especially suffered and large holes formed where the netting was knotted and where silk threads had held the netting in place. The items may have also been hung in places where draughts would put them in motion causing further abrasion on the net or other backing material.

Other items were sewn onto inappropriate backings that caused distortions over time . This is exactly what happened when the famous Star- Spangled Banner was treated in 1914. A heavy linen backing was added that tripled the weight of the 30 by 34-foot garrison flag to 150 pounds, stretching it into a rectangular shape that it had previously lost. The lining has since been removed and a highly publicized conservation effort has begun to repair the damage caused by the lining as well as other damaged areas . These previous conservation and display effort demonstrate the importance of choosing an appropriate lining material as described in the above "storage" section.

After the lining is attached a hanging system can be implemented. The two most common hanging systems are sleeve and Velcro systems. A sleeve system involves sewing a horizontal sleeve onto the back of the top edge of the textile so that a wooden or metal pole can be inserted. The pole is then hung on brackets mounted in the wall. As in all other cases wooden rods should be sealed while metal rods should be rustproof. Additionally care should be taken so that sewing does not damage the textile when the sleeve is attached. Some display specialists prefer rectangular rods instead of round rods because the latter tend to tip the item forward when hanging. In either case the sleeve should be sewn on slightly below the top edge of the textile and extend the entire length of the textile minus one centimeter on either side.

The Velcro system is easy way to display large or medium sized textiles, and it also allows for items to be moved quickly and safely. To implement a Velcro system, cut strips of Velcro (one hooked and one looped strip) that are 2 centimeters shorter than the textile being hung. Create a sleeve from folded cotton fabric and machine-sew the looped Velcro strip onto the cotton on all four edges. The mounted Velcro strip can then be hand sewn onto the textile just below the top edge of the fabric. Both the lining and the textile itself should be carefully incorporated into the stitches. The hooked strip of Velcro is then mounted onto a strip of sealed wood that is then mounted to the display wall. The textile can then be pressed onto the wooden mount beginning from the center and moving outwards.

Hanging methods using rings or tabs should be avoided because they do not provide an even distribution of weight across the length of the textile and will distort the weave of the textile. Clamp systems can also provide the necessary support but their construction is relatively complicated and they do not provide any significant benefit over simpler systems.

Finally, the hanging textile should be protected from handling and vandalism while on display. This can be particularly challenging for large items because museum quality cases are expensive and the textile needs to be ventilated and protected at the same time. A simple and inexpensive way to achieve this is to mount Plexiglas sheeting over the hung textile with spacers inserted between the sheet and the wall to prevent the textile from coming into contact with the Plexiglas. The finished system should prevent the textile from touching both the Plexiglas and the wall to prevent the transference of any harmful chemicals and to increase air circulation.

RESULT

Even with limited experience or a small budget, utilizing basic knowledge about textile structure and deterioration can effectively preserve flat textile objects. Almost all components of textile objects are organic compounds and will inevitably decompose over time, but how quickly a particular object decomposes can be controlled to some extent through proper care and handling. The underlying idea in all the systems described above is that storage systems should always be designed with the agents of deterioration in mind. Systems that protect from pollution, excessive handling, dust, heat, humidity, and especially light can slow the inevitable deterioration of textiles considerably and preserve them until professional textile conservation can be performed. Textile objects that are in relatively stable condition might also be safely displayed for short periods of time by following the same basic guidelines outlined for storage systems, although it is recommended that professional advice be sought before implementing a display system. When full professional conservation is not a possibility, preservation can ensure that little further damage is done and that the textile can be safely viewed and enjoyed.

SELVAGE

A selvage (US English) or selvedge (British English) is a self-finished edge of fabric. The selvages keep the fabric from unraveling or fraying. The selvages are a result of how the fabric is created. In woven fabric, selvages are the edges that run parallel to the warp (the longitudinal threads that run the entire length of the fabric), and are created by the weft thread looping back at the end of each row. In knitted fabrics, selvages are the unfinished yet structurally sound edges that were neither cast on nor bound off. Historically, the term selvage applied only to loom woven fabric, though now can be applied to flat-knitted fabric.

The terms *selvage* and *selvedge* are a corruption of "self-edge", and have been in use since the 16th century.

IN PRINTING AND PHILATELY

In the print industry, selvage is the excess area of a printed or perforated sheet of any material, such as the white border area of a sheet of stamps or the wide margins of an engraving etc. The term is also widely used in philately.

IN WOVEN CLOTH

In textile terminology, threads that run the length of the fabric (longitudinally) are warp ends. Threads running laterally from edge to edge, that is from left side to right side of the fabric as it emerges from the loom, are weft picks. Selvages form the extreme lateral edges of the fabric and are formed during the weaving process. The weave used to construct the selvage may be the same or different from the weave of the body of the fabric cloth. Most selvages are narrow, but some may be as wide as .75 inches (19 mm). Descriptions woven into the selvage using special jacquards, colored or fancy threads may be incorporated for identification purposes. For many end-uses the selvage is discarded. Selvages are 'finished' and will not fray because the weft threads double back on themselves and are looped under and over the warp.

Handwoven selvages vs. industrial selvages

There is a slight difference between the selvages in handweaving and in industry, because while industrial looms originally very closely mimicked handweaving looms, modern industrial looms are very different. A loom with a shuttle, such as most hand weaving looms, will produce a very different selvage from a loom without a shuttle, like some of the modern industrial looms. Also in industry sometimes the selvage is made thicker with a binding thread.

Selvages of fabrics formed on weaving machines with shuttles, such as hand looms, are formed by the weft turning at the end of each pick (pass of the weft thread) or every second pick. To prevent fraying, various selvage motions (or "styles") are used to bind the warp into the body of the cloth. Selvages are created to protect the fabric during weaving and subsequent processing (*i.e.* burnishing, dyeing and washing) but ideally should not detract from the finished cloth via ripples, contractions or waviness.

In handweaving the selvage is generally the same thickness as the rest of the cloth, and the pattern may or may not continue all the way to the edge, thus the selvage may or may not be patterned. A plain weave selvage is the other option, where the last few threads on either side are woven in plain weave.

In industry the selvage may be thicker than the rest of the fabric, and is where the main weft threads are reinforced with a tight weft back binding to prevent fraying. More simply, they "finish" the left and right-hand edges of fabric as it exits the loom, especially for the ubiquitous "criss-cross" *simple* or *tabby* weave, referred to in industry as taffeta weave. Selvages on machine-woven fabric often have little holes along their length, through the thick part, and can also have some fringe. The type or motion of selvage depends on the weaving technique or loom used. A water- or air-jet loom creates a fringed selvage that is the same weight as the rest of the cloth, as by the weft thread is drawn via a jet nozzle, which sends the weft threads through the shed with a pulse of water. The selvage is then created by a heat cutter which trims the thread at both ends close to the edge of the cloth, and then it is beaten into place. Thus it creates a firm selvage with the same thickness as the rest of the cloth.

Usability of the selvage

In the decorative embellishment of garments, especially in decorative pleat or ruffles, a selvage used as a ruffle is "self-finished", that is, it does not require additional finishing work such as hem or bias tape to prevent fraying.

Very often fabric near the selvage is unused and discarded, as it may have a different weave pattern, or may lack pile or prints that are present on the rest of the fabric, requiring that the selvage fabric be cut off or hidden in a hem. Since industrial loomed fabric often has selvages that are thicker than the rest of the fabric, the selvage reacts differently. It may shrink or "pucker" during laundering and cause the rest of the object made with it to pucker also.

Thicker selvages are also more difficult to sew through. Quilters especially tend to cut off the selvage right after washing the fabric and right before cutting it out and sewing it together.

For garments, however, the selvage can be used as a structural component as there is no need to turn under that edge to prevent fraying if a selvage is used instead. Using the selvage eliminates unnecessary work, thus the garment article can be made faster, the finished garment is less bulky and can be stitched entirely by machine. This is of major benefit for the mass-produced ready-to-wear clothing of modern society, however it is less used in homemade clothes because of the tendency of the selvage to pucker.

IN KNITTED CLOTH

Applying the term *selvage* to a hand-knitted object is still relatively new. Most books on fabric define a selvage as the edge of a woven cloth, however the term is coming into usage for hand-knitted objects. The edges of machine-knitted fabric on the other hand are rarely if ever referred to as selvages.

Selvages in knitting can either bear a special pattern worked into the first and last stitches or simply be the edge of the fabric. The two most common selvage stitches are the chain-edge selvage and the slipped-garter edge, both of which produce a nice edge. The chain-edge selvage is made by alternating rows of slipping the first stitch knitwise and knitting the last stitch, with rows of slipping the first stitch purlwise and purling the last stitch. The slipped garter edge is made by slipping the first stitch knitwise and knitting the last in every row. Other selvages include a garter stitch border one stitch wide, or a combination of the above techniques.

Knitting selvages makes the fabric easier to sew together than it would be otherwise. It also makes it easier to pick up stitches later, and is a good basis for crocheting a further decorative edge.

7

Sewing

AN OVERVIEW

Sewing or stitching or tailoring is the fastening of cloth, leather, furs, bark, or other flexible materials, using needle and thread. Its use is nearly universal among human populations and dates back to Paleolithic times (30,000 BCE). Sewing predates the weaving of cloth.

Sewing is used primarily to produce clothing and household furnishings such as curtains, bedclothes, upholstery, and table linens. It is also used for sails, bellows, skin boats, banners, and other items shaped out of flexible materials such as canvas and leather.

Most sewing in the industrial world is done by machines. Pieces of a garment are often first tacked together. The machine has a complex set of gears and arms that pierces thread through the layers of the cloth and semi-securely interlocks the thread.

Some people sew clothes for themselves and their families. More often home sewers sew to repair clothes, such as mending a torn seam or replacing a loose button. A person who sews for a living is known as a seamstress (from seams-mistress) or seamster (from seams-master), dressmaker, tailor, garment worker, or machinist.

"Plain" sewing is done for functional reasons: making or mending clothing or household linens. "Fancy" sewing is primarily decorative, including techniques such as shirring, smocking, embroidery, or quilting.

Sewing is the foundation for many needle arts and crafts, such as applique, canvas work, and patchwork.

While sewing is sometimes seen as a semi-skill job, flat sheets of fabric with holes and slits cut into the fabric can curve and fold in complex ways that require a high level of skill and experience to manipulate into a smooth, ripple-free design. Aligning and orienting patterns printed or woven into the fabric further complicates the design process. Once a clothing designer with these skills has created the initial product, the fabric can then be cut using templates and sewn by manual laborers or machines.

SEAM ALLOWANCE

Seam allowance is the area between the edge of the fabric and the line of stitching. It is usually 1.5 cm away from the edge of the fabric except for the hem, where the seam allowance is usually 2.5 cm or more. This is often the case for standard home dressmaking. Industry seam allowances vary but are usually 0.6 cm.

OCCUPATIONS REQUIRING SEWING

- Bookbinding/Bookbinder
- Shoemaking/Cobbler
- Corsetier
- Draper
- Dressmaker
- Glove/Glover
- Hatter
- Milliner
- Parachute rigger
- Quilting/Quilter
- Sailmaker
- Seamstress
- Tailor
- Taxidermist
- Upholsterer
- Leatherworker

SEWING TOOLS AND ACCESSORIES

stitching awl

The Stitch-It-Awl is a hand sewing awl for safely and quickly sewing by hand mainly medium to heavy materials such as sailcloth, canvas, leather, soft luggage, upholstery or webbing. It was developed to solve the problems associated with the use of existing hand sewing devices for heavy materials, such as the needle and leather palm used by sail makers.

Bobbin

A bobbin is the little plastic or metal wheel that fits underneath the needle plate of a sewing machine. Thread runs off the bobbin, gets picked up by the needle's thread and forms the bottom part of a seam. The bobbin is an essential part of a sewing machine. It enables the machine to form a seam from two threads, which makes a stronger, smoother, neater seam. Even the best hand sewer cannot duplicate the look of a machine seam. Bobbins, in one form or another, have been around for years. They have always been a part of the sewing machine, even in the old treadle machine days. Using the term bobbin for this implement probably dates from the days of weaving, since the weft threads of a loom are wound around bobbins. The word bobbin is of indeterminate age and origin, which in English, means it probably dates back at least to Middle English.

The sewing machine bobbin was wound by hand for early machines, but the advent of the electric motor sewing machine meant that the bobbin could be wound on the machine itself. After all, the motor turned, so why not put it to another useful purpose?

Depending on the sewing machine, a bobbin may be wound on top of the machine or underneath the needle plate. Winding a bobbin on top of the sewing machine is not difficult, but the operator must make sure to wind the thread in the right direction. Putting the bobbin on upside down will wind the thread off the bobbin.

Some sewing machines require the user to put the bobbin underneath the needle plate, hold the needle thread and turn the hand-wheel until the needle thread "catches" the bobbin thread. Other machines have a "drop-in" bobbin, so the user only needs to make certain the bobbin is oriented correctly and drop it into the correct place. The machine then picks up the bobbin thread on its own.

A sewer should make sure that the sewing machine's bobbin is of the proper size for the machine. A plastic bobbin should not have cracks or splits. Bobbins are available for most sewing machine models in stores where sewing machines and fabrics are sold, as well as online.

Bodkin

Several different implements go by the name of bodkin, including a weapon, an archaic type of hairpin, and two different sewing notions. As a weapon, a bodkin is a dagger; as a hairpin, it's an ornament with a stiletto shape; and as a sewing tool, it's either an instrument similar to an awl for piercing holes in cloth or a long needle with a large eye and blunt tip for drawing tape, elastic, or ribbon through a casing.

The needle-like bodkin is also an item of interest. It was the tool used to lace stays and bodices. A bodkin was stored in a bodkin case, which looks rather like a flat rectangular version of a lipstick container.

Modern bodkins may have several special features. Some needle-like bodkins have a ballpoint tip for ease of movement through the casing. A flat bodkin is a small metal tool that resembles a matchstick with a hole (or holes) in the end for threading. A third type of bodkin is a piece of wire with one end covered with a plastic tip and the other formed into a loop. A completely different style of bodkin is fashioned like a miniature tongs with gripping teeth on its ends to grasp the material being threaded and a neck ring to tighten and hold it in place.

It may prove useful to know that in the absence of a threading bodkin, several other tools can be substituted. A tapestry needle, with its large eye and blunt tip, can serve the purpose of a bodkin by tying the material around the eye so it can't come loose. A large safety pin can also be used by piercing the material to be drawn through the casing, closing the pin, and working it around in the same manner as the bodkin.

Dress form

A dress form is used to give a three-dimensional view on the article of clothing that is being sewed. They come in all sizes and shapes for almost every article of clothing that can be made. When a piece of clothing is made, it can be put on the dress form so one can see how the piece of clothing will turn out. Then one can make alterations upon the clothing after seeing what it looks like on a body.

Scissors

Scissors are hand-operated cutting instruments. They consist of a pair of metal blades, or tangs, connected in such a way that the sharpened edges slide against each other. Scissors are used for cutting various thin materials, such as paper, cardboard, metal foil, thin plastic, cloth, rope and wire. Scissors can also be used to cut hair and food.

Scissors and shears exist in a wide variety of forms depending on their intended uses. Children's scissors, used only on paper, have dull blades and rounded corners to ensure safety. Scissors used to cut hair or fabric must be much sharper. The largest shears used to cut metal or to trim shrubs must have very strong, sharp blades.

Specialized scissors include sewing scissors, which often have one sharp point and one blunt point for intricate cutting of fabric, and nail scissors, which have curved blades for cutting fingernails and toenails.

Special kinds of shears include pinking shears, which have notched blades that cut cloth to give it a wavy edge, and thinning shears, which have teeth that cut every second hair strand, rather than every strand giving the illusion of thinner hair.

Tape measure

A tape measure or measuring tape is a flexible form of ruler. It consists of a ribbon of cloth, plastic, fibre glass, or metal strip with linear-measurement markings. It is a common measuring tool. Its flexibility allows for a measure of great length to be easily carried in pocket or toolkit and permits one to measure around curves or corners. Today it is ubiquitous, even appearing in miniature form as a keychain fob, or novelty item. Surveyors use tape measures in lengths of over 100 m (300+ ft).

Sewing needle

A sewing needle is a long slender tool with a pointed tip. The first needles were made of bone or wood; modern ones are manufactured from high carbon steel wire, nickel- or gold plated for corrosion resistance. The highest quality embroidery needles are plated with two-thirds platinum and one-thirds titanium alloy. Traditionally, needles have been kept in needle books or needle cases which have become an object of adornment. A needle for hand sewing has a hole, called the eye, at the non-pointed end to carry thread or cord through the fabric after the pointed end pierces it. Hand sewing needles have different names depending on their purpose. Needle size is denoted by a number on the packet. The convention for sizing is that the length and thickness of a needle increases as the size number decreases. For example, a size 1 needle will be thicker and longer, while a size 10 will be shorter and finer.

Pattern (sewing)

In sewing and fashion design, a pattern is an original garment from which other garments of a similar style are copied, or the paper or cardboard templates from which the parts of a garment are traced onto fabric before cutting out and assembling (sometimes called paper patterns).

Patternmaking, pattern making or pattern cutting is the art of designing patterns. A custom-fitted basic pattern from which patterns for many different styles can be created is called a sloper or block.

Patterns for custom dressmaking

A custom dressmaker frequently employs one of three pattern creation methods. The flat-pattern method begins with the creation of a sloper or block, a basic pattern for a fitted, jewel-neck bodice and narrow skirt, made to the wearer's measurements. The sloper is usually made of lightweight cardboard or tagboard, without seam allowances or style details. Once the shape of the sloper has been refined by making a series of mock-up garments called toiles (UK) or muslins (US), the final sloper can be used in turn to create patterns for many styles of garments with varying necklines, sleeves, dart placements, and so on.

Although it is also used for women's' clothing, the drafting method is more commonly employed in men's garments and involves drafting a pattern directly onto pattern paper using a variety of straightedges and curves.

The pattern draping method is used for more elaborate and unique designs that are hard to obtain through the flat pattern method. This is because it is nearly impossible to account for the way a fabric will drape or hang on the body without an actual 3-dimensional test run. It involves creating a muslin mock-up pattern by pinning fabric directly on a dress form, then transferring the muslin outline and markings onto a paper pattern or using the muslin as the pattern itself.

Patterns for home sewing

Home sewing patterns are generally printed on tissue paper and sold in packets containing sewing instructions and suggestions for fabric and trim. Modern patterns are available in a wide range of prices, sizes, styles, and sewing skill levels, to meet the needs of consumers.

Home sewing patterns are graded, that is, redrawn to fit larger and smaller sizes than the original design. Ebenezer Butterick invented the commercially produced graded home sewing pattern in 1863 (based on grading systems used by Victorian tailors), originally selling hand-drawn patterns for men's and boys' clothing. In 1866, Butterick added patterns for women's clothing, which remains the heart of the home sewing pattern market today.

There are some applications today that enable a home sewer to customize a computerized pattern to fit her body measurements and or body shape. The 3D technology enables the home sewer to see a virtual simulation of the final garment as it will appear on her. This reduces the Time-to-Market as well as the number of muslins/test garments that are needed.

A variation on the theme was evolved by iconic British brand Clothkits. Clothkits devised ingenious cut and sew clothing kits for home sewing that avoided the need for paper patters. Rather than using conventional techniques, Clothkits pre printed fabric with both quirky designs and the pattern lines, to make dressmaking for the novice easier.

Pin

A pin is a device used for fastening objects or material together. It is usually made of steel, or on occasion copper or brass. It is formed by drawing out a thin wire, sharpening the tip, and adding a head. Nails are related, but are typically larger.

Pincushion

A pincushion (or less commonly pin cushion) is a small cushion, typically 3-5 cm across, which is used in sewing to store pins or needles with their heads protruding so as to take hold of them easily, collect them, and keep them tidy.

Pincushions are typically filled tightly with stuffing, so as to hold pins rigidly once placed. The stereotypical design – a tomato with a strawberry attached – seems to have been around forever but, most likely, was introduced during the Victorian Era.Typically, the tomato is filled with wool roving to prevent rust, and the strawberry is filled with an abrasive to clean the pins and keep them sharp.

Pincushions come in all shapes, sizes and forms. Magnetic pincushions have the advantage of being able to easily keep the work area clear of pins, but the drawback that the pins become magnetized easily and an attempt to retrieve one may pick up several.

Rotary cutter

A rotary cutter is a tool generally used by quilters to cut fabric. It consists of a handle with a circular blade that rotates, thus the tool's name. Rotary blades can be found in different sizes: usually smaller blades are used to cut around corners, while larger blades are used to cut to straight lines. Several layers of fabric can be cut simultaneously, making it easier to cut out patchwork pieces of the same shape and size than with regular scissors. In conjunction with a rotary cutting ruler – specially designed to be used with a rotary cutter, and made of thick and resistant plastic – it is also a practical tool for achieving the straight, squared lines that are so important in patchwork. The first rotary cutter was introduced by the Olfa company in 1979.

Seam ripper

A seam ripper, stitch ripper or stitch unpicker is a small tool used for unpicking stitches.

The most common form consists of a handle, shaft and head. The head is usually forked with one side of the fork flattening out and becoming a blade and the other side forming a small point. In some designs the blade side then tapers back to a point to allow easier insertion in tight stitching.

In use the blade is inserted into the seam underneath the thread to be cut. The thread is allowed to slip down into the fork and the tool is then lifted upwards allowing the blade to cut through the thread. Once the seam has been undone in this way the loose ends can be removed and the seam resewn.

Sewing table

A sewing table or work table is a table or desk used for sewing. Generally is has large amounts of space and a full set of sewing tools. Nearby there will be a chair and a waste bin. A common attachment is a dropleaf to give expanded space. Other attachments can be a cloth bag for storing sewing materials, drawers, or shelves.

Chalk

Chalk (pronounced /ˆtTˆk/) is a soft, white, porous sedimentary rock, a form of limestone composed of the mineral calcite. Calcite is calcium carbonate or $CaCO_3$. It forms under relatively deep marine conditions from the gradual accumulation of minute calcite plates (coccoliths) shed from micro-organisms called coccolithophores. It is common to find chert or flint nodules embedded in chalk. Chalk can also refer to other compounds including magnesium silicate and calcium sulfate.

Chalk is resistant to weathering and slumping compared to the clays with which it is usually associated, thus forming tall steep cliffs where chalk ridges meet the sea. Chalk hills, known as chalk downland, usually form where bands of chalk reach the surface at an angle, so forming a scarp slope. Because chalk is porous it can hold a large volume of ground water, providing a natural reservoir that releases water slowly through dry seasons.

Thimble

A thimble is a protective shield worn on the finger or thumb. It is generally used for sewing.

The earliest known thimble was Roman and was found at Pompeii. Made of bronze, its creation has been dated to the first century AD. A second Roman thimble was found at Verulamium, present day St Albans, in the UK and can be viewed in the museum there.

According to the United Kingdom Detector Finds Database , thimbles dating to the tenth century have been found in England, and thimbles were in widespread use there by the 1300s. Although there are isolated examples of thimbles made of precious metals — Elizabeth I is said to have given one of her ladies-in-waiting a thimble set with precious stones — the vast majority of metal thimbles were made of brass. Medieval thimbles were either cast brass or made from hammered sheet. Early centers of thimble production were those places known for brass-working, starting with Nuremburg in the 15th century, and moving to Holland by the 17th.

In 1693, a Dutch thimble manufacturer named John Lofting established a thimble manufactory in Islington, in London, England, expanding British thimble production to new heights. He later moved his mill to Buckinghamshire to take advantage of water-powered production, resulting in a capacity to produce more than two million thimbles per year. By the end of the 18th century, thimble making had moved to Birmingham, and shifted to the "deep drawing" method of manufacture, which alternated hammering of sheet metals with annealing, and produced a thinner-skinned thimble with a taller shape. At the same time, cheaper sources of silver from the Americas made silver thimbles a popular item for the first time.

Thimbles are usually made from metal, leather, rubber, and wood, and even glass or china. Early thimbles were sometimes made from whale bone, horn, or ivory. Natural sources were also utilized such as Connemara marble, bog oak, or mother of pearl. Rarer works from thimble makers utilized diamonds, sapphires, or rubies.

Advanced thimblemakers enhanced thimbles with semi-precious stones to decorating the apex or along the outer rim. Cabochon adornments are sometimes made of cinnabar, agate, moonstone, or amber. Thimble artists would also utilize enameling, or the Guilloché techniques advanced by Peter Carl Fabergé.

Originally, thimbles were used solely for pushing a needle through fabric or leather as it was being sewn. Since then, however, they have gained many other uses. In the 1800s they were used to measure spirits, which brought rise to the phrase "just a thimbleful". Prostitutes used them in the practice of thimble-knocking where they would tap on a window to announce their presence. Thimble-knocking also refers to the practice of Victorian schoolmistresses who would tap on the heads of unruly pupils with dames thimbles.

Before the 18th century the small dimples on the outside of a thimble were made by hand punching, but in the middle of that century, a machine was invented to do the job. If one finds a thimble with an irregular pattern of dimples, it was likely made before the 1850s. Another consequence of the mechanization of thimble production is that the shape and the thickness of the metal changed. Early thimbles tend to be quite thick and to have a pronounced dome on the top. The metal on later ones is thinner and the top is flatter.

Collecting thimbles became popular in the UK when many companies made special thimbles to commemorate the Great Exhibition held in the Crystal Palace in Hyde Park, London.

In the 19th century, many thimbles were made from silver; however, it was found that silver is too soft a metal and can be easily punctured by most needles. Charles Horner solved the problem by creating thimbles consisting of a steel core covered inside and out by silver, so that they retained their aesthetics but were now more practical and durable. He called his thimble the Dorcas, and these are now popular with collectors. There is a small display of his work in Bankfield Museum, Halifax, England.

Early American thimbles made of whale bone or tooth featuring miniature scrimshaw designs are considered valuable collectibles. Such rare thimbles are prominently featured in a number of New England Whaling Museums.

During the First World War, silver thimbles were collected from "those who had nothing to give" by the British government and melted down to buy hospital equipment. In the 1930s and 40s red-topped thimbles were used for advertising. Leaving a sandalwood thimble in a fabric store was a common practice for keeping moths away. Thimbles have also been used as love-tokens and to commemorate important events. A miniature thimble is one of the tokens in the game of Monopoly. People who collect thimbles are known as digitabulists.

Yarn

Yarn is a long continuous length of interlocked fibres, suitable for use in the production of textiles, sewing, crocheting, knitting, weaving, embroidery and ropemaking. Thread is a type of yarn intended for sewing by hand or machine. Modern manufactured sewing threads may be finished with wax or other lubricants to withstand the stresses involved in sewing. Embroidery threads are yarns specifically designed for hand or machine embroidery.

Tracing paper

Tracing paper is a type of translucent paper. It is made by immersing uncut and unloaded paper of good quality in sulphuric acid for a few seconds. The acid converts some of the cellulose into amyloid form having a gelatinous and impermeable character. When the treated paper is thoroughly washed and dried, the resultant product is much stronger than the original paper. Tracing paper is resistant to oil grease and to a large extent impervious to water and gas.

Tracing paper is named as such for its ability for an artist to trace an image onto it. When tracing paper is placed onto a picture, the picture is easily viewable through the tracing paper. Thus, it becomes easy for the artist to find edges in the picture and trace the image onto the tracing paper. Pure Cellulose fibre is translucent, and it is the air trapped between fibres, that makes paper opaque and looks white. If the fibres are refined and beaten until all the air is taken out, then the resulting sheet will be translucent. Translucent papers are dense and contain up to 10 per cent moisture at 50 per cent humidity. This type of paper is roughly 25 per cent lighter than regular paper.

The sizing in production will determine whether it is for laser printer or inkjet/offset printing.

Tracing paper may be uncoated or coated. Natural tracing paper for laser printing is usually uncoated.

Tracing wheel

A tracing wheel is an instrument with serrated teeth on a wheel attached to a handle. Used to transfer markings from patterns onto fabric with or without tracing paper, this sewing tool also makes slotted perforations. Such markings might include pleats, darts, buttonholes, notches or placement wikt:lines for appliques or pockets. There are two basic types of tracing wheels available to the modern sewer, one with a serrated edge and one with a smooth edge.

Wax

Traditionally, wax (or beeswax) is a substance secreted by bees and used in constructing their honeycombs. The term has come to refer more generally to a class of substances with properties similar to beeswax, enumerated below:

- Plastic (malleable) at normal ambient temperatures
- A melting point above approximately 45 °C (113 °F) (which differentiates waxes from fats and oils)
- A relatively low viscosity when melted (unlike many plastics)
- Insoluble in water
- Hydrophobic

Waxes may be natural secretions of plants or animals, artificially produced by purification from natural petroleum or completely synthetic. In addition to beeswax, carnauba (a plant epicuticular wax) and paraffin (a petroleum wax) are commonly encountered waxes which occur naturally. Earwax is an oily substance found in the human ear. Some artificial materials such as silicone wax that exhibit similar properties are also described as wax or waxy.

SEWING EQUIPMENT

BAR TACK

Bar tack is a series of hand or machine made stitches used for reinforcing areas of stress on a garment, such as pocket openings, bottom of a fly opening or buttonholes. It consists of a series of close-set zig-zag stitches (machine) or whip-stitches (hand), usually 1/16"-1/8" in width and 1/4"-3/8" in length. In denim jeans, it is often in a contrasting colour, such as orange or white.

BLANKET STITCH

The blanket stitch is a stitch used to reinforce the edge of thick materials. Depending on circumstances, it may also be called a whip stitch or a crochet stitch. It is defined as "A decorative stitch used to finish an unhemmed blanket. The stitch can be seen on both sides of the blanket."

This stitch has long been both an application by hand and as a machine sewn stitch. When done by hand, it is sometimes considered a crochet stitch. When done by machine, it is called a whip stitch or, sometimes, a Merrow Machine Company Crochet Stitch, after the first sewing machine that was used to sew a blanket stich. This machine was produced and patented by the Merrow Machine Company in 1877. The defining characteristic of the crochet machine is its ability to sew with yarn and stitch thick goods with a consistent overlock edge. From 1877-1925 the machine evolved dramatically, and consequently so did the capacity of manufacturers to produce goods with the whip stitch. To this day, Merrow retains a patent on the mechanism that creates the blanket stitch, renewing it with each redesign, which has occurred hundreds of times since its introduction in 1877. The most recent patent renewal occurred in 2007.

In the early 1990s Chinese manufacturers copied the design of the original crochet sewing machines; Merrow is currently involved in litigation with several Chinese firms who are accused of violating Merrow's intellectual property rights.

BUTTONHOLE STITCH

Buttonhole stitch and the related blanket stitch are hand-sewing stitches used in tailoring, embroidery, and needle lace-making.

Buttonhole stitches catch a loop of the thread on the surface of the fabric and needle is returned to the back of the fabric at a right angle to the original start of the thread. The finished stitch in some ways resembles a letter "L" depending on the spacing of the stitches. For buttonholes the stitches are tightly packed together and for blanket edges they are more spaced out. The properties of this stitch make it ideal for preventing raveling of woven fabric. Buttonhole stitches are structurally similar to featherstitches.

Applications

In addition to reinforcing buttonholes and preventing cut fabric from raveling, buttonhole stitches are used to make stems in crewel embroidery, to make sewn eyelets, to attach applique to ground fabric, and as couching stitches. Buttonhole stitch scallops, usually raised or padded by rows of straight or chain stitches, were a popular edging in the 19th century.

Buttonhole stitches are also used in cutwork, including Broderie Anglaise, and form the basis for many forms of needlelace.

Variants

Examples of buttonhole or blanket stitches include:

- Blanket stitch
- Buttonhole stitch
- Closed buttonhole stitch, in which the tops of the stitch touch to form triangles
- Crossed buttonhole stitch, in which the tops of the stitch cross
- Detached buttonhole stitch, in which rows of buttonhole stitches are worked to form a "floating" filling stitch
- Buttonhole shading, in which rows of buttonhole stitch are sewn in related colours to give a naturalistic shaded effect
- Buttonhole stitches combined with knots include:
    - Top knotted buttonhole stitch
    - German knotted buttonhole stitch
    - Tailor's buttonhole stitch
    - Armenian edging stitch

Buttonhole bars are parallel rows of thread laid across an open space in lace or cutwork and then completely covered with closely space buttonhole stitches.

CHAIN STITCH

Chain stitch is a sewing and embroidery technique in which a series of looped stitches form a chain-like pattern. Chain stitch is an ancient craft - examples of surviving Chinese chain stitch embroidery worked in silk thread have been dated to the Warring States period (5th-3rd century BC). Handmade chain stitch embroidery does not require that the needle pass through more than one layer of fabric. For this reason the stitch is an effective surface embellishment near seams on finished fabric. Because chain stitches can form flowing, curved lines, they are used in many surface embroidery styles that mimic "drawing" in thread. Chain stitches are also used in making tambour lace, needlelace, macramé and crochet.

CROSS-STITCH

Cross-stitch is a popular form of counted-thread embroidery in which X-shaped stitches in a tiled, raster-like pattern are used to form a picture. Cross-stitch is usually executed on easily countable evenweave fabric called aida cloth. The stitcher counts the threads in each direction so that the stitches are of uniform size and appearance. This form of cross-stitch is also called counted cross-stitch in order to distinguish it from other forms of cross-stitch. Sometimes cross-stitch is done on designs printed on the fabric (stamped cross-stitch); the stitcher simply stitches over the printed pattern.

DARNING

Darning is a sewing technique for repairing holes or worn areas in fabric or knitting using needle and thread alone. It is often done by hand, but it is also possible to darn with a sewing machine. Hand darning employs the darning stitch, a simple running stitch in which the thread is "woven" in rows along the grain of the fabric, with the stitcher reversing direction at the end of each row, and then filling in the framework thus created, as if weaving. Darning is a traditional method for repairing fabric damage or holes that do not run along a seam, and where patching is impractical or would create discomfort for the wearer, such as on the heel of a sock.

Darning also refers to any of several needlework techniques that are worked using darning stitches:

- Pattern darning is a type of embroidery that uses parallel rows of straight stitches of different lengths to create a geometric design.
- Net darning, also called filet lace, is a 19th century technique using stitching on a mesh foundation fabric to imitate lace.
- Needle weaving is a drawn thread work embroidery technique that involves darning patterns into barelaid warp or weft threads.

LOCKSTITCH

A lockstitch is the most common mechanical stitch made by a sewing machine.

The lockstitch uses two threads, an upper and a lower. The upper thread runs from a spool kept on a spindle on top of or next to the machine, through a tension mechanism, a take-up arm, and finally through the hole in the needle. The lower thread is wound onto a bobbin, which is inserted into a case in the lower section of the machine. To make one stitch, the machine lowers the threaded needle through the cloth into the bobbin area, where a hook catches the upper thread at the point just after it goes through the needle. The hook mechanism carries the upper thread entirely around the bobbin case, so that it has made one wrap of the bobbin thread. Then the take-up arm pulls the excess upper thread (from the bobbin area) back to the top forming the lockstitch ideally in the center of the thickness of the material, the tension mechanism prevents the thread from being pulled from the spool side, the needle is pulled out of the cloth, and the feed dogs pull the cloth back one stitch length, the cycle is repeated as the machine turns mechanically .

Lockstitch is so named because the two threads, upper and lower, "lock" together in the hole in the fabric which they pass through. The term "single needle stitching", often found on dress shirt labels, refers to lockstitch, as opposed to chain stitch which unravels easily and is usually used on lower quality garments.

OVERLOCK

An overlock stitch sews over the edge of one or two pieces of cloth for edging, hemming or seaming. Usually an overlock sewing machine will cut the edges of the cloth as they are fed through (such machines are called 'sergers'), though some are made without cutters. The inclusion of automated cutters allows overlock machines to create finished seams easily and quickly. An overlock sewing machine differs from a lockstitch sewing machine in that it uses loopers fed by multiple thread cones rather than a bobbin. Loopers serve to create thread loops that pass from the needle thread to the edges of the fabric so that the edges of the fabric are contained within the seam. Overlock sewing machines usually run at high speeds, from 1000 to 9000 rpm, and most are used in industry for edging, hemming and seaming a variety of fabrics and products. Overlock stitches are extremely versatile, as they can be used for decoration, reinforcement, or construction.

Overlocking is also referred to as "overedging", "merrowing" or "serging". Though "serging" technically refers to overlocking with cutters, in practice the four terms are used interchangeably.

PAD STITCH

Pad stitches are types of running stitches, made by taking small stitches perpendicular to the line of stitching.

Pad stitches secure two or more layers of fabric together and give the layers more firmness; smaller/denser stitches create more firmness. Pad stitches may also enforce an overall curvature of the layers. Tailors pad stitch a jacket's lapel and undercollar to give them more firmness and to help them maintain their curvature. The line of stitching usually runs parallel the direction of the most important curve of the layers. For example, pad stitches in a suit's lapel run parallel to the lapel's roll line; pad stitches in the undercollar of a tailored jacket run parallel to the collar back edge.

RUNNING STITCH

The running stitch or straight stitch is the basic stitch in hand-sewing and embroidery, on which all other forms of sewing are based. The stitch is worked by passing the needle in and out of the fabric. Running stitches may be of varying length, but typically more thread is visible on the top of the sewing than on the underside.

Running stitches are used in hand-sewing and tailoring to sew basic seams, in hand patchwork to assemble pieces, and in quilting to hold the fabric layers and batting or wadding in place. Loosely spaced rows of short running stitches are used to support padded satin stitch.

Running stitches are a component of many traditional embroidery styles, including kantha of India and Bangladesh, and Japanese sashiko quilting.

TACK (SEWING)

In sewing, to tack or baste is to make quick, temporary stitching intended to be removed. Tacking is used in a variety of ways:

- To temporarily hold a seam or trim in place until it can be permanently sewn, usually with a long running stitch made by hand or machine called a tacking stitch or basting stitch.
- To temporarily attach a lace collar, ruffles, or other trim to clothing so that the attached article may be removed easily for cleaning or to be worn with a different garment. For this purpose, tacking stitches are sewn by hand in such a way that they are almost invisible from the outside of the garment.
- To transfer pattern markings to fabric, or to otherwise mark the point where two pieces of fabric are to be joined. A special loose looped stitch used for this purpose is called a tack or tailor's tack.
- A basting stitch is essentially a straight stitch, sewn with long stitches and unfinished ends. The basting stitch is used for temporarily holding sandwiched pieces of fabric in place. The stitch is removed after the piece is finished. Often used in quilting or embroidery.

TOPSTITCH

Topstitching is a sewing technique. It is used most often on garment edges such as necklines and hems, where it helps facings to stay in place and gives a crisp edge. Decorative topstitching is designed to show, and may be done in a fancy thread or with a special type of stitch. Otherwise, topstitching is generally done using a straight stitch with a thread that matches the fashion fabric.

8

Tools

DRESS FORM

A dress form is a three-dimensional model of the torso used for fitting clothing that is being designed or sewed. When making a piece of clothing, it can be put on the dress form so one can see the fit and drape of the garment as it would appear on a body, and make adjustments or alterations. Dress forms come in all sizes and shapes for almost every article of clothing that can be made. Dress forms in standard clothing sizes are used to make patterns, while adjustable dress forms allow garments to be tailored to fit a specific individual. This is often colloquially referred to as a *Judy* for the female form and a *James* for the male.

NEEDLE THREADER

A needle threader is a small device for helping put the thread through the eye of small needles. Still popular today is the needle threader of Victorian design, consisting of a small tinned plate stamped with a profile image (usually a female figure) with a diamond-shaped steel wire attached.

PIN

A pin is a device used for fastening objects or material together. Pins often have two components: a long body and sharp tip made of steel, or occasionally copper or brass, and a larger head often made of plastic. The sharpened body penetrates the material, while the larger head provides a driving surface. It is formed by drawing out a thin wire, sharpening the tip, and adding a head. Nails are related, but are typically larger. In machines and engineering, pins are commonly used as pivots, hinges, shafts, jigs, and fixtures to locate or hold parts.

Curved sewing pins have been used for over four thousand years. Originally, they were fashioned out of iron and bone by the Sumerians and were used to hold clothes together. Later, these pins were also used to hold pages together by threading the needle through their top corner.

Many late pins were made of brass, a hard metal. Steel was used later, as it was much stronger, but there was no easy process to keep steel from rusting, so higher quality pins were plated with nickel, but the metal would start to break down and flake off in high humidity, allowing rust to form. Steel pins were not that inconvenient for homemaking uses as they were usually only used temporarily while sewing garments.

The term "pin money" dates to the 17th century; according to Oxford University Press, it refers to an allowance for decorative clasps that were worn in hair or on clothing. It was subsequently applied to money to buy clothing generally, and later to money for any minor personal expenditure.

Walter Hunt invented the safety pin by forming an eight-inch brass pin into a bent pin with a spring and guard. He sold the rights to his invention to pay a debt to a friend, not knowing that he could have made millions of dollars.

Bobby pins and other types of hairpins are used for restraining the hair. Collar pins are used to hold the collar of a dress shirt in men's fashion. Lapel pins are decorative jewelry attached to the clothing.

The push pin was invented in 1900 by Edwin Moore and quickly became a success. These pins are also called "thumbtacks". There is also a new push pin called a "paper cricket".

Thin and hardened ones can be driven into wood with a hammer aiming the goal rather not to be seen then.

A different style, somewhat thicker (1.5–2 mm diameter), 2 cm short and with rounded tip is used with grammophones to transform the waves of the groove to movement of a membrane and in this way to hearable sound.

In engineering and machine design, a pin is a machine element that secures the position of two or more parts of a machine relative to each other. A large variety of types has been known for a long time, the most commonly used are solid cylindrical pins, solid tapered pins, groove pins, slotted spring pins and spirally coiled spring pins.

- Clevis pin
- Cotter pin
- Spring pin
- Split pin

CLEVIS FASTENER

A clevis fastener is a three-piece fastener system consisting of a *clevis*, *clevis pin*, and *tang*. The clevis is a U-shaped piece that has holes at the end of the prongs to accept the clevis pin. The clevis pin is similar to a bolt, but is only partially threaded or unthreaded with a cross-hole for a split pin. The tang is a piece that fits in the space within the clevis and is held in place by the clevis pin. The combination of a simple clevis fitted with a pin is commonly called a shackle, although a clevis and pin is only one of the many forms a shackle may take.

Clevises are used in a wide variety of fasteners used in farming equipment and sailboat rigging, as well as the automotive, aircraft and construction industries. They are also widely used to attach control surfaces and other accessories to servo controls in airworthy model aircraft. As a part of a fastener, a clevis provides a method of allowing rotation in some axes while restricting rotation in others.

Clevis pin

There are two main types of clevis pins: threaded and unthreaded. Unthreaded clevis pins have a domed head at one end and a cross-hole at the other end. A cotter pin (USA usage) or split pin is used to keep the clevis pin in place. Threaded clevis pins have a partially threaded shank on one end and a formed head on the other. The formed head has a lip, which acts as a stop when threading the pin into the shackle, and a flattened tab with a cross-hole. The flattened tab allows for easy installation of the pin and the cross-hole allows the pin to be moused.

A bolt can function as a clevis pin, but a bolt is not intended to take the lateral stress that a clevis pin must handle. Normal bolts are manufactured to handle tension loads, whereas clevis pins and bolts are designed to withstand shearing forces.

Clevis pins should be closely fitted to the holes in the clevis to limit wear and reduce the failure rate of either the pin or the clevis.

Twist clevis

A twist shackle provides a loop at a right angle to the axis of rotation. Older farming implements intended to be pulled by a team of draft animals often require a twist shackle to be hitched.

Clevis hanger

A clevis hanger consists of one U-shaped clevis and a second V-shaped clevis with a hole in a flattened section at the base of the V, joined together with a bolt or pin. Clevis hangers are used as a pipe attachment providing vertical adjustment for pipes.

Clevis bracket

A clevis bracket generally takes the form of a solid metal piece with a flat rectangular base, fitted with holes for bolts or machine screws, and two rounded wings in parallel forming a clevis. Commonly used in aircraft and cars, clevis brackets allow mounting of rods to flat surfaces.

Clevis hook

A clevis hook is a hook, with or without a snap lock, with a clevis and bolt or pin at the base. The clevis is used to fasten the hook to a bracket or chain.

Clevis rod end

A clevis rod end is a folded or machined piece formed into a clevis and fitted with a hole at its base to which a rod is attached. In machined pieces, the hole is most often threaded.

Twin clevis

A twin clevis is a solid piece with two clevises directly opposite one another, each fitted with a pin. Twin clevises are commonly used to join two lengths of chain.

R-CLIP

An R-clip, also known as an R-pin, R-key, hairpin cotter pin, hairpin cotter, bridge pin, hitch pin or spring cotter pin, is a fastener made of a springy material, commonly hardened metal wire, resembling the shape of the letter "R".

R-clips are commonly used to secure the ends of round shafts such as axles and clevis pins. The straight leg of the R-clip is pushed into a hole near one end of the shaft until the semicircular "belly" in the middle of the other, bent leg of the R-clip grips one side of the shaft resisting any force removing the R-clip from its hole. To assist insertion the end of the bent leg is angled away from the straight leg. This angled end rides the side of the shaft and opens the "belly" mouth enough to pass the widest part of the shaft as the R-clip is inserted.

There is also a double loop variety when a wider range of suitable shafts is required. It also spreads out the stresses more evenly for a longer life.

This type of pin is usually made of round wire of a harder metal than is appropriate for traditional cotter pins.

R-clips are similar in function to split pins and linchpins. Compared to split pins, they are easier to remove and are re-usable.

Uses

- Hitch pin clip (vehicles, towing, caravans)
- Hood pin (Stock car racing)
- Button attachment (brass buttons on certain military, railroad, and other uniform jackets)
- Body fastener (Radio-controlled car)

SPRING PIN

A spring pin (also called tension pin or roll pin) is a mechanical fastener that secures the position of two or more parts of a machine relative to each other. Spring pins have a body diameter which is larger than the hole diameter, and a chamfer on either one or both ends to facilitate starting the pin into the hole. The spring action of the pin allows it to compress as it assumes the diameter of the hole. The radial force exerted by the pin against the hole wall retains it in the hole, therefore a spring pin is considered a self retaining fastener.

Types

There are two types of spring pins: *slotted spring pins* and *coiled spring pins*.

Coiled spring pins

A coiled spring pin, also known as a *spiral pin*, is a self retaining engineered fastener manufactured by roll forming metal strip into a spiral cross section of 2 1/4 coils. Coiled spring pins have a body diameter larger than the recommended hole diameter and chamfers on both ends to facilitate starting the pin into the hole, the spring action of the pin allows it to compress as it assumes the diameter of the hole.

When coiled spring pins are installed, the compression starts at the outer edge and moves through the coils towards the center. Coiled pins continue to flex after insertion when a load is applied to the pin thus providing excellent performance to counter fatigue in dynamic applications. Coiled spring pins were invented by Herman Koehl circa 1948.

Coiled pins are commercially available in three different duties, standard (ISO 8750), heavy (ISO 8748) and light duty (ISO 8751), which provide for a variety of combinations of strength, flexibility and diameter to suit different mating host materials and performance requirements. Typical materials for coiled spring pins include high carbon steel, stainless steel and alloy 6150.

Coiled pins are used extensively in cosmetic cases, automotive door handles and locks, and latches as hinge pins. They're also used as pivots and axles, for alignment and stopping, to fasten multiple components together—such as a gear and shaft—and even as ejector pins to remove motherboards from PCs. The automotive and electrical industries use coiled pins in such products as steering boxes and columns, pumps, electric motors and circuit breakers.

International Standards

- Standard Duty: UNE–EN-ISO 8750, NASM10971, NASM51923, NAS1407, ASME B18.8.2, ASME B18.8.3M
- Heavy Duty: UNE–EN-ISO 8748, NASM10971, NASM39086, NAS561, ASME B18.8.2, ASME B18.8.3M
- Light Duty: UNE–EN-ISO 8751, NASM10971, NASM51987, NAS1407, ASME B18.8.2, ASME B18.8.3M
- Standard duty coiled spring pins offer the best balance between flexibility and strength and are recommended for most applications.

- Heavy duty coiled spring pins are typically used in high shear strength applications and hardened host materials.
- Light duty pins are used in applications with soft metals and plastics holes where there is a high risk of enlarging or breaking the host using a traditional press fit solid pin.

Slotted spring pins

Slotted spring pins are cylindrical pins rolled from a strip of material with a slot to allow the pin to have some flexibility during insertion. Slotted spring pins are also known as *roll pins* or "C". pins.

SPLIT PIN

A split pin, also known in the United States as a cotter pin or cotter key, is a metal fastener with two tines that are bent during installation, similar to a staple or rivet. Typically made of thick wire with a half-circular cross section, split pins come in multiple sizes and types.

The British definition of "cotter pin" is equivalent to U.S. term "cotter", which can be a cause for confusion when companies of both countries work together. There are signs that manufacturers and stockists are increasingly listing both names together to avoid confusion; this led to the term split cotter sometimes being used for a split pin.

A new split pin has its flat inner surfaces touching for most of its length so that it appears to be a split cylinder. Once inserted, the two ends of the pin are bent apart, locking it in place. When they are removed they are supposed to be discarded and replaced, because of fatigue from bending.

Split pins are typically made of soft metal, making them easy to install and remove, but also making it inadvisable to use them to resist strong shear forces. Common materials include mild steel, brass, bronze, stainless steel, and aluminium.

PINCUSHION

A pincushion (or pin cushion) is a small cushion, typically 3–5 cm across, which is used in sewing to store pins or needles with their heads protruding so as to take hold of them easily, collect them, and keep them organized.

Pincushions are typically filled tightly with stuffing, so as to hold pins rigidly once placed. Magnetic pin cushions are also used, though technically they are not "cushions" they serve the same basic function of holding pins neatly.

The recorded origins of pincushions date back to the Middle Ages of Europe. In the English language, they became known by many names: "pimpilowes, pimpilos, pimplos, pimploes, pin-pillows, pin-poppets". In 1376, Jehanne de Mesnil was bequeathed a silver pin case in a French text called *Testament of Advice* written by a woman known as La Monteure, from Rouen. Other references to pin cases during the Medieval era exist. By the 16th century, these were supplanted by references to "pyn pillows". Some examples from various parts of Europe survive that have elaborate embroidery. Small porcelain baskets with a pin cushion inside were highly popular, as were small cushions such as wedding pillows or maternity pillows, embroidered with messages.

During the 18th century, weighted pincushions became popular among seamstresses. In England, seam clamps attached to a table and designed for holding hems for sewing became common, and were often in the shape of a bird (the tail would be pinched to open and close the "beak" to hold the fabric); attached to the back of the bird was a velvet pin cushion.

One popular design—a tomato with a strawberry attached—was most likely introduced during the Victorian Era. According to folklore, placing a tomato on the mantel of a new house guaranteed prosperity and repelled evil spirits. If tomatoes were out of season, families improvised by using a round ball of red fabric filled with sand or sawdust. The good-luck symbol also served a practical purpose—a place to store pins. Typically, the tomato is filled with wool roving to prevent rust, and the strawberry is filled with an abrasive to clean and sharpen the pins.

PINCUSHION DOLLS

Porcelain pincushion dolls, or half-dolls, were fashionable in late 19th-century Europe, and remain collectable today. Millions were made and sold during the 19th century, but due to their fragility, examples in excellent condition remain scarce. The form resembles a typical china figurine of a beautiful woman, but the porcelain doll ends at the waist, where holes are included in the design to allow the half-doll to be stitched to a pincushion. The pincushion half of the doll may be made of satin fabric and trimmings to resemble a skirt.

The original popularity of the pincushion dolls continued into the early 20th century, and some styles reflect Art Deco or similar 20th-century styles. The most collectable examples are from pre-World War II Dresden and Meissen, which may sell for around $500 US when the condition is perfect. More common designs in imperfect condition sell for less than $25 US. Similar dolls were produced to top the covers of tissue boxes, jewelry boxes, or tea pot cosies. Some uncommon examples are nude and have a risque style.

PINKING SHEARS

Pinking shears are scissors, the blades of which are sawtoothed instead of straight. They leave a zigzag pattern instead of a straight edge.

Pinking shears have a utilitarian function for cutting woven cloth. Cloth edges that are unfinished will easily fray, the weave becoming undone and threads pulling out easily. The sawtooth pattern does not prevent the fraying but limits the length of the frayed thread and thus minimizes damage.

These scissors can also be used for decorative cuts and a number of patterns (arches, sawtooth of different aspect ratios, or asymmetric teeth) are available. True dressmaker's pinking shears, however, should not be used for paper decoration because paper dulls the cutting edge.

The cut produced by pinking shears may have given its name to (or be derived from) the plant name pink, a flowering plant in the genus *Dianthus* (commonly called a carnation). The colour pink may have been named after these flowers, although the origins of the name are not definitively known. As the carnation has scalloped, or "pinked", edges to its petals, pinking shears can be thought to produce an edge similar to the flower.

The word "pink" can be used as a verb dating back to 1300 meaning "pierce, stab, make holes in". The French word *piquer* and Spanish word *picar* are derived from the Latin *pungere* meaning "to pierce, prick" and related to "pungent".

PATENTS

Louise Austin of Whatcom, Washington, received U.S. Patent 489,406 on January 3, 1893, for "pinking scissors." The patent describes how "pinking scissors" are superior to the existing tools at the time, "pinking irons" and "pinking cutters." The operation of the shears are described as "pinking" or "scalloping." There are references to "cut ornamental openings in the body portion of fabrics," but no references to the more utilitarian function of preventing fraying. One of the primary early uses of pinking shears was the formation of decorative edging for patchwork quilting squares.

Samuel Briskman, of Brooklyn, received U.S. Patent 1,959,190, for a method of manufacturing pinking shears, and U.S. Patent 1,965,443 and U.S. Patent 1,970,408, describing the shears themselves, in 1934. He formed the Pinking Shears Corporation and set up a factory at 102 Prince Street in Manhattan. His firm milled the teeth into the blades. Wiss made the actual shears and had the exclusive sales in the USA and through their agents abroad. Briskman was also entitled to sell abroad under the name of Pinking Shears Corp. through his agents. He had one son, Artie, who worked with him. Sales fell off after a change in the type of fabric that was popular. Briskman died in February 1967, and was known as an inventor and philanthropist at that time. Norman Wiss, Sr. was the one who pushed for and who managed the agreement with Briskman.

Benjamin Luscalzo, of Chicago, Illinois, received U.S. Patent 2,600,036 on June 10, 1952, for his improvements to pinking shears. He provided an adjustable tension means connected to one of the blades or jaws of the shears which kept the teeth in efficient cutting relationship so that the cutting plane is always perpendicular to the pivotal axis. In other words, Benjamin Luscalzo brought traditional scissors and the pinking blade together to create what we know today as pinking shears.

SCISSORS

Scissors are hand-operated shearing instruments. They consist of a pair of metal blades pivoted so that the sharpened edges slide against each other when the handles (bows) opposite to the pivot are closed. Scissors are used for cutting various thin materials, such as paper, cardboard, metal foil, thin plastic, cloth, rope, and wire. Scissors can also be used to cut hair. Hair-cutting scissors have a specific blade angle ideal for cutting hair. Using the incorrect scissors to cut hair will result in increased damage and or split ends by breaking the hair. Food scissors, also known as kitchen scissors, are for cutting and trimming foods such as meats. Hair-cutting scissors and shears are functionally equivalent, but the larger implements tend to be called shears.

A large variety of scissors and shears exist for different specialized purposes.

Modern scissors are often designed ergonomically with composite thermoplastic and rubber handles which enable the user to exert either a power grip or a precision grip.

The noun "scissors" is treated as a plural noun, and therefore takes a plural verb ("these scissors are"). Alternatively, it is also referred to as "a pair of scissors". In American English, "a pair" is singular and therefore takes a singular verb ("this pair of scissors is"). In other forms of English, "a pair" does not take the singular (so simply "these scissors are"). The word shears is used to describe similar instruments that are larger in size and for heavier cutting. Opinions vary geographically as to the size at which 'scissors' become 'shears', but this is often at between six to eight inches (about 15 to 20 cm) in length.

HISTORY

It is most likely that scissors were invented around 1500 BC in ancient Egypt. The earliest known scissors appeared in Mesopotamia 3,000 to 4,000 years ago. These were of the 'spring scissor' type comprising two bronze blades connected at the handles by a thin, flexible strip of curved bronze which served to hold the blades in alignment, to allow them to be squeezed together, and to pull them apart when released.

Spring scissors continued to be used in Europe until the 16th century. However, pivoted scissors of bronze or iron, in which the blades were pivoted at a point between the tips and the handles, the direct ancestor of modern scissors, were invented by the Romans around AD100. They entered common use not only in ancient Rome, but also in China, Japan, and Korea, and the idea is still used in almost all modern scissors.

Early manufacture

During the Middle Ages and Renaissance, spring scissors were made by heating a bar of iron or steel, then flattening and shaping its ends into blades on an anvil. The center of the bar was heated, bent to form the spring, then cooled and reheated to make it flexible.

William Whiteley and Sons (Sheffield) Ltd. was manufacturing scissors by 1760, although it is believed the business began trading even earlier. The first trade-mark, 332, was granted in 1791.

Pivoted scissors were not manufactured in large numbers until 1761, when Robert Hinchliffe produced the first pair of modern-day scissors made of hardened and polished cast steel. He lived in Cheney Square, London and was reputed to be the first person who put out a signboard proclaiming himself "fine scissor manufacturer".

During the 19th century, scissors were hand-forged with elaborately decorated handles. They were made by hammering steel on indented surfaces known as bosses to form the blades. The rings in the handles, known as bows, were made by punching a hole in the steel and enlarging it with the pointed end of an anvil.

In 1649, in Swedish-ruled Finland, an ironworks was founded in the village of Fiskars between Helsinki and Turku. In 1830, a new owner started the first cutlery works in Finland, making, among other items, scissors with the Fiskars trademark. In 1967, Fiskars Corporation introduced new methods to scissors manufacturing.

DESCRIPTION AND OPERATION

A pair of scissors consists of two pivoted blades. In lower-quality scissors the cutting edges are not particularly sharp; it is primarily the shearing action between the two blades that cuts the material. In high-quality scissors the blades can be both extremely sharp, and tension sprung - to increase the cutting and shearing tension only at the exact point where the blades meet. The hand movement (pushing with the thumb, pulling with the fingers in right handed use) can add to this tension. An ideal example is in high-quality tailor's scissors or shears, which need to be able to perfectly cut (and not simply tear apart) delicate cloths such as chiffon and silk.

Children's scissors are usually not particularly sharp, and the tips of the blades are often blunted or 'rounded' for safety.

Mechanically, scissors are a first-class double-lever with the pivot acting as the fulcrum. For cutting thick or heavy material, the mechanical advantage of a lever can be exploited by placing the material to be cut as close to the fulcrum as possible. For example, if the applied force (at the handles) is twice as far away from the fulcrum as the cutting location (*i.e.*, the point of contact between the blades), the force at the cutting location is twice that of the applied force at the handles. Scissors cut material by applying a local shear stress at the cutting location which exceeds the material's shear strength.

Some scissors have an appendage, called a finger brace or finger tang, below the index finger hole for the middle finger to rest on to provide for better control and more power in precision cutting. A finger tang can be found on many quality scissors (including inexpensive ones) and especially on scissors for cutting hair. In hair cutting, some claim the ring finger is inserted where some place their index finger, and the little finger rests on the finger tang.

For people who do not have the use of their hands, there are specially designed foot operated scissors. Some quadriplegics can use a motorized mouth-operated style of scissor.

LEFT-HANDED SCISSORS

Most scissors are best-suited for use with the right hand, but *left-handed* scissors are designed for use with the left hand. Because scissors have overlapping blades, they are not symmetric. This asymmetry is true regardless of the orientation and shape of the handles: the blade that is on top always forms the same diagonal regardless of orientation. Human hands are also asymmetric, and when closing, the thumb and fingers do not close vertically, but have a lateral component to the motion. Specifically, the thumb pushes out from the palm and the fingers pull inwards. For right-handed scissors held in the right hand, the thumb blade is closer to the user's body, so that the natural tendency of the right hand is to force the cutting blades together. Conversely, if right-handed scissors are held in the left hand, the natural tendency of the left hand would be to force the cutting blades laterally apart. Furthermore, with right-handed scissors held by the right-hand, the shearing edge is visible, but when used with the left hand, the cutting edge of the scissors is behind the top blade, and one cannot see what is being cut.

Some scissors are marketed as ambidextrous. These have symmetric handles so there is no distinction between the thumb and finger handles, and have very strong pivots so that the blades simply rotate and do not have any lateral give. However, most "ambidextrous" scissors are in fact still right-handed in that the upper blade is on the right, and hence is on the outside when held in the right hand. Even if they successfully cut, the blade orientation will block the view of the cutting line for a left-handed person. True ambidextrous scissors are possible if the blades are double-edged and one handle is swung all the way around (to almost 360 degrees) so that the back of the blades become the new cutting edges. Patents (U.S. Patent 3,978,584) have been awarded for true ambidextrous scissors.

SEAM RIPPER

A seam ripper is a small tool used for unpicking stitches. The most common form consists of a handle, shaft and head. The head is usually forked with one side of the fork flattening out and becoming a blade and the other side forming a small point. In some designs the blade side then tapers back to a point to allow easier insertion in tight stitching. In use the blade is inserted into the seam underneath the thread to be cut. The thread is allowed to slip down into the fork and the tool is then lifted upwards allowing the blade to cut through the thread. Once the seam has been undone in this way the loose ends can be removed and the seam resewn.

SEWING NEEDLE

A sewing needle is a long slender tool with a pointed tip. The first needles were made of bone or wood; modern ones are manufactured from high carbon steel wire, nickel- or 18K gold plated for corrosion resistance. The highest quality embroidery needles are plated with two-thirds platinum and one-thirds titanium alloy. Traditionally, needles have been kept in needle books or needle cases which have become an object of adornment. Sewing needles can also be kept in an etui, a small box that held needles and other items such as scissors, pencils and tweezers. A needle for hand-sewing has an eye, at the blunt end to carry thread or cord through the fabric after the pointed end pierces it.

TYPES OF HAND SEWING NEEDLES

Hand sewing needles come in a variety of types/ classes designed according to their intended use and in a variety of sizes within each type.

- Sharp Needles: used for general hand sewing; built with a sharp point, a round eye, and are of medium length. Those with a double-eyes are able to carry two strands of thread while minimizing fabric friction.
- Appliqué: These are considered another all-purpose needle for sewing, appliqué, and patch work.
- Embroidery: Also known as crewel needles; identical to sharps but have a longer eye to enable easier threading of multiple embroidery threads and thicker yarns.
- Betweens or Quilting: These needles are shorter than sharps, with a small rounded eye and are used for making fine stitches on heavy fabrics such as in tailoring, quilt making and other detailed handwork; note that some manufacturers also distinguish between quilting needles and quilting *between* needles, the latter being slightly shorter and narrower than the former.
- Milliners: A class of needles generally longer than sharps, useful for basting and pleating, normally used in millinery work.
- Easy- or Self-threading: Also called calyxeyed sharps, side threading, and spiral eye needles, these needles have an open slot into which a thread may easily be guided rather than the usual closed eye design.
- Beading: These needles are very fine, with a narrow eye to enable them to fit through the centre of beads and sequins along with a long shaft to thread and hold a number of beads at a time.
- Bodkin: Also called ballpoints, this is a long, thick needle with a ballpoint end and a large, elongated eye. They can be flat or round

and are generally used for threading elastic, ribbon or tape through casings and lace openings.

- Chenille: These are similar to tapestry needles but with large, long eyes and a very sharp point to penetrate closely woven fabrics. Useful for ribbon embroidery.
- Darning: Sometimes called finishing needles, these are designed with a blunt tip and large eye making them similar to tapestry needles but longer; yarn darners are the heaviest sub-variety.
- Doll: Not designed for hand sewing at all, these needles are made long and thin and are used for soft sculpturing on dolls, particularly facial details.
- Leather: Also known as glovers and as wedge needles, these have a triangular point designed to pierce leather without tearing it; often used on leather-like materials such as vinyl and plastic.
- Sailmaker: Similar to leather needles, but the triangular point extends further up the shaft; designed for sewing thick canvas or heavy leather.
- Tapestry: The large eye on these needles lets them to carry a heavier weight yarn than other needles, and their blunt tip—usually bent at a slight angle from the rest of the needle—allows them to pass through loosely woven fabric such as embroidery canvas or even-weave material without catching or tearing it; comes in a double-eyed version for use on a mounted frame and with two colours of thread.
- Tatting: These are built long with an even thickness for their entire length, including at the eye, to enable thread to be pulled through the double stitches used in tatting.
- Upholstery: These needles are heavy, long needles that may be straight or curved and are used for sewing heavy fabrics, upholstery work, tufting and for tying quilts; the curved variety is practical for difficult situations on furniture where a straight needle will not work Heavy duty 12" needles are used for repairing mattresses. Straight sizes: 3"-12" long, curved: 1.5"-6" long.

HISTORY

Prehistoric sewing needles

A variety of archaeological finds illustrate sewing has been present for thousands of years. A point that might be from a bone needle dates to 61,000 years ago and was discovered in Sibudu Cave, South Africa. Later Stone Age finds, such as the excavations on the island of Öland at Alby, Sweden, reveal objects such as bone needle cases dating to 60,000 years ago. Ivory needles were also found dated to 30,000 years ago at the Kostenki site in Russia. A bone needle, dated to the Aurignacian age (47000 to 41,000 years ago), was discovered in Potok Cave in the Eastern Karavanke, Slovenia.

Native Americans were known to use sewing needles from natural sources. One such source, the agave plant, provided both the needle and the "thread." The agave leaf would be soaked for an extended period of time, leaving a pulp, long, stringy fibres and a sharp tip connecting the ends of the fibres. The "needle" is essentially what was the tip end of the leaf. Once the fibres dried, the fibres and "needle" could then be used to sew together skins and other items used in a cloth-like manner.

TAILOR'S HAM

A tailor's ham or dressmaker's ham is a tightly stuffed pillow used as a curved mold when pressing curved areas of clothing, such as darts, sleeves, cuffs, collars, or waistlines. Pressing on a curved form allows a garment better to fit body contours. To accommodate tapering or garments of different sizes, it has roughly the shape of a ham.

TAPE MEASURE

A tape measure or *measuring tape* is a flexible ruler. It consists of a ribbon of cloth, plastic, fibre glass, or metal strip with linear-measurement markings. It is a common measuring tool. Its design allows for a measure of great length to be easily carried in pocket or toolkit and permits one to measure around curves or corners. Today it is ubiquitous, even appearing in miniature form as a keychain fob, or novelty item. Surveyors use tape measures in lengths of over 100 m (300+ ft).

USES

Tape measures that were intended for use in tailoring or dressmaking were made from flexible cloth or plastic.These types of tape measures were mainly used for the measuring of the human's waist line. Today, measuring tapes made for sewing are made of fiberglass, which does not tear or stretch as easily. Measuring tapes designed for carpentry or construction often use a stiff, curved metallic ribbon that can remain stiff and straight when extended, but retracts into a coil for convenient storage. This type of tape measure will have a floating tang or hook on the end to aid measuring. The tang is connected to the tape with loose rivets through oval holes, and can move a distance equal to its thickness, to provide both inside and outside measurements that are accurate. A tape measure of 25 or even 100 feet can wind into a relatively small container. The self-marking tape measure allows the user an accurate one hand measure.

HISTORY

The first record of a people using a measuring device was by the Romans using marked strips of leather, but this was more like a regular ruler than a tape measure. On 3 January 1922, Hiram Farrand received the patent he filed in 1919. Sometime between 1922 and December 1926, Farrand experimented with the help of The Brown Company in Berlin, New Hampshire. It is there Hiram and W.W. Brown began mass-producing the tape measure. Their product was later sold to Stanley Works.

DESIGN

The design on which most modern spring tape measures are built was patented by a New Haven, Connecticut resident named Alvin J. Fellows on 14 July 1868. According to the text of his patent, Fellows' tape measure was an improvement on other versions previously designed. The spring tape measure has existed since Fellows' patent in 1868, but did not come into wide usage until the early 1900s, when carpenters began slowly adopting H. A. Farrand's design as the one more commonly used, which is the design all modern tape measures use today. With the mass production of the integrated circuit (IC) the tape measure has also entered into the digital age with the digital tape measure. Some incorporate a digital screen to give measurement readouts in multiple formats. There are also other styles of tape measures that have incorporated lasers and ultrasonic technology to measure the distance of an object with fairly reliable accuracy.

United States

Justus Roe, a surveyor and tape-maker by trade, made the longest tape measure in 1956, at 600 feet (180 m). The Northern Virginia Surveyors Association presented the 600-foot, gold-plated surveyor's tape measure to Mickey Mantle in 1956. Some tapes sold in the United States have additional marks in the shape of small black diamonds, which appear every 19.2 inches (49 cm). These are known as "metric layouts" markings, and are used to mark out equal truss lengths for roofing materials (five trusses per standard 8-foot (2.4 m) length of building material). Many tapes also have special markings every 16 inches (41 cm), which is a standard interval for studs in housing.

THIMBLE

A thimble is a small hard pitted cup worn for protection on the finger that pushes the needle in sewing. Usually, thimbles with a closed top are used by dressmakers but special thimbles with an opening at the end are used by tailors as this allows them to manipulate the cloth more easily. Finger guards differ from tailors' thimbles in that they often have a top but are open on one side. Some finger guards are little more that a finger shield attached to a ring to maintain the guard in place. The Old English word *þ3^mel*, the ancestor of thimble, is derived from Old English *þûma*, the ancestor of our word thumb.A single steel needle from the time of the Han Dynasty ancient China (206BC - 202AD) was found in a tomb in Jiangling, and it could conceivably be assumed that thimbles were in use at this time also although no thimble seems to have been discovered with the needle.

The earliest known thimble — in the form of a simple ring — dates back to the Han Dynasty ancient China also and was discovered during the Cultural Revolution of the People's Republic of China (PRC) in a lesser dignitary's tomb. Oddly, neither the Romans nor the Greeks before them appear to have used metal thimbles. It may be that leather or cloth finger guards proved sufficiently robust for their purposes. There are so-called Roman thimbles in museum collections, but the provenance of these metal thimbles is, in fact, not certain, and many have been removed from display. No well-documented archeological data link metal thimbles to any Roman site. According to the United Kingdom Detector Finds Database, thimbles dating to the 10th century have been found in England, and thimbles were in widespread use there by the 14th century.

Although there are isolated examples of thimbles made of precious metals—Elizabeth I is said to have given one of her ladies-in-waiting a thimble set with precious stones—the vast majority of metal thimbles were made of brass. Medieval thimbles were either cast brass or made from hammered sheet.Early centers of thimble production were those places known for brass-working, starting with Nuremberg in the 15th century, and moving to Holland by the 17th.

In 1693, a Dutch thimble manufacturer named John Lofting established a thimble manufactory in Islington, in London, England, expanding British thimble production to new heights. He later moved his mill to Buckinghamshire to take advantage of water-powered production, resulting in a capacity to produce more than two million thimbles per year. By the end of the 18th century, thimble making had moved to Birmingham, and shifted to the "deep drawing" method of manufacture, which alternated hammering of sheet metals with annealing, and produced a thinner-skinned thimble with a taller shape. At the same time, cheaper sources of silver from the Americas made silver thimbles a popular item for the first time.

Thimbles are usually made from metal, leather, rubber, and wood, and even glass or china. Early thimbles were sometimes made from whale bone, horn, or ivory. Natural sources were also utilized such as Connemara marble, bog oak, or mother of pearl. Rarer works from thimble makers utilized diamonds, sapphires, or rubies.

Advanced thimblemakers enhanced thimbles with semi-precious stones to adorn the apex or along the outer rim. Cabochon adornments are sometimes made of cinnabar, agate, moonstone, or amber. Thimble artists would also utilize enameling, or the Guilloché techniques advanced by Peter Carl Fabergé.

Originally, thimbles were used simply solely for pushing a needle through fabric or leather as it was being sewn. Since then, however, they have gained many other uses. From the 16th century onwards silver thimbles were regarded as an ideal gift for ladies.

Early Meissen porcelain and elaborate, decorated gold thimbles were also given as 'keepsakes' and were usually quite unsuitable for sewing. This tradition has continued to the present day. In the early modern period, thimbles were used to measure spirits, and gunpowder, which brought rise to the phrase "just a thimbleful". Prostitutes used them in the practice of thimble-knocking where they would tap on a window to announce their presence. Thimble-knocking also refers to the practice of Victorian schoolmistresses who would tap on the heads of unruly pupils with dames thimbles.

Before the 18th century the small dimples on the outside of a thimble were made by hand punching, but in the middle of that century, a machine was invented to do the job. If one finds a thimble with an irregular pattern of dimples, it was likely made before the 1850s. Another consequence of the mechanization of thimble production is that the shape and the thickness of the metal changed. Early thimbles tend to be quite thick and to have a pronounced dome on the top. The metal on later ones is thinner and the top is flatter.

Collecting thimbles became popular in the UK when many companies made special thimbles to commemorate the Great Exhibition held in the Crystal Palace in Hyde Park, London. In the 19th century, many thimbles were made from silver; however, it was found that silver is too soft a metal and can be easily punctured by most needles.

Charles Horner solved the problem by creating thimbles consisting of a steel core covered inside and out by silver, so that they retained their aesthetics but were now more practical and durable. He called his thimble the Dorcas, and these are now popular with collectors. There is a small display of his work in Bankfield Museum, Halifax, England.

Early American thimbles made of whale bone or tooth featuring miniature scrimshaw designs are considered valuable collectibles. Such rare thimbles are prominently featured in a number of New England Whaling Museums.

During the First World War, silver thimbles were collected from "those who had nothing to give" by the British government and melted down to buy hospital equipment. In the 1930s and 40s glass-topped thimbles were used for advertising. Leaving a sandalwood thimble in a fabric store was a common practice for keeping moths away. Thimbles have also been used as love-tokens and to commemorate important events. People who collect thimbles are known as digitabulists. One superstition about thimbles says that if you have three thimbles given to you, you will never be married.

KNOWN THIMBLE MAKERS

Most of these thimble makers are no longer in existence.

- Avon Fashion Thimbles
- Wicks (Inventor USA)
- A Feaù and René Lorillon (French)
- Charles Horner (UK) (1837–1896)

- Charles Iles (UK)
- Charles May
- Anthony Stavrianoudakis (GR)
- Gabler Bros (German)
- Henry Fidkin (UK)
- Henry Griffith (UK)
- James Fenton (UK)
- James Swann (UK)
- Jean Levy (France)
- Johan Caspar Rumpe (Germany)
- Ketcham and McDougall (USA)
- Meissen (German)
- Roger Lenain (French)
- Samuel |Foskett (UK)
- Simons Bros Co (USA)
- Stern Bros and Co (USA)
- Waite-Thresher (USA)
- Webster (USA)
- William Prym (Germany)

THIMBLETTE

Thimblettes (also known as rubber finger, rubber thimbles and finger cones) are soft thimbles, made predominately of rubber, used primarily for leafing through or counting documents, bank notes, tickets, or forms. They also protect against paper cuts as a secondary function. Unlike thimbles, the softer thimblettes become worn over time. They are considered disposable and sold in boxes. The surface is dimpled with the dimples inverted to provide better grip. Thimblettes are sized from 00 through to 3.

TRACING PAPER

Tracing paper is paper made to have low opacity, allowing light to pass through. It was originally developed for architects and design engineers to create drawings which could be copied precisely using the diazo copy process, it then found many other uses. The original use for drawing and tracing was largely superseded by technologies which do not require diazo copying or manual copying (by tracing) of drawings.

The transparency of the paper is achieved by careful selection of the raw materials and the process used to create transparency. Cellulose fibre forms the basis of the paper, usually from wood species but also from cotton fibre. Often, paper contains other filler materials to enhance opacity and print quality. For tracing or translucent paper it is necessary to remove any material which obstructs the transmission of light.

There are three main processes to manufacture this type of paper, as follows:-

1. Through mechanical 'refining' of the cellulose fibre to create a fibre which is highly fibrillated and gelatinous, so that in forming the sheet of paper, virtually all air is excluded from the internal structure of the paper. This method produces a very translucent and even looking paper over a range of grammages from 42 to over 280 gsm.
2. By making a 'normal' sheet of paper and then filling the spaces between the fibres with a material that has the same refractive index as the cellulose. This was a common process adopted in the USA. The product was frequently called Vellum, although this terminology can refer to a wider range of special papers. Due to the relatively high cost, this method of manufacture has largely disappeared.
3. As with 2, by making a normal sheet of paper, which is followed by immersing uncut and unloaded paper of good quality in sulfuric acid for a few seconds. The acid converts some of the cellulose into amyloid form having a gelatinous and impermeable character. When the treated paper is thoroughly washed and dried, the resultant product is much stronger than the original paper. Tracing paper is resistant to oil, grease and to a large extent impervious to water and gas.

Tracing paper is named as such for its ability for an artist to trace an image onto it. When tracing paper is placed onto a picture, the picture is easily viewable through the tracing paper. Thus, it becomes easy for the artist to find edges in the picture and trace the image onto the tracing paper. Pure cellulose fibre is translucent, and it is the air trapped between fibres, that makes paper opaque and looks white. If the fibres are refined and beaten until all the air is taken out, then the resulting sheet will be translucent. Translucent papers are dense and contain up to 10 per cent moisture at 50 per cent humidity. This type of paper is roughly 25 per cent lighter than regular paper.

The sizing in production will determine whether it is for laser printer or inkjet/offset printing. Tracing paper may be uncoated or coated. Natural tracing paper for laser printing is usually uncoated. The HS code for tracing paper is 4806. Tracing paper is usually made from sulfite pulp by reducing the fibres to a state of fine subdivision and hydrolysing them by very prolonged beating in water.

TRACING WHEEL

A tracing wheel is an instrument with multiple teeth on a wheel attached to a handle. The teeth can be either serrated or smooth. It is used to transfer markings from patterns onto fabric with or without tracing paper, this sewing tool also makes slotted perforations. Such markings might include pleats, darts, buttonholes, notches or placement lines for appliques or pockets.

BOBBIN

A bobbin is a spindle or cylinder, with or without flanges, on which wire, yarn, thread or film is wound. Bobbins are typically found in sewing machines, cameras, and within electronic equipment. In non-electrical applications the bobbin is used for tidy storage without tangles. In electrical applications a coil of wire carrying a current has important magnetic properties. As used in spinning, weaving, knitting, sewing, or lacemaking, the bobbin provides temporary or permanent storage for yarn and may be made of plastic, metal, bone or wood.

Bobbin lacemaking is a handcraft which requires the winding of yarn onto a temporary storage spindle made of wood, previously bone, often turned on a lathe. Many lace designs use dozens of bobbins at any one time. Exotic woods are extremely popular with contemporary lacemakers. Both traditional and contemporary bobbins may be decorated with designs, inscriptions, or pewter or wire inlays. Often, the bobbins are 'spangled' to provide additional weight to keep the thread in tension. A hole is drilled near the base to enable glass beads and other ornaments to be attached by a loop of wire. Again, in the modern context of the hobby of bobbin lacemaking, these spangles provide a means of self-expression in the decoration of a tool of the craft. Both antique and unique bobbins, sometimes spangled, have become highly sought after by collectors.

In the case of an electrical transformer, inductor or relay, the bobbin is a permanent container for the wire, acting to form the shape of the coil (and ease assembly of the windings into or onto the magnetic core). The bobbin may be made of thermoplastic or thermosetting (for example, phenolic) materials. This plastic often has to have a TÜV, UL or other regulatory agency flammability rating for safety reasons.

9

Trades and Suppliers

DRAPER

Draper was originally a term for a retailer or wholesaler of cloth that was mainly for clothing. A draper may additionally operate as a cloth merchant or a haberdasher. Drapers were an important trade guild during the medieval period, when the sellers of cloth operated out of draper's shops. However the original meaning of the term has now largely fallen out of use. In 1724, Jonathan Swift wrote a series of satirical pamphlets in the guise of a draper called the *Drapier's Letters*.

HISTORICAL DRAPERS

A number of notable people who have at one time or another worked as drapers include:

- William Barley
- Norman Birkett
- Margaret Bondfield
- Eleanor Coade (1733-1821), successful businesswoman with Coade stone
- Antonie van Leeuwenhoek
- John Spedan Lewis
- Anthony Munday
- H. G. Wells
- Edward Whalley, regicide, cousin of Oliver Cromwell
- George Williams, founder of the YMCA
- John Woodward, geologist and physician to King Charles II

CURRENT USAGE

A draper is now defined as a highly skilled role within the fashion industry. The term is used within a fashion design or costume design studio for people tasked with creating garments or patterns by draping fabric over a dress form; draping uses a human form to physically position the cloth into a desired pattern.

This is an alternative method to drafting, when the garment is initially worked out from measurements on paper.

A fashion draper may also be known as a "first hand" because they are often the most skilled creator in the workshop and the "first" to work with the cloth for a garment. However a First Hand in a costume studio is often an assistant to the draper. They are responsible for cutting the fabric with the patterns and assisting in costume fittings.

DRESSMAKER

A dressmaker is a person who makes custom clothing for women, such as dresses, blouses, and evening gowns. Also called a mantua-maker (historically) or a modiste.

Notable dressmakers:

- Cristóbal Balenciaga
- Pierre Balmain
- Coco Chanel
- Christian Dior
- David Emanuel
- Jean Muir, fashion designer (though she herself preferred to be called a dressmaker)
- Isabel Toledo
- Madeleine Vionnet
- Charles Frederick Worth

HABERDASHER

A haberdasher is a person who sells small articles for sewing, such as buttons, ribbons, zips, and other notions (in the United Kingdom) or a men's outfitter (American English). A haberdasher's shop or the items sold therein are called *haberdashery*.

The word appears in Chaucer's *Canterbury Tales*. Haberdashers were initially peddlers, sellers of small items such as needles and buttons. It should be noted that the word is thought to have no connection with an Old Norse word akin to the Icelandic *haprtask*, which means *peddlers' wares* or the sack in which the peddler carried them. If that had been the case, a *haberdasher* (in its hypothetical Scandinavian meaning) would be very close to a *mercer* (French). Since the word has no recorded use in Scandinavia, it is most likely derived from the Anglo-Norman *hapertas*, meaning *small ware*. A haberdasher would retail small wares, the goods of the peddler, while a mercer would specialize in "linens, silks, fustian, worsted piece-goods and bedding".

Saint Louis IX, King of France 1226–70, is the patron saint of French haberdashers. In Belgium and elsewhere in Continental Europe, Saint Nicholas remains their patron saint, while Saint Catherine was adopted by the Worshipful Company of Haberdashers in the City of London.

Notable haberdashers:

- William Adams – a 17th-century London haberdasher who founded Adams' Grammar School in 1656
- Robert Aske – a philanthropist and London Alderman
- William Baldwin – an actor who became interested in haberdashery on the set of *Double Bang* in 2001
- Captain James Cook RN FRS – the 18th-century British navigator and explorer, apprenticed to this job in his youth
- Daniel Defoe – the author of *Robinson Crusoe*
- Richard Goldthorpe – a Yorkshire based haberdasher and real estate owner
- John Graunt – one of the first demographers
- John Jarrold - Norwich retailer
- Jerome Knapp Junior – an English barrister and businessman
- Wayne Knight – an actor, *e.g.* "Newman" from *Seinfeld*
- Christopher Lloyd – an actor, *e.g.* "Dr. Emmett Brown" in the *Back to the Future* trilogy
- Joseph Merrick – "the Elephant Man", worked as a haberdasher's assistant before being a freak show act
- George Newnes – founder of the *Tit-Bits* newspaper and the popular *The Strand Magazine*, of Sherlock Holmes-fame
- Paavo Nurmi – a Finnish distance runner
- Charles Taze Russell – founder of the Bible Student Movement which evolved into Jehovah's Witnesses
- Harry S. Truman – President of the United States from 1945 to 1953

MERCERY

Mercery (from French *mercerie*, the notions trade) initially referred to silk, linen, and fustian textiles imported to England in the 12th century. The term mercery later extended to goods made of these and the sellers of those goods. The term mercer for cloth merchants (from French *mercier*, "notions dealer") is now largely obsolete. Mercers were formerly merchants or traders who dealt in cloth, typically fine cloth that was not produced locally. However inventories of mercers in small towns suggest that many were shopkeepers who dealt in various other dry commodities, not only cloth. Related occupations include haberdasher, draper and cloth merchant, while clothier historically referred to someone who manufactured cloth, often under the domestic system. By the 21st century the word *mercer* was primarily used in connection with the Worshipful Company of Mercers, one of the twelve great Liveries Companies of the City of London.

Prominent mercers:

- Wynne Ellis, 19th century British mercer
- Geoffrey Boleyn, 15th century English mercer.

- Richard le Lacer, 14th century English mercer.
- Charles Woodmason

CORDWAINER

A cordwainer is a shoemaker who makes fine soft leather shoes and other luxury footwear articles. The word is derived from "cordwain", or "cordovan", the leather produced in Córdoba, Spain. The term *cordwainer* (also "Corviser") was used as early as 1100 in England. Historically, there was a distinction between a *cordwainer*, who made luxury shoes and boots out of the finest leathers, and a *cobbler*, who repaired them. This distinction gradually weakened, particularly during the twentieth century, when there was a predominance of shoe retailers who neither made nor repaired shoes.

In London, the occupation of cordwainer was historically controlled by the guild of the Worshipful Company of Cordwainers. The ward of the City of London named Cordwainer is historically where most cordwainers lived and worked.

Until 2000, a Cordwainers' Technical College existed in London. For over a hundred years, the College had been recognised as one of the world's leading establishments for training shoemakers and leather workers. It produced some of the leading fashion designers, including Jimmy Choo and Patrick Cox. In 2000, Cordwainers' College was absorbed into the London College of Fashion, the shoe-design and accessories departments of which are now called "Cordwainer's at London College of Fashion".

CORDWAINERS IN AMERICA

Cordwainers were among those who sailed to Virginia in 1610 to settle in Jamestown. By 1616, the secretary of Virginia reported that the leather and shoe trades were flourishing. Christopher Nelme, of England, was the earliest shoemaker in America whose name has been recorded; he sailed to Virginia from Bristol, England, in 1619. In 1620, the Pilgrims landed in Massachusetts. Nine years later, in 1629, the first shoemakers arrived, bringing their skills with them.

CORSET

A corset is a garment worn to hold and shape the torso into a desired shape for aesthetic or medical purposes (either for the duration of wearing it or with a more lasting effect). Both men and women are known to wear corsets, though women are more common wearers.

In recent years, the fashion industry has also borrowed the term "corset" to refer to tops which, to varying degrees, mimic the look of traditional corsets without actually acting as them. While these modern corsets and corset tops often feature lacing or boning and generally imitate a historical style of corsets, they have very little, if any, effect on the shape of the wearer's body. Genuine corsets are usually made by a corsetmaker and are frequently fitted to the individual wearer.

The word *corset* is derived from the Old French word *corps* and the diminutive of *body*, which itself derives from *corpus*—Latin for body. The craft of corset construction is known as *corsetry*, as is the general wearing of them. (The word *corsetry* is sometimes also used as a collective plural form of corset.) Someone who makes corsets is a *corsetier* or *corsetière* (French terms for a man and for a woman, respectively), or sometimes simply a *corsetmaker*.

In 1828, the word corset came into general use in the English language. The word was used in *The Ladies Magazine* to describe a "quilted waistcoat" called *un corset* by the French. It was used to differentiate the lighter corset from the heavier stays of the period.

USES

Fashion

The most common and well-known use of corsets is to slim the body and make it conform to a fashionable silhouette. For women, this most frequently emphasizes a curvy figure by reducing the waist and thereby exaggerating the bust and hips. However, in some periods, corsets have been worn to achieve a tubular straight-up-and-down shape, which involves minimizing the bust and hips.

However, there was a period from around 1820 to 1835 when a wasp-waisted figure (a small, nipped-in look to the waist) was also desirable for men; this was sometimes achieved by wearing a corset.

An "overbust corset" encloses the torso, extending from just under the arms towards the hips. An "underbust corset" begins just under the breasts and extends down towards the hips. A "longline corset" – either overbust or underbust – extends past the iliac crest, or the hip bone. A longline corset is ideal for those who want increased stability, have longer torsos or want to smooth out their hips. A "standard" length corset will stop short of the iliac crest and is ideal for those who want increased flexibility or have a shorter torso. Some corsets, in very rare instances, reach the knees. A shorter kind of corset, which covers the waist area (from low on the ribs to just above the hips), is called a *waist cincher*. A corset may also include garters to hold up stockings (alternatively a separate garter belt may be worn for that).

Traditionally, a corset supports the visible dress and spreads the pressure from large dresses, such as the crinoline and bustle. At times, a corset cover is used to protect outer clothes from the corset and to smooth the lines of the corset. The original corset cover was worn under the corset to provide a layer between it and the body. Corsets were not worn next to the skin, possibly due to difficulties with laundering these items during the 19th century, as they had steel boning and metal eyelets which would rust. The corset cover was in the form of a light chemise, made from cotton lawn or silk. Modern corset wearers also wear corset liners for many of the same reasons, but tight-lacers also use them to prevent rope burn from the corset's laces.

Medical

People with spinal problems, such as scoliosis, or with internal injuries may be fitted with a form of corset to immobilize and protect the torso. Andy Warhol was shot in 1968 and never fully recovered; he wore a corset for the rest of his life.

Fetish

Aside from fashion and medical uses, corsets are also used in sexual fetishism, most notably in BDSM activities. In BDSM, a submissive can be forced to wear a corset which would be laced very tight and give some degree of restriction to the wearer. A dominant can also wear a corset, often black, but for entirely different reasons, such as aesthetics. A specially designed corset, in which the breasts and vulva are exposed, can be worn during vanilla sex or BDSM activities.

CONSTRUCTION

Corsets are typically constructed of a flexible material (like cloth, particularly coutil, or leather) stiffened with boning (also called ribs or stays) inserted into channels in the cloth or leather. In the 19th century, bones of whale were favoured for the boning. Plastic is now the most commonly used material for lightweight, faux corsets and the majority of poor quality corsets. Spring and/or spiral steel is preferred for stronger, and generally better quality, corsets. Other materials used for boning have included ivory, wood, and cane. (By contrast, a girdle is usually made of elasticized fabric, without boning.)

Corsets are held together by lacing, usually (though not always) at the back. Tightening or loosening the lacing produces corresponding changes in the firmness of the corset. Depending on the desired effect and time period, corsets can be laced from the top down, from the bottom up, or both up from the bottom and down from the top, using two laces that meet in the middle. In the Victorian heyday of corsets, a well-to-do woman's corset laces would be tightened by her maid, and a gentleman's by his valet. However, Victorian corsets also had a buttoned or hooked front opening called a busk. If the corset was worn loosely, it was possible to leave the lacing as adjusted and take the corset on and off using the front opening (if the corset is worn snugly, this method will damage the busk if the lacing is not significantly loosened beforehand). Self-lacing is also incredibly difficult with tightlacing - also called waist training - which strives for the utmost possible reduction of the waist. Corset and bodice lacing became a mark of class, front laced bodices being worn by women who could not afford servants.

WAIST REDUCTION

By wearing a tightly-laced corset for extended periods, known as tightlacing, men and women can learn to tolerate extreme waist constriction and eventually reduce their natural waist size. Though petite women are often able to get down to a smaller waist in absolute numbers, women with more fat are typically able to reduce their waists by a larger percentage. Many tightlacers dream of waists that are 16 inches (41 cm) and 17 inches (43 cm). Some went so far that they could only breathe with the top part of their lungs. This caused the bottom part of their lungs to fill with mucus. Symptoms of this include a slight but persistent cough, as well as heavy breathing, causing a heaving appearance of the bosom. Until 1998, the Guinness Book of World Records listed Ethel Granger as having the smallest waist on record at 13 inches (33 cm). After 1998, the category changed to "smallest waist on a living person" and Cathie Jung took the title with a waist measuring 15 inches (38 cm). Other women, such as Polaire, also have achieved such reductions (14 inches (36 cm) in her case). However, these are extreme cases. Corsets were and are still usually designed for support, with freedom of body movement an important consideration in their design.

COMFORT

In the past, a woman's corset was usually worn over a chemise, a sleeveless low-necked gown made of washable material (usually cotton or linen). It absorbed perspiration and kept the corset and the gown clean. In modern times, an undershirt or corset liner may be worn. Moderate lacing is not incompatible with vigourous activity. Indeed, during the second half of the 19th century, when corset wearing was common, there were sport corsets specifically designed to wear while bicycling, playing tennis, or horseback riding, as well as for maternity wear.

HISTORY

The corset has been erroneously attributed to Catherine de' Medici, wife of King Henry II of France. She enforced a ban on thick waists at court attendance during the 1550s. For nearly 350 years, women's primary means of support was the corset, with laces and stays made of whalebone or metal. Other researchers have found evidence of the use of corsets in early Crete.

The corset has undergone many changes. Originally, it was known as "a pair of bodys" in the late 16th century. It was a simple bodice, stiffened with boning of reed or whalebone. The central front was further reinforced by a busk made of wood, horn, whalebone, metal or ivory. It was most often laced in the back, and was, at first, a garment reserved for the aristocracy. The term "stays" was generally used during the 17th and 18th centuries. In the 17th century, tabs (called "fingers") at the waist were added.

Stays evolved in the 18th century when whalebone began to be used more, and there was more boning used in the garment. The shape of the stays changed as well. The stays were low and wide in the front, while in the back they could reach as high as the upper shoulder. Stays could be strapless or use shoulder straps. The straps of the stays were generally attached in the back and tied at the front sides.

The purpose of 18th century stays was to support the bust and confer the fashionable conical shape while drawing the shoulders back. At this time, the eyelets were reinforced with stitches and were not placed across from one another, but instead staggered. This allowed the stays to be spiral laced. One end of the stay lace was inserted and knotted in the bottom eyelet; the other end was wound through the stays' eyelets and tightened on the top. Tight-lacing was not the purpose of stays in this time period. Stays were worn by women in all societal levels, from ladies of the court to street vendors. During this time period, there is evidence of a variant of stays, called "jumps", which were looser than stays with attached sleeves, like a jacket.

Corsets were originally quilted waistcoats, worn by French women as an alternative to stiff corsets. They were only quilted linen, laced in the front, and un-boned. This garment was meant to be worn on informal occasions, while stays were worn for court dress. In the 1790s, stays began to fall out of fashion. This development coincided with the French Revolution, and the adoption of neo-classical styles of dress. Interestingly, it was the men, Dandies, who began to wear corsets. The fashion persisted thorough the 1840s, though after 1850 men who wore corsets claimed they needed them for "back pain".

In the early 19th century, when gussets were added for room for the bust, stays became known as corsets. They also lengthened to the hip and the lower tabs were replaced by gussets at the hip and had less boning. The shoulder straps disappeared in the 1840s for normal wear.

In the 1820s, fashion changed again, with the waistline lowered back to almost the natural position. Corsets began to be made with some padding and more boning. Some women made their own, while others bought their corsets. Corsets were one of the first mass-produced garments for women. Corsets began to be more heavily boned in the 1840s. By 1850, steel boning became popular.

With the advent of metal eyelets, tight lacing became possible. The position of the eyelets changed. They were now situated across from one another at the back. The front was fastened with a metal busk in front. Corsets were mostly white. The corsets of the 1850s–1860s were shorter than the corsets of the 19th century through 1840s. This was because of a change in the silhouette of women's fashion. The 1850s and 60s emphasized the hoopskirt. After the 1860s, when the hoop fell out of style, the corset became longer to mold the abdomen, exposed by the new lines of the princess or cuirass style.

During the Edwardian period, the straight front corset (also known as the S-Curve corset) was introduced. This corset was straight in front with a pronounced curve at the back that forced the upper body forward and the derrière out. This style was worn from 1900 to 1908.

The corset reached its longest length in the early 20th century. The longline corset at first reached from the bust down to the upper thigh. There was also a style of longline corset that started under the bust, and necessitated the wearing of a brassiere. This style was meant to complement the new silhouette. It was a boneless style, much closer to a modern girdle than the traditional corset. The longline style was abandoned during World War I.

The corset fell from fashion in the 1920s in Europe and North America, replaced by girdles and elastic brassieres, but survived as an article of costume. Originally an item of lingerie, the corset has become a popular item of outerwear in the fetish, BDSM and goth subcultures. In the fetish and BDSM literature, there is often much emphasis on tightlacing, and many corset makers cater to the fetish market.Outside the fetish community, living history re-enactors and historic costume enthusiasts still wear stays and corsets according to their original purpose to give the proper shape to the figure when wearing historic fashions. In this case, the corset is underwear rather than outerwear. Skilled corset makers are available to make reproductions of historic corset shapes or to design new styles.

There was a brief revival of the corset in the late 1940s and early 1950s in the form of the waist cincher sometimes called a "waspie". This was used to give the hourglass figure as dictated by Christian Dior's "New Look". However, use of the waist cincher was restricted to haute couture, and most women continued to use girdles. This revival was brief, as the New Look gave way to a less dramatically-shaped silhouette.

Since the late 1980s, the corset has experienced periodic revivals, all which have usually originated in haute couture and have occasionally trickled through to mainstream fashion. These revivals focus on the corset as an item of outerwear rather than underwear. The strongest of these revivals was seen in the Autumn 2001 fashion collections and coincided with the release of the film *Moulin Rouge!*; the costumes featured many corsets as characteristic of the era. Another fashion movement, which has renewed interest in the corset, is the Steampunk culture that utilizes late-Victorian fashion shapes in new ways.

SPECIAL TYPES

There are some special types of corsets and corset-like devices which incorporate boning.

Corset dress

A corset dress (also known as hobble corset because it produces similar restrictive effects to a hobble skirt) is a long corset. It is like an ordinary corset, but it is long enough to cover the legs, partially or totally. It thus looks like a dress, hence the name. A person wearing a corset dress can have great difficulty in walking up and down the stairs (especially if wearing high-heeled footwear) and may be unable to sit down if the boning is too stiff. Other types of corset dresses are created for unique high fashion looks by a few modern corset makers. These modern styles are functional as well as fashionable and are designed to be worn with comfort for a dramatic look.

Neck corset

A neck corset is a type of posture collar incorporating stays and it is generally not considered to be a true corset.

GLOVE

A glove (Middle English from Old English *glof*) is a garment covering the whole hand. Gloves have separate sheaths or openings for each finger and the thumb; if there is an opening but no covering sheath for each finger they are called "fingerless gloves". Fingerless gloves with one large opening rather than individual openings for each finger are sometimes called gauntlets. Gloves which cover the entire hand or fist but do not have separate finger openings or sheaths are called mittens. Mittens are warmer than gloves made of the same material because fingers maintain their warmth better when they are in contact with each other. Reduced surface area reduces heat loss.

A hybrid of glove and mitten also exists, which contains open-ended sheaths for the four fingers (as in a fingerless glove, but not the thumb) and also an additional compartment encapsulating the four fingers as a mitten would. This compartment can be lifted off the fingers and folded back to allow the individual fingers ease of movement and access while the hand remains covered. The usual design is for the mitten cavity to be stitched onto the back of the fingerless glove only, allowing it to be flipped over (normally held back by Velcro or a button) to transform the garment from a mitten to a glove. These hybrids are called convertible mittens or glittens, a combination of "glove" and "mittens".

Gloves protect and comfort hands against cold or heat, damage by friction, abrasion or chemicals, and disease; or in turn to provide a guard for what a bare hand should not touch. Latex, nitrile rubber or vinyl disposable gloves are often worn by health care professionals as hygiene and contamination protection measures. Police officers often wear them to work in crime scenes to prevent destroying evidence in the scene. Many criminals wear gloves to avoid leaving fingerprints, which makes the crime investigation more difficult. However, the gloves themselves can leave prints that are just as unique as human fingerprints. After collecting glove prints, law enforcement can then match them to gloves that they have collected as evidence. In many jurisdictions the act of wearing gloves itself while committing a crime can be prosecuted as an inchoate offence.

Fingerless gloves are useful where dexterity is required that gloves would restrict. Cigarette smokers and church organists use fingerless gloves. Some gloves include a gauntlet that extends partway up the arm. Cycling gloves for road racing or touring are usually fingerless. Guitar players often use fingerless gloves in circumstances when weather is much too cold to play with an uncovered hand.

Gloves are made of materials including cloth, knitted or felted wool, leather, rubber, latex, neo-prene, and metal (as in mail). Gloves of kevlar protect the wearer from cuts. Gloves and gauntlets are integral components of pressure suits and spacesuits such as the Apollo/Skylab A7L which went to the moon. Spacesuit gloves combine toughness and environmental protection with a degree of sensitivity and flexibility.

Gloves appear to be of great antiquity. According to some translations of Homer's *The Odyssey*, Laërtes is described as wearing gloves while walking in his garden so as to avoid the brambles. (Other translations, however, insist that Laertes pulled his long sleeves over his hands.) Herodotus, in *The History of Herodotus* (440 BC), tells how Leotychides was incriminated by a glove (gauntlet) full of silver that he received as a bribe. There are also occasional references to the use of gloves among the Romans as well. Pliny the Younger (c. 100), his uncle's shorthand writer wore gloves in winter so as not to impede the elder Pliny's work.

A gauntlet, which could be a glove made of leather or some kind of metal armour, was a strategic part of a soldier's defence throughout the Middle Ages, but the advent of firearms made hand-to-hand combat rare. As a result, the need for gauntlets also disappeared.

During the 13th century, gloves began to be worn by ladies as a fashion ornament. They were made of linen and silk, and sometimes reached to the elbow. Such worldly accoutrements were not for holy women, according to the early 13th century *Ancrene Wisse*, written for their guidance. Sumptuary laws were promulgated to restrain this vanity: against samite gloves in Bologna, 1294, against perfumed gloves in Rome, 1560.

A Paris *corporation* or guild of glovers (*gantiers*) existed from the thirteenth century. They made them in skin or in fur.

It was not until the 16th century that gloves reached their greatest elaboration, however, when Queen Elizabeth I set the fashion for wearing them richly embroidered and jewelled, and for putting them on and taking them off during audiences, to draw attention to her beautiful hands. The 1592 "Ditchley" portrait of her features her holding leather gloves in her left hand. In Paris, the *gantiers* became *gantiers parfumeurs*, for the scented oils, musk, ambergris and civet, that perfumed leather gloves, but their trade, which was an introduction at the court of Catherine de Medici, was not specifically recognised until 1656, in a royal *brevet*. Makers of knitted gloves, which did not retain perfume and had less social cachet, were organised in a separate guild, of *bonnetiers* who might knit silk as well as wool. Such workers were already organised in the fourteenth century. Knitted gloves were a refined handiwork that required five years of apprenticeship; defective work was subject to confiscation and burning. In the 17th century, gloves made of soft chicken skin became fashionable. The craze for gloves called "limericks" also took hold. This particular glove-fad was the product of a manufacturer in Limerick, Ireland, who fashioned the gloves from the skin of unborn calves.

Embroidered and jeweled gloves also formed part of the insignia of emperors and kings. Thus Matthew of Paris, in recording the burial of Henry II of England in 1189, mentions that he was buried in his coronation robes with a golden crown on his head and gloves on his hands. Gloves were also found on the hands of King John when his tomb was opened in 1797 and on those of King Edward I when his tomb was opened in 1774.

Pontifical gloves are liturgical ornaments used primarily by the pope, the cardinals, and bishops. They may be worn only at the celebration of mass. The liturgical use of gloves has not been traced beyond the beginning of the 10th century, and their introduction may have been due to a simple desire to keep the hands clean for the holy mysteries, but others suggest that they were adopted as part of the increasing pomp with which the Carolingian bishops were surrounding themselves. From the Frankish kingdom the custom spread to Rome, where liturgical gloves are first heard of in the earlier half of the 11th century.

When short sleeves came into fashion in the 1700s, women began to wear long gloves, reaching half-way up the forearm. By the 1870s, buttoned kid, silk, or velvet gloves were worn with evening or dinner dress, but long suede gloves were also worn during the day and when having tea.

In 1905 *The Law Times* made one of the first references to the use of gloves by criminals to hide fingerprints, stating: *For the future... when the burglar goes a-burgling, a pair of gloves will form a necessary part of his outfit.*

Early Formula One race cars used steering wheels taken directly from road cars. They were normally made from wood, necessitating the use of driving gloves.

Latex gloves were developed by the Australian company Ansell. Ansell also launched the ActivArmr line, which is dedicated to producing protective gloves for construction, plumbing, HVAC, and military applications.

More recently in history, Tommie Smith and John Carlos held up their leather glove-clad fists at the awards ceremony of the 1968 Summer Olympics. Their actions were intended to symbolize Black Power, but they were banned from the Olympics for life as a result of the incident. Yet another of the more infamous episodes involving a leather glove came during the 1995 O.J. Simpson murder case in which Simpson demonstrated that the glove purportedly used in the alleged murder was too small to fit his hand. The glove and its impact on the case have caused the term O.J. Gloves to become a popular nickname for black or brown leather gloves. Rappers 50 Cent and Kanye West have referenced these infamous gloves in their songs.

TYPES OF GLOVE

Commercial and industrial

- Aircrew gloves: fire resistant
- Barbed wire handler's gloves
- Chainmail gloves are used by butchers, scuba divers, woodcutters and police
- Chainsaw gloves
- Cut-resistant gloves
- Disposable gloves can be used by anyone from doctors making examination to caregivers changing diapers.
- Fireman's gauntlets
- Food service gloves
- Gardening gloves
- Impact protection gloves
- Medical gloves
- Military gloves
- Rubber gloves
- Sandblasting gloves
- Welder's gloves

Sport and recreational

- American football various position gloves
- Archer's glove
- Baseball glove or *catcher's mitt*: in baseball, the players in the field wear gloves to help them catch the ball and prevent injury to their hands.
- Billiards glove
- Boxing gloves: a specialized padded mitten
- Cricket gloves
  - The batsmen wear gloves with heavy padding on the back, to protect the fingers in case of being struck with the ball.
  - The wicket keeper wears large webbed gloves.
- Cycling gloves
- Driving gloves intended to improve the grip on the steering wheel. Driving gloves have external seams, open knuckles, open backs, ventilation holes, short cuffs, and wrist snaps. The most luxurious are made from Peccary gloving leather.

- Falconry glove
- fencing glove
- Football - Goalkeeper glove
- Gardening glove
- Golf glove
- Ice hockey glove
- Riding gloves
- Lacrosse gloves
- Kendo Kote
- LED gloves
- Motorcycling gloves
- Oven gloves - or Oven mitts, used when cooking
- Paintball Glove
- Racing drivers gloves with long cuffs, intended for protection against heat and flame for drivers in automobile competitions.
- Scuba diving gloves :
  - Cotton gloves; good abrasion but no thermal protection
  - Dry gloves; made of rubber with a latex wrist seal to prevent water entry
  - Wet gloves; made of neo-prene and allowing water entry
- Shooting glove
  - Biathlon glove - an articulated padded combination of a skiing glove and a shooting glove, offers cold temperature protection outside in winter, as well as padding to support the .22lr ammunition single-action/Fortner-action biathlon rifle, and is suitable for using with poles in cross country skiing.
  - Pistol glove - used in competition pistol shooting to improve performance and cushion the shooting hand.
  - Target rifle glove - open-fingered heavily padded one-hand (non-shooting) glove with non-skid surfaces, used to support the rifle in prone shooting position. Also may be used in kneeling, sitting and standing positions. The glove cushions and distributes the weight of the rifle, which varies from 3 kilograms (6.6 lb) to 7 kilograms (15 lb), depending on type of rifle stock used.
- Skiing gloves are padded and reinforced to protect from the cold, and also from injury by skis.
- Touchscreen gloves - made with conductive material to enable the wearer's natural electric capacitance to interact with capacitive touchscreen devices without the need to remove one's gloves
  - finger tip conductivity; where conductive yarns or a conductive patch is found only on the tips of the fingers (typically the index finger and thumb) thus allowing for basic touch response

    - full hand conductivity; where the entire glove is made from conductive materials allowing for robust tactile touch and dexterity good for accurate typing and multi-touch response
- Underwater Hockey gloves - with protective padding, usually of silicone rubber or latex, across the back of the fingers and knuckles to protect from impact with the puck; usually only one, either left- or right-hand, is worn depending on which is the playing hand.
- Washing mitt or Washing glove: a tool for washing the body (one's own, or of a child, a patient, a lover).
- Webbed gloves - a swim training device or swimming aid.
- Weightlifting gloves
- Wired glove
    - Power Glove - an alternate controller for use with the Nintendo Entertainment System
- Wheelchair gloves - for users of manual Wheelchairs

Fashion

Western lady's gloves for formal and semi-formal wear come in three lengths: wrist ("matinee"), elbow, and opera or full-length (over the elbow, reaching to the biceps). Satin and stretch satin are popular and mass-produced. Some women wear gloves as part of "dressy" outfits, such as for church and weddings. Long white gloves are common accessories for teenage girls attending formal events such as prom, quinceañera, cotillion, or formal ceremonies at church such as confirmation.

In Japan, white gloves are worn frequently. Work-oriented white gloves are worn for activities such as gardening and cleanup; "dress" white gloves are worn by professionals who want a clean public appearance, such as taxi drivers, police, politicians and elevator operators. However white gloves are not recommended for touching old books and similar antiquities.

FINGERLESS GLOVES

Fingerless gloves or "glovelettes" are garments worn on the hands which resemble regular gloves in most ways, except that the finger columns are half-length and opened, allowing the top-half of the wearer's fingers to be shown.

Fingerless gloves are often padded in the palm area, to provide protection to the hand, and the exposed fingers do not interfere with sensation or gripping. In contrast to traditional full gloves, often worn for warmth, fingerless gloves will often have a ventilated back to allow the hands to cool; this is commonly seen in weightlifting gloves.

Fingerless gloves are also worn by bikers as a means to better grip the handlebars, as well as by skateboarders and rollerbladers, to protect the palms of the hands and add grip in the event of a fall. Some anglers, particularly fly fishermen, favour fingerless gloves to allow manipulation of line and tackle in cooler conditions. Fingerless gloves are common among marching band members. The lack of fabric on the fingertips allows for better use of touchscreens, as on smartphones and tablet computers. Professional MMA fighters are also required to wear fingerless gloves in fights.

LEATHER GLOVES

A leather glove is a fitted covering for the hand with a separate sheath for each finger and the thumb. This covering is composed of the tanned hide of an animal (with the hair removed), though in recent years it is more common for the leather to be synthetic.

Common uses

Leather gloves have been worn by people for thousands of years. The unique properties of leather allow for both a comfortable fit and useful grip for the wearer. The grain present on the leather and the pores present in the leather gives the gloves the unique ability to assist the wearer as he or she grips an object. As soft as a leather glove may be, its pores and grain provide a level of friction when "gripped" against an item or surface.

A common use for leather gloves is sporting events. In baseball, a baseball glove is an oversized leather glove with a web used for fielding the ball. Leather gloves also factor into playing handball. Cyclists also use leather gloves. Leather gloves are also used frequently by football players so that they can more easily grip the ball.Early Formula One racing drivers used steering wheels taken directly from road cars. They were normally made from wood necessitating the use of driving gloves.Leather gloves also provide protection from occupational hazards. For example, beekeepers use leather gloves to avoid being stung by bees. Construction workers might also use leather gloves for added grip and for protecting their hands. Welders use gloves too for protection against electrical shocks, extreme heat, ultraviolet and infrared.

Criminals have also been known to wear leather gloves during the commission of their crimes. These gloves are worn by criminals because the tactile properties of the leather allow for good grip and dexterity. These same properties are the result of their being a grain present on the surface of the leather. This understandably makes the surface of the leather as random as human skin since the leather itself is skin, usually from livestock. Investigators are able to dust for the glove prints left behind from the leather the same way in which they dust for fingerprints.

Leather dress gloves

Main types of gloving leather

Leather is a natural product with special characteristics that make it comfortable to wear, and give it great strength and flexibility. Because it is a natural product, with its own unique variations, every piece has its own individual characteristics. As they are worn and used, leather gloves (especially if they fit snugly) will conform to the wearer's hand. As this occurs the leather of the glove will become more malleable, and thus softer and more supple. This process is known as 'breaking-in' the glove. Overtime wear spots may appear on certain parts of the palm and fingertips, due to the constant use of those areas of the glove. Creases and wrinkles will also appear on the palm side of the leather glove and will generally correspond to the locations of the hinge joints of the wearer's hands, including the interphalangeal articulations of hand, metacarpophalangeal joints, intercarpal articulations, and wrists.

Because the leather is natural as well as delicate, the wearer must take precaution as to not damage them. The constant handling of damp or wet surfaces will discolor lighter-colored gloves and stiffen the leather of any glove. The wearer will often unknowingly damage or stain their gloves while doing such tasks as twisting a wet door knob or wiping a running nose with a gloved hand. Leather dress gloves that are worn very tight and possess very short, elasticized wrists, are most often referred to as *cop gloves* or *law enforcement gloves* because of their prevalence as issued duty gloves for many law enforcement agencies. It is also common attire in leather subculture and BDSM communities.

- Cowhide is often used for lower-priced gloves. This leather is generally considered too thick and bulky for the majority of glove styles, particularly finer dress gloves. It is, however, used for some casual styles of glove.
- Deerskin has the benefit of great strength and elasticity, but has a more rugged appearance, with more grain on the surface, than "hairsheep". It is very hard-wearing and heavier in weight.
- Goatskin is occasionally used for gloves. It is hard-wearing but coarser than other leathers and is normally used for cheaper gloves.
- Hairsheep originates from sheep that grow hair, not wool. Hairsheep leather is finer and less bulky than other leathers. Its major benefits are softness of touch, suppleness, strength, and lasting comfort. It is very durable and is particularly suited for the manufacture of dress gloves.
- Peccary is the world's rarest and most luxurious gloving leather. Peccary leather is very soft, difficult to sew, and hard-wearing.
- Sheepskin, also called shearling, is widely used for casual and country gloves. It is very warm in cold weather, and as a leather reversed, it has still attached wool on the inside.
- Slink lamb is used only in the most expensive lambskin gloves. Some of the finest lambskin comes from New Zealand.

Leather glove linings

- Cashmere is warm, light in weight, and very comfortable to wear. Cashmere yarn comes from the hair of mountain goats, whose fleece allows them to survive the extreme weather conditions they are exposed to.
- Silk is warm in winter and cool in summer and is used both in men's and women's gloves, but is more popular in women's.
- Wool is well known for its natural warmth and comfort, as well as having a natural elasticity.
- Other linings, which include wool mixtures and acrylics.

Component parts

The component parts that may be found in a leather dress glove are one pair of tranks, one pair of thumbs, four whole fourchettes, four half fourchettes, two gussets, and six quirks. Depending on the style of the glove there may also be roller pieces, straps, rollers, eyelets, studs, sockets and domes. Finally, linings will themselves consist of tranks, thumbs and fourchettes.

Stitching

The most popular types of leather glove sewing stitches used today are:

- Hand stitched, which is most popular in men's gloves and some women's styles. Hand stitching is a very time-consuming and skilled process.
- Inseam, which is mainly used on women's gloves, but occasionally on men's dress gloves.

Some glove terms

- *Button length* is the measurement in inches that is used to determine the length/measurement from the base of the glove thumb to the cuff of the glove.
- *Fourchettes* are the inside panels on the fingers of some glove styles.
- *Perforations* are small holes that are punched in the leather. They are often added for better ventilation, grip, or aesthetics and can be as fine as a pin hole.
- *Points* are the three, or sometimes single, line of decorative stitching on the back of the glove.
- *Quirks* are found on only the most expensive hand sewn gloves. They are small diamond shaped pieces of leather sewn at the base of the fingers, where they are attached to the hand of the glove to improve the fit.
- A *strap and roller* is used to adjust the closeness of the fit around the wrist.

- A *Vent* is the 'V' shaped cut out of the glove, sometimes at the back, but more often on the palm, to give the glove an easier fit around the wrist.

Driving gloves

Driving gloves are designed for holding a steering wheel and transmitting the feeling of the road to the driver. They provide a good feel and protect the hands. They are designed to be worn tight and to not interfere with hand movements. The increased grip allows for more control and increased safety at speed. True driver's gloves offer tactile advantages to drivers frequently handling a car near the limits of adhesion. Made of soft leather, drivers gloves are unlined with external seams.

MITTENS

loves which cover the entire hand but do not have separate finger openings or sheaths are called mittens. Generally, mittens still separate the thumb from the other four fingers. They have different colours and designs. Mittens have a higher thermal efficiency than gloves as they have a smaller surface area exposed to the cold.

The earliest mittens known to archeologists date to around 1000 A.D. in Latvia. Mittens continue to be part of Latvian national costume today. Wool biodegrades quickly, so it is likely that earlier mittens, possibly in other countries, may have existed but were not preserved. An exception is the specimen found during the excavations of the Early Medieval trading town of Dorestad in the Netherlands. In the harbour area a mitten of wool was discovered dating from the 8th or early 9th century.

Many people around the Arctic Circle have used mittens, including other Baltic peoples, Native Americans and Vikings. Mittens are a common sight on ski slopes, as they not only provide extra warmth but extra protection from injury.

Idiot mittens are two mittens connected by a length of yarn, string or lace, threaded through the sleeves of a coat. This arrangement is typically provided for small children to prevent the mittens becoming discarded and lost; when removed, the mittens simply dangle from the string just beyond the cuff of the sleeve.

- Gunner's Mittens - In the 1930s, special fingerless mittens were introduced that have a flap located in the palm of the mitten so a hunter or soldier could have his finger free to fire his weapon. Originally developed for hunters in the frigid zones of the US and Canada, eventually most military organizations copied them.
- Scratch mitts do not separate the thumb, and are designed to prevent babies— who do not yet have fine motor control — from scratching their faces.

SAFETY STANDARDS

Several European standards relate to gloves. These include:

- BS EN388- Mechanical hazards including Abrasion, cut, tear and puncture.
- BS EN388:2003 - Protective Against Mechanical Rist (Abrasion/Blade Cut Resistance/Tear Resistance/Abrasion Resistance)
- BS EN374-1:2003 Protective Against Chemical And Micro-Organisms
- BS EN374-2- Micro-organisms
- BS EN374-3- Chemicals
- BS EN420- General requirements for gloves includes sizing and a number of health and safety aspects including latex protein and chromium levels.
- BS EN60903- Electric shock
- BS EN407- Heat resistance
- BS EN511- Cold resistance
- BS EN1149- Antistatic

These exist to fulfill Personal protective equipment (PPE) requirements. PPE places gloves into three categories:

- Minimal risk - End user can easily identify risk. Risk is low.
- Complex design- Used in situations that can cause serious injury or death.
- Intermediate - Gloves that don't fit into minimal risk or complex design categories.

HATMAKING

Hatmaking is the manufacture of hats and headwear. Millinery is the designing and manufacture of hats. A millinery shop is a store that sells those goods. A milliner or hatter designs, makes, trims, or sells hats. Millinery is sold to women, men and children, though some definitions limit the term to women's hats. Historically, milliners, typically female shopkeepers, produced or imported an inventory of garments for men, women, and children, including hats, shirts, cloaks, shifts, caps, neckerchiefs, and undergarments, and sold these garments in their millinery shop.

More recently, the term *milliner* has evolved to describe a person who designs, makes, sells or trims hats primarily for a female clientele. The origin of the term is likely the Middle English *milener*, an inhabitant of Milan or one who deals in items from this Italian city known for its fashion and clothing. Many styles of headgear have been popular through history and worn for different functions and events. They can be part of uniforms or worn to indicate social status. Styles include the top hat, hats worn as part of military uniforms, cowboy hat, and cocktail hat.

NOTABLE HATTERS AND MILLINERS

Hatters:

- John Cavanagh, an American hatter whose innovations included manufacturing regular, long and wide-oval fitting hats to enable customers to find better-fitting ready-to-wear hats.
- James Lock and Co. of London (founded 1676), is credited with the introduction of the bowler hat in 1849.
- John Batterson Stetson, credited with inventing the classic cowboy hat
- Giuseppe Borsalino, with the famous "Borsalino" Fedora hat.

Milliners:

- Anna Ben-Yusuf wrote *The Art of Millinery* (1909), one of the first reference books on millinery technique.
- Rose Bertin, milliner and modiste to Marie Antoinette, is often described as the world's first celebrity fashion designer.
- John Boyd is one of London's most respected milliners and is known for the famous pink tricorn hat worn by Diana, Princess of Wales.
- Lilly Daché was a famous American milliner of the mid-20th century.
- Mr. John was an American milliner considered by some to be the millinery equivalent of Dior in the 1940s and 50s.
- Stephen Jones of London, is considered one of the world's most radical and important milliners of the late 20th and early 21st centuries.
- Simone Mirman was known for her designs for Elizabeth II and other members of the British Royal Family.
- Caroline Rèboux was a renowned milliner of the 19th and early 20th centuries.
- David Shilling is a renowned milliner, artist and designer based in Monaco.
- Philip Treacy of London is an award-winning milliner.

LEATHER CRAFTING

Leather crafting or simply Leathercraft is the practice of making leather into craft objects or works of art, using shaping techniques, coloring techniques or both.

TECHNIQUES

Dyeing

Leather dyeing usually involves the use of spirit- or alcohol-based dyes where alcohol quickly gets absorbed into moistened leather, carrying the pigment deep into the surface. "Hi-liters" and "Antiquing" stains can be used to add more definition to patterns. These have pigments that will break away from the higher points of a tooled piece and so pooling in the background areas give nice contrasts. Leaving parts unstained also provides a type of contrast.

Alternatives to spirit stains might include a number of options. Shoe polish can be used to dye and preserve leather. Oils such as neatsfoot or linseed can be applied to preserve leather but darkens them. A wax paste more often than not serves as the final coat.

Sweat and grime will also stain and 'antique' leather over time. Gun holsters, saddlebags, wallets and canteens used by cowboys and buckaroos were rarely colored in the Old West. The red, brown, and black tones develop naturally through handling and as the oiled leathers absorb the rays of the desert sun.

Due to changing environmental laws, alcohol-based dyes are soon to be unavailable. There are currently water-based alternatives available, although they tend not to work as well.

Painting

Leather painting differs from leather dyeing in that paint remains only on the surface while dyes are absorbed into the leather. Due to this difference, leather painting techniques are generally not used on items that can or must bend nor on items that receive friction, such as belts and wallets because under these conditions, the paint is likely to crack and flake off. However, latex paints can be used to paint such flexible leather items. In the main though, a flat piece of leather, backed with a stiff board is ideal and common, though three-dimensional forms are possible so long as the painted surface remains secured.

Acrylic paint is a common medium, often painted on tooled leather pictures, backed with wood or cardboard, and then framed. Unlike photographs, leather paintings are displayed without a glass cover, to prevent mold.

Carving

Leather carving entails using metal implements to compress moistened leather in such a way as to give a three-dimensional appearance to a two-dimensional surface. The surface of the leather is not intended to be cut through, as would be done in filigree.

The main tools used to "carve" leather include: swivel knife, veiner, beveler, pear shader, seeder, cam, and background tool. The swivel knife is held similar to pencil and drawn along the leather to outline patterns. The other tools are punch-type implements struck with a wooden, nylon or rawhide mallet. The object is to add further definition with them to the cut lines made by the swivel knife.

In the United States and Mexico, the western floral style, known as "Sheridan Style", of carving leather predominates. Usually, these are stylized pictures of acanthis or roses. California, Texas, and a few other styles are common. By far the most preeminent carver in the United States was Al

Stohlman. His patterns and methods have been embraced by many hobbyists, scout troops, reenacters, and craftsmen.

Stamping

Leather stamping involves the use of shaped implements (stamps) to create an imprint onto a leather surface, often by striking the stamps with a mallet.

Commercial stamps are available in various designs, typically geometric or representative of animals. Most stamping is performed on vegetable tanned leather that has been dampened with water, as the water makes the leather softer and able to be compressed by the design being pressed or stamped into it. After the leather has been stamped, the design stays on the leather as it dries out, but it can fade if the leather becomes wet and is flexed. To make the impressions last longer, the leather is conditioned with oils and fats to make it waterproof and prevent the fibres from deforming.

Molding/Shaping

Leather shaping or molding consists of soaking a piece of leather in hot or room temperature water to greatly increase pliability and then shaping it by hand or with the use of objects or even molds as forms. As the leather dries it stiffens and holds its shape. Carving and stamping may be done prior to molding. Dying however must take place after molding, as the water soak will remove much of the colour. Molding has become popular among hobbyists whose crafts are related to fantasy, goth/steampunk culture and cosplay.

PARACHUTE RIGGER

A parachute rigger is a person who is trained or licensed to pack, maintain or repair parachutes. A rigger is required to understand fabrics, hardware, webbing, regulations, sewing, packing, and other aspects related to the building, packing, repair, and maintenance of parachutes.Militaries around the world train their own parachute riggers to support their airborne or paratrooper forces. These military riggers also pack parachutes for aerial delivery operations, through which military supplies and equipment are delivered by aircraft to combat zones.

Parachute riggers in the Australian Army are responsible for the preparation, maintenance and supply of parachutes and other aerial delivery components.

Prior to commencing the parachute rigger course, all trainees must be static-line parachute qualified. Parachute riggers frequently make parachute jumps, and at any time may be required to jump with any parachute they have packed. This is to help them better understand how the equipment they prepare and maintain works, and to help ensure that each parachute is professionally packed to a safe standard.

Riggers in the Canadian Forces train at the Canadian Army Advanced Warfare Centre at CFB Trenton in Trenton, Canada.

When Canada entered the airborne world with the creation of two airborne battalions in 1942, all the would-be jumpers were trained at Fort Benning, Georgia or Ringway, UK. Later, however, the flow of reinforcements for the parachute battalions posed an acute problem and it was decided to remedy this situation by training paratroopers in Canada. In May 1943, a Canadian Parachute Training Centre was formed in Shilo, Manitoba. With background knowledge in American and British parachuting techniques, Canadian trainers were able to develop a truly Canadian method of parachuting by incorporating the best features of both the American and British systems. Following several name and location changes, the school was moved to Edmonton in 1970 as the Canadian Airborne Centre (CABC) and then moved to Trenton in August 1996, becoming the Canadian Parachute Centre (CPC). On 1 April 1998 the former Canadian Forces Parachute Maintenance Depot (CFPMD) was amalgamated into CPC as Support Company.

On 1 April 2006 the renaming of CPC to CFLAWC began a transformation that was more than just another name change. CFLAWC became the Centre of Excellence (CoE) for Land Advanced Warfare, in addition to its previous focus on delivery of training. To meet the new challenges and added responsibilities, CFLAWC is currently organized with a Command team, Training Company, Support Company with the Canadian Forces Parachute Team (CFPT - the SkyHawks) and a Headquarters Company that includes the Standards Section, the Airborne Trials and Evaluation Section (ATES) and the Unit Orderly Room (UOR). Training Company is organized into four subject matter expert (SME) platoons for the conduct of the majority of the courses at CFLAWC. Support Company is based on the old CFPMD structure and provides the CF with parachute packing and maintenance services including the major repair of parachutes and associated aerial delivery equipment. Support Company is also responsible for training all parachute rigger specialists in the CF. It traces its roots to 1943 as part of the Canadian Army Parachute Training Centre. In those early days, parachute trainees were taught to pack their own parachutes, but this system was soon discarded as impractical and the packing and maintenance of parachutes became a centralized operation. Since its formation, Support Company has changed its name from 28 Central Ordnance Depot to 28 Canadian Forces Supply Depot in 1968, and upon the move from Camp Shilo, MB to Edmonton, AB in 1970, was given the name Canadian Forces Parachute Maintenance Depot. All riggers are jumpers and can be asked at any time to jump with a parachute they have packed.

CFLAWC currently delivers, as part of the Army National Individual Training Calendar, the following courses: the Arctic Operations Advisor Course, Drop Zone/ Landing Zone Controller, Aerial Delivery, Basic Helicopter Operations, Basic Parachuting, Jump Master, Parachute Instructor, Static Line

Square Parachuting, the three different phases of Parachute Rigger training, the Advanced Mountain Operations Course, the Helicopter Insertion Instructor Course, Military Freefall Parachuting, Military Freefall Jump Master, Military Freefall Parachute Instructor and a revised Patrol Pathfinder Course.

Parachute Riggers/Packers in training attend the 15 day Basic Parachute qualification course at CFB Trenton, and then for approx 2.5-3yrs undertake 3 different 45 day courses, that cover Maintaining parachutes, packing parachutes, and quality control of parachutes.

Riggers have played an important role in the American military since the advent of the use of the parachute for aerial insertion of troops, supplies, and equipment into combat zones. In addition to the maroon beret worn by paratroopers in airborne units, riggers are authorized the wear of a distinctive red baseball cap as their military headgear when on rigger duties.

When the U.S. Army formed its first paratrooper unit in 1940, a parachute test platoon, the paratroopers themselves prepared and took care of their own parachutes. The test platoon had only 3 men, two enlisted soldiers and one warrant officer, from the Army Air Corps serving as the precursors of the U.S. Army's parachute riggers.

When the U.S. Army created five Airborne divisions for World War II, the U.S. Army stopped training paratroopers on how to pack their own chutes and started support organizations for parachute packing and rigging. The first riggers received their training at Fort Benning, Georgia.

After 1950, the U.S. Army assigned the Quartermaster Corps with the mission of aerial delivery, including parachute rigging. A parachute rigger course was established at the U.S. Army Quartermaster School at Fort Lee, Virginia in 1951, and has continued since then.

Airborne Orientation Course. For students completing basic training at Fort Jackson, South Carolina, preparation for Airborne and rigger training begins before even departing for Fort Lee with attendance at the post's Airborne Orientation Course. According to an *Army News Service* story, "while most of the course involves physical training, soldiers are also familiarized with such Airborne operations as parachute landing falls, rigging equipment and actions in the aircraft." The AOC has raised the success rate for soldiers subsequently attending Airborne training from 60 percent to 89 percent.

From AOC, rigger recruits go to Airborne School at Fort Benning, Georgia. If a rigger recruit does not pass Airborne School, that soldier is reclassified.

The U.S. Army MOS (Military Occupational Specialty) designation for parachute riggers is graded in 5 skill levels, from 92R1P to 92R5P. Prior to fiscal year 2003, it was 43E2P. Recruits are designated 92R0P.

After Airborne School, 92R0P recruits head to Fort Lee to attend the 13 week Parachute Rigger Course. The course provides training on inspecting, packing, rigging, recovering, storing, and maintaining air item equipment. It is divided into three phases. Air Drop Phase - Includes instruction in cargo parachute packing, rigging supplies and equipment for airdrop, types and limitations of aircraft. Students become proficient in the use of the various technical manuals for rigging airdrop loads. At the conclusion of the instruction, the students participate in an airdrop exercise. They pack the cargo parachutes, rig the loads to be dropped and place the loads in the aircraft. After the airdrop, the students recover the loads and equipment. Aerial Equipment Repair Phase - Trains fundamentals and procedures of inspection, classification, and repair of maintenance of personnel, cargo, extraction parachutes and airdrop equipment to include the service of High Altitude Low Opening (HALO) Automatic Ripcord Release (ARR). Parachute Pack Phase - Is designed to equip students with the working knowledge of inspection and packing procedures relative to personnel, light cargo and extraction parachutes. The student receives concentrated instruction on the troop back parachute. The student is required to jump the parachute he/she packed during the examination. Throughout the course, the student is constantly reminded of the fact that all parachutes must be packed with meticulous care to insure proper functioning. Any malfunction could result in death or in equipment loss.

All U.S. Army parachute riggers are required to be Airborne qualified, and by tradition are required to be prepared to jump any parachute packed by any U.S. Army parachute rigger, without checking the log book for the name of the rigger who last prepared it. The official motto of the U.S. Army parachute rigger is: "I will be sure always."

Service members from other branches of the U.S. Armed Forces also attend parachute rigger courses at Fort Lee.

United States Air Force parachute riggers are trained at Sheppard AFB in Texas. The career field is classified under "Aircrew Flight Equipment" Airmen attend a 3½-month course learning to inspect, pack, and repair emergency parachutes, as well as a wide variety of other types of aircrew equipment. Once graduated from this technical school, students are assigned to a duty location where they are further instructed using on the job training. USAF aerial delivery riggers (2T2X1) packed training airdrop loads for airlift units in peacetime; wartime airdrop missions would be rigged by Army riggers. In recent years, uniformed Air Force riggers have been replaced by contract civilian employees since the mission does not require deployment overseas, and instead consists of supporting training missions at home station.

In the U.S. Air National Guard, two-week rigger courses have been organized to teach "packing, inspecting and repairing minor damage", as described on page 9 of the January–February 2005 *Guard Times*.

In mid-2009, the U.S. Air Force's 98th Virtual Uniform Board announced "Airmen earning and awarded the Army Parachute Riggers badge are authorized permanent wear on all uniform combinations. For the airman battle uniform and the battle dress uniform, the badge will be blue. On the desert combat uniform the approved colour is brown." Previous guidance had limited the wear of the badge to airmen attached to U.S. Army rigger units.

The Parachute Materials School was first established in 1924 at Lakehurst, New Jersey by two U.S. Navy chief petty officers. Parachute Rigger, or "PR", became an enlisted job rating in 1942, but the name changed during the 1960s to Aircrew Survival Equipmentman. The United States Navy parachute riggers are now trained at Naval Air Station Pensacola during a 12 week (55 training day) school. When they graduate, they do become PRs, but the rating is called Aircrew Survival Equipmentman. While in school they go through 9 courses, 3 courses of "Common Core" skills over 19 days, 3 courses of Organizational-Level skills for 17 days, and finishing with 3 courses of Intermediate-Level skills for 19 days. The first week is a course taught on materials manufacturing using the Consew 206RB-5 industrial sewing machine, dubbed by students and instructors alike as "Combat Rigger Sewing" or simply "Combat Sewing". Students will manufacture a "rigger bag" completely from scratch and will learn about tool control. The next course is NB-8 parachutes, where students will learn basics of parachute rigging, inspection cycles and nomenclature. This is followed by a course of general survival equipment named ESE. Then "O" strand begins with Survival I Fixed Wing, followed by Survival II Rotary Wing, where students learn inspection and maintenance concepts unique to squadron level work. The final "O" level subject is Survival Radios. "I" strand will start with NES-12s, the Navy's most complicated parachute system, for advanced rigging concepts. Seat Survival Kits and Life Preservers finish out the entire course of instruction, where they will graduate upon completion. The PR "A" School House graduates one class every 7 training days.

During the entire time of study students will undergo physical training at least three times a week, be subjected to rigorous inspections every Monday, and will march to and from the building, being accountable for showing up on time, cleanliness, and homework. No student is allowed to continue in the course if their grade average falls below an 90, making it one of the most challenging courses at the Naval Aviation Technical Training Center.

Special Operations Parachute Riggers assist Naval Special Warfare (NSW), US Navy SEALs, and Explosive Ordnance Disposal (EOD) units throughout the world. They inspect, maintain, pack, and use specialized premeditated personnel static-line and free-fall parachute systems. They use and maintain specialized aerial delivery and re-supply systems, and helicopter insertion and extraction systems unique to NSW and EOD units. They function as Parachute Jump (P.J.) and Helicopter Rope Suspension Techniques (HRST) masters. They also perform paraloft management, administrative functions, ordnance handling functions, and Quality Assurance (Q.A.) inspections.

The Navy Enlisted Classification Code (NEC) of Special Operations Parachute Rigger is awardable upon completion of Army courses 431 F3 PARA

NAVY or 860 43E10. Special Operations Parachute Rigger NEC OJT is awardable if personnel attached to a rigger unit of EOD for 1 year and observed by Army/Navy school graduate and qualified prior to 1 July 1990. Personnel other than Parachute Riggers must hold NEC 53XX to be assigned this NEC

There are a select few who perform duties as a Special Operations Parachute Rigger. The minimum prerequisite qualifications are graduating the Basic Airborne course at Fort Benning, GA and the EOD Rigger course at Fort Lee, Virginia. Although their primary duty is to maintain parachuting equipment, many go on to achieve greater qualifications such as Static Line Jumpmaster, Military Free Fall Parachutist, Military Free Fall Jumpmaster, Air Load Planner, Hazardous Cargo Certifier, FAA Senior Rigger, Rappel Master, and Fast Rope Master.

Riggers who work on the parachutes of sport parachutists are usually certified by the local aviation authorities, a local aviation association, or a parachuting association. The licensing system varies from country to country, but usually there are several levels of licenses, the higher licenses giving the rigger more privileges in the field. In the US, former and active duty military parachute riggers are allowed credit for FAA certification upon recommendation of commanding officer or providing officials with documentation of recorded parachute packs.In Canada, parachute rigger ratings are issued by the Canadian Sport Parachuting Association's Technical Committee. CSPA issues two levels of rigger ratings: A and B. Entry level includes packing ten reserves under supervision then attending a one-week course given by a CSPA Rigger Instructor. Canadian Rigger As are limited to assembling and packing parachutes. They can replace components and do simple hand-sewing, but are not trained to use sewing machines. At the end of the Rigger A Course candidates can choose to be tested on round or square parachutes and they can chose which type of container for their practical test (one-pin sport, two-pin sport, Pop-Top or chest). Certification for packing Pilot Emergency Parachutes (PEP) can be obtained only after passing practical tests on all other types. Two more years of experience, including learning sewing machine operation, is needed before riggers can challenge for Rigger B ratings. The SOLO programme includes sewing a bag of samples and submitting them to CSPA's Technical Committee. CSPA Rigger Bs enjoy the same privileges as American Master Riggers and are allowed to do most major repairs that can be done outside of a factory.

In the United Kingdom, sport parachute rigger ratings are issued by the British Parachute Association's Rigging Committee, itself a subcommittee of the Safety and Training Committee. The BPA issues two working levels of riggers rating: Parachute Rigger and Advanced Rigger.

- A Parachute Rigger is authorised to manufacture new components as listed in the BPA Parachute Rigger Manufacturing Syllabus. They are not cleared for harness manufacture or harness work, nor repair work or modifications to reserve canopies, reserve containers or reserve component parts.

- An Advanced Rigger is cleared for all work on all sport parachute assemblies.

To become a BPA Parachute Rigger, first a candidate must become a BPA Advanced Packer, which certifies them to pack reserve parachutes. Following this they become a BPA Basic Rigger. This is their apprenticeship whereby they work under the supervision of an appropriately qualified rigger and where they gain experience manufacturing parachute component parts and repairing damaged parachutes and systems.Following this and a successful attendance on a BPA Parachute Rigger exam course, the candidate becomes a BPA Parachute Rigger.

The next level is a BPA Advanced Rigger,whereby Parachute Riggers of at least 2 years standing can attend a BPA Advanced Rigger course. This involves major repairs to canopies and container systems including harness work. In advance of the course, the candidate must also manufacture a full skydiving container system including component parts for assessment at the course.

Such courses are administered by at least two BPA Rigger Examiners. A Rigger Examiner is a BPA Advanced Rigger who has been successfully assessed on his ability to run Advanced Packer courses, Basic Rigger courses and Parachute Rigger courses.

The following documents record the criteria for examination and work limitations for each BPA Rigger rating –

- Form 199 Basic Riggers Course syllabus
- Form 200 Parachute Riggers Course Syllabus
- Form 201 Advanced Riggers Course Syllabus
- Form 202 Rigger Examiner Course Syllabus

Also of relevance are the following documents

- Form 238 List of all BPA Rigging Related Documents
- Form 169 Advanced Packing Course Syllabus

In the U.S., the Federal Aviation Administration (FAA) licenses civilian riggers. The FAA issues two levels of civilian parachute rigger ratings: senior and master. Entry-level riggers start by apprenticing under another licensed rigger, then test for the Senior Rigger rating. The Senior Rigger test involves three parts: written, oral and practical. The written test is usually done at a computerized learning center and results are available immediately.

The oral and practical exams include questions about common rigging practices. The practical test consists of inspecting and repacking 20 reserves, along with hand sewing and a simple machine-sewn patch on a canopy. Candidates have the option of testing on back, chest, seat or lap type parachutes. The FAA does not distinguish between round and (modern) square parachutes.

After three years experience — including packing at least 200 reserves, 100 each of two different types — Senior Riggers can test for the Master Rigger rating which allows them to do most major repairs. There is no written test for Master Riggers, but the oral exam is far more extensive, including identifying dozens of material samples. The Master practical exam starts with assembling and adjusting a sewing machine, then doing a major canopy repair that includes a seam, reinforcing tape and line attachment. Master candidates are usually asked to demonstrate a harness repair also. FAA riggers are tested by Parachute Rigger Examiners (government employees) or Designated Parachute Rigger Examiners (independent civilians, usually highly-experienced Master Riggers). U.S. military riggers only need a letter from their commanding officer and the written test to earn FAA rigger ratings.

QUILTING

Quilting can refer either to the process of creating a quilt or to the sewing of two or more layers of material together to make a thicker padded material. "Quilting" as the process of creating a quilt uses "quilting" as the joining of layers as one of its steps, often along with designing, piecing, appliqué, binding and other steps. A quilter is the name given to someone who works at quilting. Quilting can be done by hand, by sewing machine, or by a specialized longarm quilting system.

The process of quilting uses a needle and thread to join two or more layers of material to make a quilt. Typical quilting is done with 3 layers: the top fabric or quilt top, batting or insulating material and backing material. The quilter's hand or sewing machine passes the needle and thread through all layers and then brings the needle back up. The process is repeated across the entire area where quilting is wanted. A rocking, straight or running stitch is commonly used and these stitches can be purely functional, or decorative and elaborate. Quilting is done to create bed spreads, art quilt wall hangings, clothing, and a variety of textile products. Quilting can make a project thick, or with dense quilting, can raise one area so that another stands out.

Quilt stores often sell fabric, thread, patterns and other goods that are used for quilting. They often have group sewing and quilting classes, where one can learn how to sew or quilt and work with others to exchange skills. Quilt stores often have quilting machines that can be rented out for use, or customers can drop off their quilts and have them professionally quilted.

The word "quilt" comes from the Latin *culcita* meaning a stuffed sack, but it came into the English language from the French word *cuilte*. The origins of quilting remain unknown, but sewing techniques of piecing, applique, and quilting have been used for clothing and furnishings in diverse parts of the world for several millennia.

The earliest known quilted garment is depicted on the carved ivory figure of a Pharaoh of the Egyptian First Dynasty, about 3400 B.C.

In 1924 archaeologists discovered a quilted floor covering in Mongolia. They estimated its date as between 100 BC to 200 AD. There are numerous references to quilts in literature and inventories of estates. Crusaders brought quilted objects from the Middle East to Europe in the late 11th century. Quilted garments known as gambesons were popular in the European Middle Ages. Knights wore them under their armor for comfort and sometimes as an outer garment to protect the metal armor from the weather. The earliest known surviving European bed quilt is from late 14th century Sicily. It is made of linen and padded with wool. The blocks across the center are scenes from the legend of Tristan. The quilt is 122" by 106" and is in the Victoria and Albert Museum in London.

Quilting has been part of the needlework tradition in Europe from about the 5th century CE. Early objects contain Egyptian cotton, which may indicate that Egyptian and Mediterranean trade provided a conduit for the technique. Quilted objects were relatively rare in Europe until approximately the 12th century, when quilted bedding and other items appeared after the return of the Crusaders from the Middle East. The medieval quilted gambeson, aketon and arming doublet were garments worn under, or instead of, armor of maille or plate armor. These developed into the later quilted doublet worn as part of fashionable European male clothing from the 14th to 17th century. Quilting clothing began to be generally used in the 14th century, with quilted doublets and armor worn in France, Germany, and England and quilted tunics in Italy.

In American Colonial times, most women were busy spinning, weaving, and making clothing. Meanwhile, women of the wealthier classes prided themselves on their fine quilting of wholecloth quilts with fine needlework. Quilts made during the early 19th century were not constructed of pieced blocks but were instead whole cloth quilts. Broderie perse quilts and medallion quilts were made. Some antique quilts made in North America have worn-out blankets or older quilts as the internal batting layer, quilted between new layers of fabric and thereby extending the usefulness of old material.

During American pioneer days, "paper" quilting became popular. Paper was used as a pattern and each individual piece of cut fabric was basted around the paper pattern. Paper was a scarce commodity in the early American west, and women would save letters from home, newspaper clippings, and catalogs to use as patterns. The paper not only served as a pattern but as an insulator. The paper found between the old quilts has become a primary source of information about pioneer life.

Quilts made without any insulation or batting were referred to as summer quilts. They were not made for warmth, only to keep the chill off during cooler summer evenings.

African-American women developed a distinctive style of quilting, notably different from the style most strongly associated with the Amish. Harriet Powers, a slave-born African American woman, made two famous story quilts.

She was just one of the many African American quilters who contributed to the evolution of quilting. The Gee's Bend quilting community was celebrated in an exhibition that travelled to museums including the Smithsonian. The contributions made by her and other quilters of Gee's Bend, Alabama has been recognized by the US Postal Service with a series of stamps. The *communal* nature of the quilting process (and how it can bring together women of varied races and backgrounds) was honored in the series of stamps.

During the American Civil War, slaves used quilts as a means to share and transmit secret messages to escape slavery and travel the Underground Railroad. A lack of written record on the topic has created debate among historians and scholars. However, an oral history has been told and preserved.

"Hawaiian quilting was well established by the beginning of the twentieth century. Hawaiian women learned to quilt from the wives of missionaries from New England in the 1820s. Though they learned both pieced work and applique; by the 1870s they had adapted applique techniques to create a uniquely Hawaiian mode of expression. The classic Hawaiian quilt design is a large, bold, curvilinear appliqué pattern that covers much of the surface of the quilt, and the symmetrical design is cut from only one piece of fabric."During the late 20th century, art quilts became popular for their aesthetic and artistic qualities rather than for functionality (they are displayed on a wall or table rather than spread on a bed). "It is believed that decorative quilting came to Europe and Asia during the Crusades (A.D. 1100-1300), a likely idea because textile arts were more developed in China and India than in the West."Unusual quilting designs have increasingly become popular as decorative textiles. Industrial sewing technology has become more precise and flexible, and quilting using exotic fabrics and embroidery began to appear in home furnishings in the early 21st century.

The quilt block is traditionally a patterned square of fabric that is repeated with plain blocks to form the overall design of a quilt. There are a variety of different designs for quilt blocks including the Nine-Patch, Shoo Fly, Churn Dash, and the Prairie Queen.

A Nine Patch is made by sewing five patterned or dark pieces (patches) to four light square pieces in alternating order. These nine sewn squares make one block.

The Shoo Fly varies from the Nine Patch by dividing each of the four corner pieces into a light and dark triangle.

Another variation develops when one square piece is divided into two equal rectangles in the basic Nine Patch design. The Churn Dash block combines the triangles and rectangle to expand the Nine Patch.

The Prairie Queen block combines two large scale triangles in the corner section with the middle section using four squares. The center piece is one full size square. Each of the nine sections does have the same overall measurement and fits together.

TYPES AND EQUIPMENT

Many types of quilting exist today. The two most widely used are hand-quilting and machine quilting.

Hand quilting

Hand quilting is the process of using a needle and thread to sew a running stitch by hand across the entire area to be quilted. This binds the layers together. A quilting frame or hoop is often used to assist in holding the piece being quilted off the quilter's lap. A quilter can make one stitch at a time by first driving the needle through the fabric from the right side, then pushing it back up through the material from the wrong side to complete the stitch; this is called a stab stitch. Another option is called a rocking stitch, where the quilter has one hand, usually with a finger wearing a thimble, on top of the quilt, while the other hand is located beneath the piece to push the needle back up. A third option is called "loading the needle" and involves doing four or more stitches before pulling the needle through the cloth. Hand quilting is still practiced by the Amish and Mennonites within the United States and Canada, and is enjoying a resurgence worldwide.

Machine quilting

Machine quilting is the process of using a home sewing machine or a longarm machine to sew the layers together. With the home sewing machine, the layers are tacked together before quilting. This involves laying the top, batting, and backing out on a flat surface and either pinning (using large safety pins) or tacking the layers together. Longarm Quilting involves placing the layers to be quilted on a special frame. The frame has bars on which the layers are rolled, keeping these together without the need for tacking or pinning. These frames are used with a professional sewing machine mounted on a platform. The platform rides along tracks so that the machine can be moved across the layers on the frame. A Longarm machine is moved across the fabric. In contrast, the fabric is moved through a home sewing machine.

Tying

Tying is another technique of fastening the three layers together (and is not a form of quilting at all). This is done primarily on quilts that are made to be used and are needed quickly. The process of tying the quilt is done with yarns or multiple strands of thread. Square knots are used to finish off the ties so that the quilt may be washed and used without fear of the knots coming undone. This technique is commonly called "tacking." In the Midwest, tacked bed covers are referred to as comforters.

Quilting is now taught in some American schools. It is also taught at senior centers around the U.S., but quilters of all ages attend classes. These forms of workshop or classes are also available in other countries in guilds and community colleges.

Contemporary quilters use a wide range of quilting designs and styles, from ancient and ethnic to post-modern futuristic patterns. There is no one single school or style that dominates the quilt-making world. Regardless of skill level, all quilters know the importance of having the right tools when quilting. Having the right tools increases the fluid process of making a quilt and can even be improved over time with practice. Having the right tools will maximize efficiency and make the quilting experience one to remember. Hand quilters spend much more time on making the quilts compared to machine quilters because of all of the tools that are incorporated into the machine compared to the hand quilters' ability to only use their hands. There are many other tools and machines to use to make quilts. Below is a list of the different tools and tips that can be used to make a quilt by hand or machine:

- Fabric Markers: When making a quilt it is important to mark the fabric that you are cutting in order to have some kind of guidance when cutting the fabric. When marking the fabric it is advised that you use a "fabric marker" which is a marker that washes out when the quilt is washed or will fade away after repeated washes.
- Long Arm Quilting Machines: The long arm quilting machine is something that every quilter would love to work with. This machine makes it easier to make larger quilts because of the extended arm that is used. Being able to leverage the larger machine and not having to hold the material that is being used while quilting helps the process move along much faster and makes it easier on the quilter.
- Machine Quilting Needles: When quilting the most important tool that is used is the needle. Regardless of if you are quilting by hand or by machine, the needle that is being used is critical to the final result. Using the wrong needle can lead to puckering, bumps, or even the material being torn. There are many different styles of needles and looking at Sewing Needles will be a good guide.
- Pins and Thimbles: Understanding how Pins and Thimbles work is also very important in the process of making quilts. Many different combinations of pins and thimbles can be used in order get similar results and the exciting part is figuring out existing combinations as well as coming up with new combinations. Thimbles are not required but are always seen as good practice.
- Quilting Hoops and Quilting Frames: Many different options are available for quilting hoops and frames and the quilter has the option of which one they want to use. Looking at quilting hoops or quilting frames will be beneficial in making that decision.
- How to Choose Threads and Cottons for Quilting: Choosing the right types of threads for a quilt can be difficult and beginners may need

some assistance from an expert or more advanced quilter. The colour, composition, and type of thread that is used will have a pivotal role in the outcome of the final quilt.

- Rotary Cutting – Cutters and Boards: What a quilter uses to cut the fabric is a vital step in the quilting process. It is very important each piece is perfectly aligned in order to prevent an uneven or sloppy appearance and to prevent rework. A rotary cutter offers even the shakiest of hands the ability to produce perfect even slices and minimizes the chance of error.
- Quilting Templates: Quilts can have many different templates and they can have a large impact on the final result. There are a number of mediums that can be used and depending on the usage, size and style they will give your quilt a varied look. Templates are generally considered the basis of the structure of the quilt, like a blueprint for a house. If used properly it can help quilters produce a quilt of their liking and give them a sense of satisfaction and vision for future quilts they want to make.

PROCESSES AND DEFINITIONS

The Basics of Quilt Assembly

Disclaimer: This section describes basic information about the assembly of quilts using machine quilting techniques. There are many different ways to make quilts and it would be impractical to attempt to cover all of these methods. It is, however, worth noting that many cultures and groups in different parts of the world have their own unique approaches, methods and styles of quilting which are not addressed below.

Assembling the Quilt Top

Selecting Fabric

The top-most layer of a quilt is usually made from cotton quilting fabric. Selecting the fabric can be a challenging exercise, and the number of different fabrics required depends on the quilting pattern selected. A good way to coordinate the colours of a quilt is to start by choosing a patterned or 'focus' fabric with the colour scheme that you like. Using this patterned fabric, you can pick out certain colours from within the pattern and select other fabrics with complimentary colours and tones. It is important to note that complimentary fabrics do not necessarily have identical colours, rather that each additional fabric selected draws out one or more shade, tone or aspect of the original focus fabric. Many quilters will also make use of fabrics from home, incorporating fabrics with a particular sentimental importance. When making use of these re-purposed fabrics, it is important to select fabrics that are not too worn in order to create a quilt which will last.

Fabric Preparation

Newly bought fabric is often washed before being cut or sewn. If not pre-washed, there is a risk of the fabric dyes bleeding into each other when the final product is washed. Many fabric manufacturers take this into account and have taken steps to prevent colour-bleeding. However, the only way to be absolutely certain is to wash the fabrics yourself. Washing, and subsequently drying, the fabric will also shrink some fabrics, so it is best to do this before cutting the fabric into the shapes and sizes needed. Whether you choose to wash your fabric or not, it must be ironed flat before cutting to prevent creases or wrinkles from altering your measurements.

Cutting Fabric

With large scale projects like quilts, it is often advantageous to have a rotary cutter and mat. A rotary cutter is a cutting tool with a round blade, making it easy to cut a smooth, continuous line. Rotary cutters come with different sized blades: a larger blade is useful for large projects with straight lines, while a smaller blade is helpful for small areas or curved lines. A rotary mat works to protect your tables and surfaces from the blade, while also protecting your cutting edge from damage. A quilting ruler will also be useful to help ensure that all pieces are cut to consistent sizes. Quilting rulers are made of clear plastic and possess marked grid-lines across the surface of the ruler. This type of ruler makes it possible to cut a piece of fabric in the correct width or length without having to use a measuring tape and fabric chalk. As you measure and cut, it is important to make sure that your measurements account for the seam allowance that you will be using when assembling your quilt.

Sewing the Pattern

Accurate seam allowances are especially important when it comes to quilting. With dozens, sometimes hundreds of different seams, if each seam is off by even 0.5 cm you will find it hard to make all of the components fit together evenly. When sewing a large quilt it is advantageous to use an assembly line method to maximize speed: pin together all similar fabric sections and sew the pieces together one after another without breaking the threads. Once all of the sections are sewn, clip the threads between them to separate before pressing flat. Always press the seams flat before attaching further segments.

Quilting the "Sandwich:"

- Layers of Quilts: There are generally three layers in a quilt: the quilt top, the middle layer of batting, and the fabric backing. The quilt top is the design layer. The cotton or polyester batting in the middle layer is what determines the warmth of the quilt. Batting comes in different thicknesses depending on the purpose of the final quilt, and multiple layers of batting can be combined to increase the warmth of the final product. The bottom layer is often a simple layer of cotton fabric, in a neutral or complimentary colour and design scheme, though some quilters use the extra or spare fabric from the quilt top to make a secondary design for the backing.

- Basting the Layers: Before actually quilting your fabrics, it is important to baste them together. Basting is the practice of making long, loose stitches in a grid format across the surface of the quilt to hold the layers of the quilt together and to prevent them from shifting during the quilting process. Basting can also be done using large curved safety pins rather than machine or hand basting.
- Quilting: Once the quilt has been basted, it is possible to quilt the layers together, either by hand or through the use of a sewing machine. One method of quilting involves the use of an outline or stencil applied to the surface of the quilt using fabric chalk, washable marker or iron-on pattern. The quilter will then sew along the applied pattern, washing or wiping the stencil off after the quilt is complete. Some quilters choose not to make use of a pattern. Free-motion quilting is the process of quilting without the use of a stencil or other guide, requiring a steady hand and a great deal of practice.
- Binding: Once the layers have been quilted, the edges must be finished and bound. There are many different ways to bind a quilt, one of the simplest involves sewing one side of a strip of fabric to the front side of the quilt, through all of the layers of fabric, then folding the strip over to the back side of the fabric and hand stitching the binding closed.
- Note: If the quilt will be hung on the wall, there is an additional step: making and attaching the hanging sleeve.

In China

Throughout China, a simple method of producing quilts is employed. It involves setting up a temporary roadside site. A frame is assembled within which a lattice work of cotton thread is made. Cotton batting, either new or retrieved from discarded quilts, is prepared in a mobile carding machine. The mechanism of the carding machine is powered by a small, petrol motor. The batting is then added, layer by layer, to the area within the frame. Between each layer, a new lattice of thread is created with a wooden disk used to tamp down the layer.

Definitions

- Piecing: Sewing small pieces of cloth into patterns, called blocks, that are then sewn together to make a finished quilt top. These blocks may be sewn together, edge to edge, or separated by strips of cloth called sashing. Note: Whole cloth quilts typically are not pieced, but are made using a single piece of cloth for the quilt top.

Pieced Quilt- Pieced quilts are also known as Patchworks. They consist of geometric shapes taken from different fabrics and are sewn together. After that process, it is referred to as a quilt top. The quilting patterns generally follow the design of the geometric patterns. The quilt ends up being a mixture of different fabrics and geometric designs and shapes that are organized in some fashion.

- Borders: Typically strips of fabric of various widths added to the perimeter of the pieced blocks to complete the quilt top. Note: borders may also be made up of simple or patterned blocks that are stitched together into a row, before being added to the quilt top.
- Layering: Placing the quilt top over the batting and the backing.
- Quilting: Stitching through all three layers of the quilt (the quilt top, the batting, and the quilt back), typically in decorative patterns, which serves three purposes:
  1. To secure the layers to each other,
  2. To add to the beauty and design of the finished quilt, and
  3. To trap air within the quilted sections, making the quilt as a whole much warmer than its parts.
- Binding: Long fabric strips cut on the bias that are attached to the borders of the quilt. Binding is typically machine sewn to the front side of the edge of the quilt, folded over twice, and hand sewn to the back side of the quilt.

Quilting is usually completed by starting from the middle, and moving outward towards the edges of the quilt.

Quilting can be elaborately decorative, comprising stitching fashioned into complex designs and patterns, simple or complex geometric grids, "motifs" traced from published quilting patterns or traced pictures, freehand, or complex repeated designs called tessellations. The quilter may choose to emphasize these designs by using threads that are multicolored or metallic, or that contrast highly to the fabric. Conversely, the quilter may choose to make the quilting disappear, using "invisible" nylon or polyester thread,thread that matches the quilt top, or stitching within the patchwork seams themselves (commonly known as "stitch in the ditch"). Some quilters draw the quilting design on the quilt top before stitching, while others prefer to stitch "freehand."

Quilting is often combined with embroidery, patchwork, applique, and other forms of needlework.

SPECIALTY STYLES

- Foundation piecing – also known as paper-piecing – sewing pieces of fabric onto a temporary or permanent foundation
- Shadow or Echo Quilting – Hawaiian Quilting, where quilting is done around an appliquéd piece on the quilt top, then the quilting is echoed again and again around the previous quilting line.

- Ralli Quilting – Pakistani and Indian quilting, often associated with the Sindh (Pakistan) and Gujarat (India) regions.
- Sashiko stitching – Basic running stitch worked in heavy, white cotton thread usually on dark indigo colored fabric. It was originally used by the working classes to stitch layers together for warmth.
- Trapunto quilting – stuffed quilting, often associated with Italy.
- Machine Trapunto quilting – a process of using water soluble thread and an extra layer of batting to achieve trapunto design and then sandwiching the quilt and re-sewing the design with regular cotton thread.
- Shadow trapunto – This involves quilting a design in fine Lawn and filling some of the spaces in the pattern with small lengths of colored wool.
- Tivaevae or tifaifai – A distinct art from the Cook Islands.
- Watercolor Quilting – A sophisticated form of scrap quilting whereby uniform sizes of various prints are arranged and sewn to create a picture or design.
- Thread Art – A custom style of sewing where thread is layered to create the picture on the quilt.

SAILMAKER

A sailmaker makes and repairs sails for sailboats, kites, hang gliders, wind art, architectural sails, or other structures using sails. A sailmaker typically works on shore in a sail loft. The sail loft has other sailmakers. Large ocean-going sailing ships often had sailmakers in the crew. The sailmaker maintained and repaired sails. This required knowledge of the sailmaker's craft and the tools of the sailmakers loft on shore.

Today, one of a sailmaker's important jobs is to teach people how to set and trim their sails to get the most out of them. Sometimes a sailmaker will accompany the client out on the water and adjust the sails. The modern sailmaker uses computer-aided design and manufacturing tools. Computer graphics allow the sailmaker to produce a "lines drawing" of the sail. Once the design is complete, the sailmaker can now use a low-power laser to cut the material to the exact shape.

The shape or depth of a sail is put into the sail by the sailmaker through the use of curved seams or *broadseams*.

A simple way of describing this is if you lay two pieces of paper next to each other with one overlapping the other by a fixed amount and stick them together, the result will be a larger flat piece of paper. However, if you cut a curve into one or both of the edges and maintain the same overlap, the result will be a curved surface.

Traditionally this was done "on the floor" by the sailmaker and was called *lofting*, but modern sailmakers now use computers to perform the calculations, the output files are then sent to a CAD cutting/plotting machine, which will reproduce the individual panels to micrometre accuracy.

This advance in technology has vastly improved the accuracy of sails and has reduced the space needed by the modern sailmaker to construct a sail.

Sailmakers have recently started using Computational fluid dynamics (CFD), the study of the flow of fluids over or through physical objects, in order to create more efficient sail or foil shapes in the design process.

After CFD analysis is run, complex data sets can be rendered graphically to enhance understanding of the design's likely results, before sails are ever cut.

Sailmaker's tools:

- Fid, used to stretch grommets before inserting reinforcement
- Sail palm, an oversized thimble used to drive needles through heavy canvas
- beeswax, used on thread
- Bench hook, to provide a "third hand" to hold sailcloth taut
- Seam rubber, to press folds in to fabric
- sailmaker's needles
- Sewing machine

SHOEMAKING

Shoemaking is the process of making footwear. Originally, shoes were made one at a time by hand. Traditional handicraft shoemaking has now been largely superseded in volume of shoes produced by industrial mass production of footwear, but not necessarily in quality, attention to detail, or craftsmanship.

Shoemakers or cordwainers (cobblers being those who *repair* shoes) may produce a range of footwear items, including shoes, boots, sandals, clogs and moccasins. Such items are generally made of leather, wood, rubber, plastic, jute or other plant material, and often consist of multiple parts for better durability of the sole, stitched to a leather upper.

For most of history, shoemaking has been a handicraft, limited to time consuming manufacture by hand. Traditional shoemakers used more than 15 different techniques of making shoes. Some of these were: pegged construction, English welted (machine-made versions are referred to as "Goodyear welted" after the inventor of the technique), goyser welted, Norwegian, stitchdown, turnout, German sewn, moccasin, bolognese stitched, and blake-stitched.

The most basic foot protection, used since ancient times in the Mediterranean area, was the sandal, which consisted of a protective sole, attached to the foot with leather thongs. Similar footwear worn in the Far East was made from plaited grass or palm fronds. In climates that required a full foot covering, a single piece of untanned hide was laced with a thong, providing full protection for the footand so made a complete covering.

The production of wooden shoes, was widespread in medieval Europe. They were made from a single piece of wood roughly cut into shoe form. A variant of this form was the clog, which were wooden soles to which a leather upper was attached. The sole and heel were made from one piece of maple or ash two inches thick, and a little longer and broader than the desired size of shoe. The outer side of the sole and heel was fashioned with a long chisel-edged implement, called the clogger's knife or stock; while a second implement, called the groover, made a groove around the side of the sole. With the use of a 'hollower', the inner sole's contours were adapted to the shape of the foot. The leather uppers were then fitted closely to the groove around the sole. Clogs were of great advantage to workers in muddy and damp conditions, keeping the feet dry and comfortable.

By the 1600s, leather shoes came in two main types. 'Turn shoes' consisted of one thin flexible sole, which was sewed to the upper while outside in and turned over when completed. This type was used for making slippers and similar shoes. The second type united the upper with an insole, which was subsequently attached to an out-sole with a raised heel. This was the main variety, and was used for most footwear, including standard shoes and riding boots.

The traditional shoemaker would measure the feet and cut out upper leathers according to the required size. These parts were fitted and stitched together. The sole was next assembled, consisting of a pair of inner soles of soft leather, a pair of outer soles of firmer texture, a pair of welts or bands about one inch broad, of flexible leather, and lifts and top-pieces for the heels. The insole was then attached to a last made of wood, which was used to form the shoe. Some lasts were straight, while curved lasts came in pairs: one for left shoes, the other for right shoes. The 'lasting' procedure then secured the leather upper to the sole with tacks. The soles were then hammered into shape; the heel lifts were then attached with wooden pegs and the worn out-sole was nailed down to the lifts. The finishing operation included paring, rasping, scraping, smoothing, blacking, and burnishing the edges of soles and heels, scraping, sand-papering, and burnishing the soles, withdrawing the lasts, and cleaning out any pegs which may have pierced through the inner sole.

Other types of ancient and traditionally made shoes included furs wrapped around feet, and sandals wrapped over them: used by Romans fighting in northern Europe, and moccasins - simple shoes without the durability of joined shoes. Shoemaking became more commercialized in the mid-18th century, as it expanded as a cottage industry. Large warehouses began to stock footwear in warehouses, made by many small manufacturers from the area.

Until the 19th century, shoemaking was a traditional handicraft, but by the century's end, the process had been almost completely mechanized, with production occurring in large factories. Despite the obvious economic gains of mass-production, the factory system produced shoes without the individual differentiation that the traditional shoemaker was able to provide.

The first steps towards mechanisation were taken during the Napoleonic Wars by the engineer, Marc Brunel. He developed machinery for the mass-production of boots for the soldiers of the British Army. In 1812 he devised a scheme for making nailed-boot-making machinery that automatically fastened soles to uppers by means of metallic pins or nails. With the support of the Duke of York, the shoes were manufactured, and, due to their strength, cheapness, and durability, were introduced for the use of the army. In the same year, the use of screws and staples was patented by Richard Woodman. Brunel's system was described by Sir Richard Phillips as a visitor to his factory in Battersea as follows:

- "In another building I was shown his manufactory of shoes, which, like the other, is full of ingenuity, and, in regard to subdivision of labour, brings this fabric on a level with the oft-admired manufactory of pins. Every step in it is effected by the most elegant and precise machinery; while, as each operation is performed by one hand, so each shoe passes through twenty-five hands, who complete from the hide, as supplied by the currier, a hundred pairs of strong and well-finished shoes per day. All the details are performed by the ingenious application of the mechanic powers; and all the parts are characterised by precision, uniformity, and accuracy. As each man performs but one step in the process, which implies no knowledge of what is done by those who go before or follow him, so the persons employed are not shoemakers, but wounded soldiers, who are able to learn their respective duties in a few hours. The contract at which these shoes are delivered to Government is 6s. 6d. per pair, being at least 2s. less than what was paid previously for an unequal and cobbled article."

However, when the war ended in 1815, manual labour became much cheaper, and the demand for military equipment subsided. As a consequence, Brunel's system was no longer profitable and it soon ceased business.

Similar exigencies at the time of the Crimean War stimulated a renewed interest in methods of mechanization and mass-production, which proved longer lasting. A shoemaker in Leicester, Tomas Crick, patented the design for a riveting machine in 1853. His machine used an iron plate to push iron rivets into the sole. The process greatly increased the speed and efficiency of production. He also introduced the use of steam-powered rolling-machines for hardening leather and cutting-machines, in the mid-1850s.

The sewing machine was introduced in 1846, and provided an alternative method for the mechanization of shoemaking. By the late 1850s, the industry was beginning to shift towards the modern factory, mainly in the US and areas of England. A shoe stitching machine was invented by the American Lyman Blake in 1856 and perfected by 1864. Entering in to partnership with McKay, his device became known as the McKay stitching machine and was quickly adopted by manufacturers throughout New England. As bottlenecks opened up in the production line due to these innovations, more and more of the manufacturing stages, such as pegging and finishing, became automated. By the 1890s, the process of mechanisation was largely complete.

Traditional shoemakers still exist today, especially in poorer parts of the world, and create custom shoes. Current crafters, in developing regions or supply constrained areas may use surplus car or truck tire tread sections as an inexpensive and plentiful material resource to make strong shoe soles or sandals with.

The shoemaking profession makes a number of appearances in popular culture, such as in stories about shoemaker's elves, and the proverb "The shoemaker's children go barefoot". The patron saint of shoemakers is Saint Crispin.

Chefs and cooks sometimes use the term "shoemaker" as an insult to others who have prepared sub-standard food, possibly by overcooking, implying that the chef in question has made his or her food as tough as shoe leather or hard leather shoe soles, and thus may be in the wrong profession. Similarly, to "cobble" can mean not only to make or mend shoes, but "to put together clumsily; to bungle."

UPHOLSTERY

Upholstery is the work of providing furniture, especially seats, with padding, springs, webbing, and fabric or leather covers. The word *upholstery* comes from the Middle English word *upholder*, which referred to a tradesman who held up his goods. The term is equally applicable to domestic, automobile, airplane and boat furniture, and can be applied to mattresses, particularly the upper layers, though these often differ significantly in design. A person who works with upholstery is called an upholsterer; an apprentice upholsterer is sometimes called an outsider or trimmer. Traditional upholstery uses materials like coil springs (post-1850), animal hair (horse, hog and cow), coir, straw and hay, hessians, linen scrims, wadding, etc., and is done by hand, building each layer up. In contrast, modern upholsterers employ synthetic materials like dacron and vinyl, serpentine springs, and so on.

Upholder is an archaic term used for 'upholsterer' in the past, although it appears to have a connotation of *repairing* furniture rather than creating new upholstered pieces from scratch (c.f. cobbler vs. cordwainer).

In 18th-century London, upholders frequently served as interior decorators responsible for all aspects of a room's decor. These individuals were members of the Worshipful Company of Upholders, whose traditional role, prior to the 18th century, was to provide upholstery and textiles and the fittings for funerals. In the great London furniture-making partnerships of the 18th century, a cabinet-maker usually paired with an upholder: Vile and Cobb, Ince and Mayhew, Chippendale and Rannie or Haig.

In the USA, Grand Rapids, Michigan is a centre for furniture manufacture, and many of the best upholsterers can still be found there. These craftsmen continue to create or recreate many antique and modern pieces of furniture.

TYPES OF UPHOLSTERY

Traditional upholstery

Traditional upholstery is a craft which evolved over centuries for padding and covering chairs, seats and sofas, before the development of sewing machines synthetic fabrics and plastic foam. Using a solid wood or webbed platform, it can involve the use of springs, lashings, stuffings of animal hair, grasses and coir, wools, hessians, scrims, bridle ties, stuffing ties, blind stitching, top stitching, flocks and wadding all built up by hand.

In the Middle Ages, domestic interiors were becoming more comfortable and upholstery was playing an important part in interior decoration. The decorations consisted mainly of what we would now consider as "soft furnishings", though there were simple platforms of webbing, canvas or leather for stools, chairs and elaborately decorated coverings that already demonstrated the rudimentary beginnings of upholstered furniture. By the beginning of the 17th century chair seats were being padded, but this form of upholstery was still fairly basic. All sorts of stuffings from sawdust, grass, feathers, to deer goat or horsehair were used, although in England the Livery Company forbade the use of goat and deer hair and imposed fines for misdemeanors. The stuffing was heaped on a wooden platform and held in place with a decorative top fabric and nails. This produced a simple dome shape sloping towards the seat. Only towards the end of the 17th century did upholsterers start to develop the techniques that would distribute and shape the stuffing into more controlled shapes. Curled horsehair was being used more consistently for stuffing that was easier to hold in place with stitches in twine that were developed from saddlery techniques. Thus layers of stuffing could be distributed evenly and secured to stay in place. On a basic level, squab cushions were made more stable by using tufting ties. Stuffed edge rolls appeared on seat fronts providing support for cushions to be retained and later for deeper stuffing to be held in place under a fixed top cover.

What we now think of as 'classic' upholstery shapes and techniques flourished in the 18th century. Frames of elegant line and proportion were sympathetically matched by expertly executed upholstery. By now, the upholsterers' technical knowledge meant that stuffings could be controlled along upright and sloping lines, giving new levels of comfort and a simply stated elegance. Later in the century, the border was replaced by a single piece of linen or scrim taken over the stuffed seat and tacked to the frame. At the same time the locked blind stitch and top-stitching combination (pulling the side and top surfaces together and bringing the stuffing up to make a firm top edge) had evolved.

In the Victorian era, fashions of opulence and comfort gave rise to excesses of stuffing and padding. Mass production techniques made upholstered furniture available in large quantity to all sections of society. The availability of better-quality steel springs and the development of lashing techniques enabled upholstery to be built up on seats, backs and arms quite independently of the frame shape. Stuffings became even more complex, edges became elaborately shaped into rolls and scrolls and fabrics were folded into soft padded shapes by means of buttoning.

Automobile upholstery

An automotive upholsterer, also known as a trimmer, coachtrimmer or motor trimmer, shares many of the skills required in upholstery, in addition to being able to work with carpet.

The term coachtrimmer derives from the days when car frames were produced by manufacturers and delivered to coachbuilders to add a car body and interior trimmings. Trimmers would produce soft furnishings, carpets, soft tops and roof linings often to order to customer specifications. Later, trim shops were often an in-house part of the production line as the production process was broken down into smaller parts manageable by semi-skilled labour.

Many automotive trimmers now work either in automotive design or with aftermarket trim shops carrying out repairs, restorations or conversions for customers directly. A few high-quality motor car manufacturers still employ trimmers, for example, Aston Martin.

Commercial upholstery

This is the type of upholstery work offered to businesses. Examples would be restaurant seating consisting of booth seats, dining room chairs, bar stools, etc. Also churches, including but not limited to pews and chairs for the congregation, hospitals and clinics consisting of medical tables, chiropractic tables, dental chairs, etc. Also common to this type of upholstery would be lobby and waiting-area seating. Upholstered walls are found in some retail premises.

Marine upholstery

Marine upholstery differs in that one has to consider dampness, sunlight and hard usage. A vinyl or material that is UV and cold-cracking resistant is the choice. Stainless-steel hardware such as staples, screws must be used for a quality job that will last. Any wood used must be of marine quality. Usually a high-resiliency, high-density plastic foam with a thin film of plastic over it is used to keep out water that might get by the seams. Closed-cell foam is used on smaller cushions which can double as flotation devices. Dacron thread must be used in any sewing work. Zippers should be of nylon.

TAILOR

A tailor is a person who makes, repairs, or alters clothing professionally, especially suits and men's clothing.

Although the term dates to the thirteenth century, *tailor* took on its modern sense in the late eighteenth century, and now refers to makers of men's and women's suits, coats, trousers, and similar garments, usually of wool, linen, or silk.

The term refers to a set of specific hand and machine sewing and pressing techniques that are unique to the construction of traditional jackets. Retailers of tailored suits often take their services internationally, traveling to various cities, allowing the client to be measured locally.

Traditional tailoring is called "bespoke tailoring" in the United Kingdom, where the heart of the trade is London's Savile Row tailoring, and "custom tailoring" in the United States and Hong Kong. This is unlike made to measure which uses pre-existing patterns. A bespoke garment or suit is completely original and unique to each customer.

Famous fictional tailors include the tailor in *The Tailor of Gloucester*, *The Emperor's New Clothes* and *The Valiant Little Tailor*. A more recent example is John le Carré's *The Tailor of Panama*.

TYPES OF TAILORING

As the tailoring profession has evolved, so too have the methods of tailoring. There are a number of distinctive business models which modern tailors may practice. While some may practice many, there are others who will practice only one or two.

Local tailoring

Local tailoring is as the name implies. Typically the tailor is met locally and the garment produced locally. This method enables the tailor to take professional measurements, assess posture and body shape to make unique modifications to the garment. Local tailors will typically have a showroom or shopfront allowing clients to choose fabrics from samples or return the garment easily should it require further modification. This is the most traditional form of tailoring. Hong Kong and London are the most famous for high quality bespoke tailoring, in average it takes about 2 to 3 fittings and about 3 to 4 days to handmake one suit.

Distance tailoring

Distance tailoring involves ordering a garment from an out-of-town tailor enabling cheaper labour to be used. In practice this can now be done on a global scale via e-commerce web sites. Unlike local tailoring, customers must take their own measurements, fabric selection must be made from a photo and if further alterations are required the garment must be shipped. Today, the most common platform for distance tailoring is via online tailors.

Online tailors sometimes offer to pay for needed alterations at a local tailor. Another new option is the concept where a free test suit is made to the provided measurements and shipped to the customer first. The test suit can be tried on and worn to see where any adjustments are wanted. The final suit is then tailored to the new specifications provided by the test suit fitting.

Traveling tailor

Unlike tailors who do distance tailoring, traveling tailors provide a more personal service to their customers and give the customers an opportunity to see the fabric samples and meet the tailor in person. Traveling tailors travel between cities and station in a local luxury hotel for a short period of time to meet and provide the same tailoring services they would provide in their local store. In the hotel, the customer will be able to select the fabric from samples and the tailor will take the measurements himself. The order then will be shipped to the customer within 3-4 weeks time. Unlike local tailoring, if further alterations are required the garment must be shipped. Today, most traveling tailors are from Hong Kong, traveling to USA, UK, France, Australia and Japan.

RELATED TERMS

- A *tailor-made* is a man's suit consisting of a (usually) woolen or tweed coat and pants; the name arose during the Edwardian period.
- As an adjective, *tailor-made* (from the second half of the twentieth century usually simplified to *tailored*) refers to clothing made by or in the style of clothes made by a tailor, characterized by simplicity of cut and trim and fine (often hand) finishing; as a women's clothing style *tailored* as opposed to dressmaker.
- *Rodeo tailor* is a term for a creator of the flamboyant costumes typical of country and western musicians, characterized by extensive hand embroidery, an abundance of rhinestones, and cowboy details such as pearl snaps and arrowhead pockets.
- In some documents, tailor means *adjust*, and tailoring means *adjusting*.

Sewing professional is the most general term for those who make their living by sewing, teaching, writing about sewing, or retailing sewing supplies. They may work out of their home, a studio, or retail shop, and may work part-time or full-time. They may be any or all or the following sub-specialties:

- A custom clothier makes custom garments one at a time, to order, to meet an individual customer's needs and preferences.
- A custom dressmaker specializes in women's custom apparel, including day dresses, suits, evening or bridal wear, sportswear, or lingerie.

- A tailor makes custom menswear-style jackets and trousers.
- A cutter cuts out, from lengths of cloth, the panels that make up a suit. In bespoke tailoring, the cutter may also measure the client, advise them on style choices, and commission craftsmen to sew the suit.
- An alterations specialist, or alterationist adjusts the fit of completed garments, usually ready-to-wear, or restyles them. Note that while all tailors can do alterations, not all alterationists can do tailoring.
- Designers conceive combinations of line, proportion, colour, and texture for intended garments. They may or may not have sewing or patternmaking skills, and may only sketch or conceptualize garments. They work with people who know how to actually construct the garment.
- Patternmakers flat draft the shapes and sizes of the numerous pieces of a garment by hand, using paper and measuring tools or by computer using AutoCAD based software, or by draping muslin onto a dressform. The resulting pattern pieces must comprise the intended design of the garment and they must fit the intended wearer.
- A wardrobe consultant, fashion advisor, or stylist recommends styles and colours that are flattering to a client.
- A seamstress is someone who sews seams or a machine operator in a factory who may not have the skills to make garments 'from scratch' or to fit them onto a real body. This term is not a synonym for *dressmaker*. *Seamstress* is also an unkind and archaic euphemism for prostitute.

10

Sewing Machine and Manufacturers

BARTHÉLEMY THIMONNIER

Barthélemy Thimonnier, (August 19, 1793 in L'Arbresle, Rhône - July 5, 1857 in Amplepuis), was a French inventor, who invented the first sewing machine that replicated sewing by hand. In 1795, his family moved to Amplepuis. Thimonnier was the oldest of seven children. He studied for a while in Lyon, before going to work as a tailor in Panissières. Barthelemy Thimonnier married an embroideress in January 1822. In 1823, he settled in a suburb (or called a commutety) of Saint-Étienne and worked as a tailor there.

INVENTION OF THE SEWING MACHINE

In 1829, he invented the sewing machine and in 1830 he signed a contract with Auguste Ferrand, a mining engineer, who made the requisite drawings and submitted a patent application. The patent for his machine was issued on 17 July 1830 in the names of both men, supported by the French government. Thc same year, he opened (with partners) the first machine-based clothing manufacturing company in the world. It was supposed to create army uniforms. However, the factory was burned down, reportedly by workers fearful of losing work following the issuing of the patent. A model of the machine is exhibited at the London Science Museum. The machine is made of wood and uses a barbed needle which passes downward through the cloth to grab the thread and pull it up to form a loop to be locked by the next loop.

The earliest sewing machine was actually patented by Thomas Saint in 1790. So Thimonnier's machine was not the first. Saint's contribution was not made public until 1874 when William Newton Wilson, himself a sewing machine manufacturer, found the drawings in the London Patent Office and built a machine which worked following some adjustments to the looper. So, in 1790 Thomas Saint had invented a machine with an overhanging arm, a feed mechanism (adequate for the short lengths of leather he intended it for), a vertical needle bar and a looper. The London Science Museum has the model that Wilson built from Saint's drawings.

LATER LIFE

Thimonnier then returned to Amplepuis and supported himself as a tailor again, while searching for improvements to his machine. He obtained new patents in 1841, 1845, and 1847 for new models of sewing machine. However, despite having won prizes at World Fairs, and being praised by the press, use of the machine did not spread. Thimonnier's financial situation remained difficult, and he died in poverty at the age of 64. The *Thimonnier* sewing machine company, created after his death, existed up to the 20th century.

BERNINA INTERNATIONAL

Bernina International AG is an international manufacturer of sewing and embroidery systems. The company was founded in 1893 in Steckborn, Switzerland. The origins of the company lie in the invention of the hemstitch sewing machine by Karl Friedrich Gegauf in 1893. Currently, the company's products include sewing machines, embroidery machines, serger/overlocker machines, and computer software for embroidery design.

COMPANY

- Company Profile: Bernina International AG is a privately owned company which develops, manufactures and sells goods and services for the textile market, primarily household sewing-related products in the fields of embroidery, quilting, home textiles, garment sewing and crafting.
- Target group: Primarily creative hobby sewers, but also designers implement their creative ideas with Bernina sewing and embroidery systems.
- Sales: Bernina International AG supplies 80 markets worldwide via business-to-business connections. Accordingly, there is a strong international network of independent dealers and no direct company-to-customer sales. Bernina products are sold in multi-brand stores as well as in mono-brand stores—the latter, which are devoted exclusively to the sale of the 300-or-so products manufactured mainly in Switzerland by the Bernina Textile Group.
- Production sites: The Bernina factories are located in Steckborn, Switzerland, as well as in Lamphun, Thailand.
- Subsidiaries: The Bernina Textile Group is a group of 15 companies doing business in 80 countries. Subsidiaries are established in Australia, Austria, Belgium, Germany, Japan, the Netherlands, New Zealand, Switzerland and the USA. The subsidiary Benartex, headquartered in the US, sells printed textiles, and quilting fabrics in particular. OESD, another subsidiary, develops and sells embroidery designs.

EMBROIDERY SOFTWARE

Bernina offers embroidery design editing and full digitizing software branded under its own name and written by industrial digitizing software manufacturer Wilcom International Pty Ltd.

BROTHER INDUSTRIES

Brother Industries, Ltd. is a Japanese multinational electronics and electrical equipment company headquartered in Nagoya, Japan. Its products include printers, multifunction printers, sewing machines, large machine tools, label printers, typewriters, fax machines, and other computer-related electronics. Brother distributes its products both under its own name and under OEM agreements with other companies.

Brother's history began in 1908 when the Yasui Sewing Machine Co. was established in Nagoya, Japan. In 1954, Brother International Corporation (US) was established as their first overseas sales affiliate. In 1958 a European regional sales company was established in Dublin. The corporate name was finally changed to Brother Industries, Ltd. in 1962. Brother entered the printer market during its long association with Centronics.

In 1968 the company moved its UK headquarters to Audenshaw, Manchester, after acquiring the Jones Sewing Machine Company, a long established British sewing machine maker.

In December 2011, Brother diversified its offerings by acquiring Nefsis, an innovator in web-based remote collaboration and conferencing software.

In November 2012, Brother announced that it had built the last UK-made typewriter at its north Wales factory. Saying that it had made 5.9 million typewriters in its factory in Wrexham since it opened in 1985, the firm donated the last machine to London's Science Museum.

As of March 2013, Brother's annual turnover was ¥516 billion ($5.05 billion USD (at April 2014 exchange rates)).

SEWING AND EMBROIDERY MACHINES

In 2010, the sewing divisions of Brother Industries around Europe were consolidated into one larger company called "Brother Sewing Machines Europe GmbH". With a turnover in excess of □80 million, it is the 4th biggest company under the Brother Industries Ltd umbrella of organisations.

Brother Industries manufacture mechanical sewing machines in Zhuhai, China and computerised sewing and embroidery machines in Taiwan. A new sewing machine factory was opened in 2012 in Dong Nai Province, Vietnam, which is the largest single brand sewing machine factory in the world.

In September 2012, Brother Industries manufactured their 50 millionth home sewing machine.

ADVERTISING AND SPONSORSHIP

Brother sponsored Manchester City Football Club from 1989 until 1999, one of the longest unbroken sponsorship deals of any English football club. Brother launched their first integrated, pan European advertising campaign in Autumn 2010 for their A3 printer range. Titled '141 per cent', referring to the ratio between paper sizes A3 and A4.

ELIAS HOWE

Elias Howe, Jr. (July 9, 1819 – October 3, 1867) was an American inventor and sewing machine pioneer. Elias Howe was born on July 9, 1819 to Dr. Elias Howe, Sr. and Polly (Bemis) Howe in Spencer, Massachusetts. Howe spent his childhood and early adult years in Massachusetts where he apprenticed in a textile factory in Lowell beginning in 1835. After mill closings due to the Panic of 1837, he moved to Cambridge, Massachusetts to work as a mechanic with carding machinery, apprenticing along with his cousin Nathaniel P. Banks. Beginning in 1838, he apprenticed in the shop of Ari Davis, a master mechanic in Cambridge who specialized in the manufacture and repair of chronometers and other precision instruments. It was in the employ of Davis that Howe seized upon the idea of the sewing machine. He married Elizabeth Jennings Ames, daughter of Simon Ames and Jane B. Ames on 3 Mar 1841 in Cambridge. They had three children: Jane Robinson Howe, Simon Ames Howe, and Julia Maria Howe.

INVENTION OF SEWING MACHINE AND CAREER

Contrary to popular belief, Howe was not the first to conceive of the idea of a sewing machine. Many other people had formulated the idea of such a machine before him, one as early as 1790, and some had even patented their designs and produced working machines, in one case at least 80 of them. However, Howe originated significant refinements to the design concepts of his predecessors, and on September 10, 1846, he was awarded the first United States patent (U.S. Patent 4,750) for a sewing machine using a lockstitch design. His machine contained the three essential features common to most modern machines:

- A needle with the eye at the point,
- A shuttle operating beneath the cloth to form the lock stitch, and
- An automatic feed.

A possibly apocryphal account of how he came up with the idea for placing the eye of the needle at the point is recorded in a family history of his mother's family: "He almost beggared himself before he discovered where the eye of the needle of the sewing machine should be located. It is probable that there are very few people who know how it came about. His original idea was to follow the model of the ordinary needle, and have the eye at the heel. It never occurred to him that it should be placed near the point, and he might have failed altogether if he had not dreamed he was building a sewing machine for a savage king in a strange country. Just as in his actual working experience, he was perplexed about the needle's eye. He thought the king gave him twenty-four hours in which to complete the machine and make it sew. If not finished in that time death was to be the punishment. Howe worked and worked, and puzzled, and finally gave it up. Then he thought he was taken out to be executed. He noticed that the warriors carried spears that were pierced near the head. Instantly came the solution of the difficulty, and while the inventor was begging for time, he awoke. It was 4 o'clock in the morning. He jumped out of bed, ran to his workshop, and by 9, a needle with an eye at the point had been rudely modeled. After that it was easy. That is the true story of an important incident in the invention of the sewing machine."

Despite securing his patent, Howe had considerable difficulty finding investors in the United States to finance production of his invention, so his elder brother Amasa Bemis Howe traveled to England in October 1846 to seek financing. Amasa was able to sell his first machine for £250 to William Thomas of Cheapside, London, who owned a factory for the manufacture of corsets, umbrellas and valises. Elias and his family joined Amasa in London in 1848, but after business disputes with Thomas and failing health of his wife, Howe returned nearly penniless to the United States. His wife Elizabeth, who preceded Elias back to the United States, died in Cambridge, Massachusetts shortly after his return in 1849.

Despite his efforts to sell his machine, other entrepreneurs began manufacturing sewing machines. Howe was forced to defend his patent in a court case that lasted from 1849 to 1854 because he found that Isaac Singer with cooperation from Walter Hunt had perfected a facsimile of his machine and was selling it with the same lockstitch that Howe had invented and patented. He won the dispute and earned considerable royalties from Singer and others for sales of his invention.

Howe contributed much of the money he earned to providing equipment for the 17th Connecticut Volunteer Infantry of the Union Army during the Civil War, in which Howe served during the Civil War as a private in Company D and regimental postmaster from August 14, 1862, to July 19, 1865.

INVOLVEMENT IN INVENTING THE ZIPPER

Howe received a patent in 1851 for an "Automatic, Continuous Clothing Closure". Perhaps because of the success of his sewing machine, he did not try to seriously market it, missing recognition he might otherwise have received.

LATER LIFE AND LEGACY

In 1865, Elias established the Howe Machine Company of Bridgeport, Connecticut that was operated by the Stockwell brothers, his brothers-in-law, from 1867 until about 1885. Between 1867 and 1870, Elias's brother Amasa operated a factory in New York City manufacturing sewing machines under the brand name of A. B. Howe. Elias's sewing machine won the gold medal at the Paris Exhibition of 1867, and that same year he was awarded the Légion d'honneur by Napoleon III for his invention.

Howe died at age 48, on October 3, 1867. He was buried in Green-Wood Cemetery in Brooklyn, New York. His second wife, Rose Halladay, who died on October 10, 1890 is buried with him. Both Singer and Howe ended their days as multi-millionaires.

Howe was commemorated with a 5-cent stamp in the Famous American Inventors series issued October 14, 1940. In 2004 he was inducted into the United States National Inventors Hall of Fame.

GENEALOGY

Howe was a direct descendant of John Howe (1602-1680) who arrived in Massachusetts Bay Colony in 1630 from Brinklow, Warwickshire, England and settled in Sudbury, Massachusetts. Howe was also a descendant of Edmund Rice another early immigrant to Massachusetts Bay Colony as follows:

- Elias Howe, Jr, son of
  - Elias Howe (1792 – 1872), son of
  - Elijah Howe, Jr. (1768 – 1816), son of
  - Elijah Howe (1731 – 1808), son of
  - Jaazaniah Howe (1704 – 1762), son of
  - Deliverance Rice (1681 – 1723), daughter of
  - John Rice (1659 – 1719), son of
  - Deacon Edward Rice (1622 – 1712), son of
  - Deacon Edmund Rice (1594 – 1663)

JANOME

Janome is a Japanese company that is one of the leading manufacturers of sewing machines worldwide, with manufacturing plants in Japan, Taiwan and Thailand. The Pine Sewing Machine factory was founded on October 16, 1921. In 1935, the Janome trademark was established, and the company was renamed to Janome Sewing Machine Co., Ltd. in 1954. As one of many manufacturers selling in the USA, its subsidiary Janome America based in Mahwah, New Jersey, also owns Swiss-brand Elna. The company manufactures all of its machines in the same factories. Around 1862, William Barker and Andrew J. Clark began producing the "The Pride of the West" machine, later calling it the "New England Single Thread Hand Sewing Machine" after moving the plant to Orange, Massachusetts in 1867. Over the next few years, the New England machine and the "Home Shuttle" were their two most significant products. In 1882, the company reformed under the name New Home (a combination of the labels New England and Home Shuttle). The company ran into financial difficulties in the 1920s and was taken over by The Free Sewing Machine Company in 1930, after they temporarily ran the business for two years. In 1960, New Home and the "New Home" brand were purchased by the Janome Sewing Machine Company of Tokyo, Japan. Janome was the first to develop a computerized machine for home use (the Memory 7, in 1979), the first to offer professional style embroidery to the home market (the Memory Craft 8000, in 1990) and the first to offer a long-arm quilting machine for home use (the Memory Craft 6500P, in 2003). The name "Janome" literally means "snake's eye" and was taken from the appearance of the latest bobbin design. At the time of brand establishment in 1935, the round bobbin system was the more advanced technology replacing the traditional long shuttle type. As the new round bobbin looks like a snake's eye, Janome was chosen as the company's name. Janome is also the name of the traditional Japanese bull's-eye umbrella design.

MACHINES

Machines produced by Janome range from basic sewing, quilting and embroidery through to specialty machines for overlocking (serging), embroidery, fibre arts, machines for schools and compact models. The range includes electronic and computerised machines that can be updated via software and there is a treadle for power-free sewing. Brands manufactured by Janome include Kenmore for Sears, Artistic Sewing Suite, Elna and all Janome brands. Additionally, brands like Necchi are made in their Asian factories.

JONES SEWING MACHINE COMPANY

The Jones Sewing Machine Co. was a British manufacturer of sewing machines founded in 1860 by William Jones and Thomas Chadwick under the name Chadwick and Jones that later become known as The Jones Sewing Machine Co.William Jones started making sewing machines in 1859 and in 1860 formed a partnership with Thomas Chadwick. As Chadwick and Jones they manufactured sewing machines at Ashton-under-Lyne until 1863. Their machines used designs from Howe and Wilson produced under licence. Thomas Chadwick later joined Bradbury and Co. William Jones opened a factory in Guide Bridge, Manchester in 1869. In 1893 a Jones advertising sheet claimed that this factory was the "Largest Factory in England Exclusively Making First Class Sewing Machines". The firm was renamed as the Jones Sewing Machine Co. Ltd and was later acquired by Brother Industries of Japan, in 1968. The Jones patent for his popular Serpent Neck model appeared in 1879. These were manufactured until 1909. The machines pictured employ a transverse boat shuttle mechanism forming a lock stitch. The CS Family model has "As Supplied to Her Majesty Queen Alexandra" written along the shoulder and like many Jones machines displayed very ornamental decoration ensuring that many are still kept in good condition as decorative items.

JUKI

Juki is a Japanese manufacturer of industrial sewing machines and recently domestic machines. It also produces sewing machines for the home or hobbyist market, and surface mount component placement machines for electronics manufacturing. Juki claims to have sold industrial sewing machines to customers "in about 170 nations." As of March 2013, the company had a market cap of ¥18 billion. In February 2013, Juki and Sony Corp. entered negotiations to discuss a merger of their SMT equipment businesses.

MERROW SEWING MACHINE COMPANY

The Merrow Sewing Machine Company is a manufacturer of sewing machines, established in 1838 as the Merrow Company by J. Makens Merrow. Originally a gunpowder manufacturer, in 1837 the company built a knitting mill, and in 1887 evolved to design, build and market sewing machines exclusively. Best known for inventing the overlock sewing machine, it was renamed J. B. Merrow and Sons in 1888, then The Merrow Machine Company in 1893. Originally all of its manufacturing was done at facilities in Merrow, Connecticut and then in Hartford, Connecticut. The company is currently based in Fall River, Massachusetts.

HISTORY

From gunpowder to knitting mills

In the early 19th Century Mr. Joseph Makens Merrow became interested in the manufacture of gunpowder and established a powder mill 24 miles from Hartford Connecticut. When the Mill was destroyed by explosion in 1837 it was decided to build a knitting factory on the same site using water power from an adjacent river.

At first the knitted goods were made largely of native wool which was sorted, scoured and dyed, picked, carded and spun into yarn and knitted into hosiery. The product was sold through commission merchants in New York and delivered to retail stores throughout New England by two-horse wagons. Following the gold rush of 1849 shipments of goods began to sail to San Francisco. As business increased, a small machine shop was started to support the equipment in the factory.

The first overlock machine

In Conjunction with the knitting business, the first Crochet Machines were constructed for finishing around the tops of men's socks in place of handwork. The Merrow machine as it is now known, was an invention of Mr. Joseph M. Merrow, who was president of the company until his death in 1947 at age 98.

The Merrow Machines were constructed under his direction prior to 1876 with numerous patents granted. The machines were so useful that business was undertaken to introduce the equipment to other textile manufacturers. In 1887 the knitting mill was destroyed by fire and the company moved to Hartford and reorganized concentrating on the manufacture of overlock sewing machines.

The Merrow Machine Company

In Hartford the company focused on building lines of industrial overlock sewing machines that were used to overedge fabric, add decorative edging and support the fabric processing trade by joining fabrics.

Between 1893 when the company was renamed the Merrow Machine Company, and 1932 when a line of "A Class" machines was introduced, Merrow had a significant impact on the textile industry. The technology and rate of innovation in this time, spearheaded by Joseph M. Merrow was unequaled in the industry. As a consequence there were several high profile legal confrontations, including Merrow v. Wilcox and Gibbs in 1897.

Sales for overlock sewing machines were strong and Merrow grew to employ more than 500 people in Hartford Ct. The company also excelled developing international distribution and by 1905 had agents in 35 countries and printed manuals in at least 12 languages. In 1955, Merrow patented the Merrow MG-3U Emblem Machine. In the mid 1960s Merrow opened a manufacturing facility in Lavonia GA to reduce costs and maintain proximity to an American textile market that was moving from New York City to the American South East. In the 1990s Merrow developed a new overlock machine called the Delta Class, but was never able to gain traction with the new model. In 2004 shareholders of the Merrow Machine Co. agreed to a buyout of the company by Charlie Merrow, and it was renamed the Merrow Sewing Machine Company.

The Merrow Machine Company today

After the reorganization in Massachusetts, the company released notice that it would continue supporting most models of sewing machines manufactured after 1925, and would re-release to market new versions of its most popular models. The company has capitalized on the trademarks "merrowed" and "merrowing", working with manufacturers who use Merrow Machines to brand and market "merrow" stitching.

In 2008 Merrow developed a social network for stitching named merrowing.com, and introduced series of rich media web based tools to help people research and understand the myriad of stitches produced by Merrow Machines.

Present day

the Merrow Machine Company is now based in Fall River, Massachusetts, and is managed by Charlie Merrow and Owen Merrow great great nephews of Joseph M. Merrow. The company continues to build many models of overlock sewing machines. In addition to being one of the most recognized brands of textile equipment in the world, it remains the oldest manufacturer of sewing machines still made in the United States.

TIMELINE

- 1822: J. Makens Merrow purchases a powder mill in Mansfield, Connecticut for the manufacture of gunpowder. The mill is destroyed in 1837 by a gunpowder explosion.
- 1838: J. Makins Merrow founds the first knitting mill in American in partnership with his son, Joseph B. Merrow, under the name J. M. Merrow and Son. This knitting mill is located on the site of the old gunpowder mill in Mansfield, Connecticut.
- 1840s: A machine shop is established at the Merrow mill to develop specialized machinery for the knitting operations.
- 1877: The world's first crochet machine is invented and patented by Joseph M. Merrow, then-president of the company.
- 1892: A need for expansion leads Merrow to rebuild its plant in Hartford, Connecticut.
- 1894–1947: Over 100 patents are issued in the Merrow name, most notably for the first shell-stitching machine and the first butt-seaming machine.
- 1964: Merrow expands operations in the South by opening Franklin Industries, a wholly owned subsidiary, in Lavonia, Georgia.
- 1972: Merrow acquires the Arrow Tool Company of Wethersfield, Connecticut, a machining subcontractor.
- 1982: Merrow moves its location from Hartford to Newington, Connecticut.
- 2004: Charlie Merrow and Owen Merrow great grandsons of Lena Bryant (and great nephews to Joseph B. Merrow) with their father Robert Merrow, organize a buyout of the Merrow Machine Company and move its headquarters to Wareham, Massachusetts, changing its name to The Merrow Sewing Machine Company.
- 2005: Merrow announces it will continue production of all major lines of sewing machines.
- 2008: Merrow develops and releases a dozen web based tools including video, interactive application and stitch finders, the social network for stitching named merrowing.com, a "needle configurator" and a new online store
- 2010: After moving to Fall River, MA into the historic Granite Mill Buildings, Merrow opens the only Sample Room for industrial sewing machines in the USA, and adds custom built industrial sewing machines to its standard product line.

WILLIAM L. GROUT

William L. Grout (1833-1908) was an U. S. industrialist and pioneer manufacturer of sewing machines and automobiles. In 1876 he started, together with Thomas H. White, the White Sewing Machine Company in Cleveland, Ohio. The company was founded with a joint capital of $400 which he brought in. Later, White began building steam and gasoline automobiles and became a leading truck manufacturer for decades. Later, they separated, and Grout founded the New Home Sewing Machine Company in Orange, Massachusetts. Business flourished, and in 1892, its best year, New Home sold 1.2 million sewing machines.

Alas, he had to learn that his sons Carl, Fred, and C.B. were not interested in his business. So, he set them up in the automobile business in 1900 as manufacturers of both steam and gasoline powered cars. At first, they were sold under the *Grout New Home* label. Although quite successful with up to 18 steam vehicles built per week by 1904, William L. Grout disagreed more and more with the way his sons handled business. Severe family struggle arose as he ended serving a $200,000 attachment on the factory - and faced, aged 74, a lawsuit to install a conservator for him because of his age. Finally, he succeeded and took control of the company. His sons resigned and left town. William L. Grout died on April 20th, 1908, leaving the company in trouble again. Production closed in 1912, after a reorganization.

PFAFF

Pfaff (German: *Pfaff Industriesysteme und Maschinen AG, Pfaff Industrial*) is a manufacturer of sewing machines and is now owned by the SGSB Co.LTD. Pfaff was founded in Kaiserslautern Germany in 1862 by instrument maker Georg-Michael Pfaff (1823 – 1893). Pfaff's first machine was handmade, and designed to sew leather in the manufacture of shoes.

SEWMOR

Sewmor sewing machines were designed and manufactured in post-World War II Japan (mainly using parts from miscellaneous Asian countries, though the 900 series motors are said to be manufactured in Belgium) and imported/ badged by the [[Consolidated Sewing Machine Corporation]] in New York City, New York. Numerous models existed, with many units proving to have a fairly reliable track record over the years. A few models have recently been becoming a more sought after collectors item, generally due to their aesthetic similarities to American automobiles of the same era.

SINGER CORPORATION

Singer Corporation is an American manufacturer of sewing machines, first established as I. M. Singer and Co. in 1851 by Isaac Merritt Singer with New

York lawyer Edward Clark. Best known for its sewing machines, it was renamed Singer Manufacturing Company in 1865, then The Singer Company in 1963. It is based in La Vergne, Tennessee near Nashville. Its first large factory for mass production was built in Elizabeth, New Jersey in 1863.

Presidents:

- Isaac Singer (1851–1863)
- Inslee Hopper (1863–1875)
- Edward S. Clark (1875–1882)
- George Ross McKenzie (1882–1889)
- Frederick Gilbert Bourne (1889–1905)
- Sir Douglas Alexander (1905–1949)
- Milton C. Lightner (1949–1958)
- Donald P. Kircher (1958–1975)
- Joseph Bernard Flavin (1975–1987)
- Paul Bilzerian (1987–1989)
- James H. Ting (1989–1997)
- Stephen H. Goodman (1998–2004)

THE COMPANY

In 1885 Singer produced its first "vibrating shuttle" sewing machine, an improvement over contemporary transverse shuttle designs; (see bobbin drivers).

The 11,000 workers at the largest factory of Singer, in Clydebank, went on strike in March–April 1911, ceasing to work in solidarity with 12 female colleagues protesting against work process reorganization. Following the end of the strike, Singer fired 400 workers, including all strike leaders and purported members of the IWGB, among whom was Arthur McManus, who later went on to become the first chairman of the Communist Party of Great Britain between 1920 and 1922.

During World War II, the company suspended sewing machine production to take on government contracts for weapons manufacturing. Factories in the US supplied the American forces with Norden bomb sights and M1 Garand rifle receivers, while factories in Germany provided their armed forces with weapons.

In 1939, the company was given a production study by the government to draw plans and develop standard raw material sizes for building M1911A1 pistols.

The following April 17, Singer was given an educational order of 500 units with serial numbers S800001 - S800500. The educational order was a programme set up by the US Ordnance Board to teach companies without gun-making experience to manufacture weapons.

After the 500 units were delivered to the government, the management decided to produce artillery and bomb sights. The pistol tooling and manufacturing machines were transferred to Remington Rand whilst some went to the Ithaca Gun Company. Original Singer pistols are collectable, and in excellent condition sell for $25,000 to $60,000 with the highest paid $80,000 at auction in 2002.

MARKETING

The Singer sewing machine was the first complex standardized technology to be mass marketed. It was not the first sewing machine, and its patent in 1851 led to a patent battle with Elias Howe, inventor of the lockstitch machine. This eventually resulted in a patent sharing accord among the major firms. Marketing strategies included focusing on the manufacturing industry, gender identity, credit plans, and "hire purchases."

Singer's marketing emphasized the role of women and their relationship to the home, evoking ideals of virtue, modesty, and diligence. Though the sewing machine represented liberation from arduous hand sewing, it chiefly benefited those sewing for their families and themselves. Tradespeople relying on sewing as a livelihood still suffered from poor wages, which dropped further in response to the time savings gained by machine sewing. Singer offered credit purchases and rent-to-own arrangements, allowing people to rent a machine with the rental payments applied to the eventual purchase of the machine, and sold globally through the use of direct-sales door-to-door canvassers to demonstrate and sell the machines.

Diversification

In the 1960s the company diversified, acquiring the Friden calculator company in 1965, Packard Bell Electronics in 1966 and General Precision Equipment Corporation in 1968. GPE included Librascope, The Kearfott Company, Inc, and Link Flight Simulation. In the 1968 also Singer bought out GPS Systems and added it to the Link Simulations Systems Division (LSSD). This unit produced nuclear power plant control center simulators in Silver Spring, MD; while flight simulators were produced in Binghamton, New York.

In 1987, corporate raider, Paul Bilzerian, made a "greenmail" run at Singer, and ended up owning the company when no "White Knight" rescuer appeared. To recover his money, Bilzerian sold off parts of the company. Kearfott was split, the Kearfott Guidance and Navigation Corporation was sold to the Astronautics Corporation of America in 1988. The Electronic Systems Division was purchased by GEC-Marconi in 1990, renamed GEC-Marconi Electronic Systems (and later incorporated into BAE Systems) while the Sewing Machine Division was sold in 1989 to Semi-Tech Microelectronics, a publicly traded Toronto-based company.

For several years in the 1970s, Singer set up a national sales force for CAT phototypesetting machines made by another Massachusetts company,

Graphic Systems Inc. This division was purchased by Wang Laboratories in 1978.

21st Century

The Singer Corporation produces a range of consumer products, including electronic sewing machines. It is now part of SVP Worldwide, which also owns the Pfaff and Husqvarna Viking brands, which is in turn owned by Kohlberg and Company, which bought Singer in 2004.

Its main competitors are Brother Industries, Janome and Aisin Seiki - a Toyota Group company that manufactures Toyota, Necchi and E&R Classic Sewing Machines.

SINGER BUILDINGS

Singer was heavily involved in Manhattan real estate in the 1800s through Edward Clark, a founder of the company. Clark had built The Dakota apartments and other Manhattan buildings in the 1880s. In 1900, the Singer company retained Ernest Flagg to build a 12-story loft building at Broadway and Prince Street in Lower Manhattan. The building is now considered architecturally notable, and has been restored.

The 47-story Singer Building, completed in 1908, was also designed by Flagg, who designed two landmark residences for Bourne. Constructed during Bourne's tenure, the Singer Building (demolished in 1968) was then the tallest building in the world. Singer built the largest clock face in the world, the Singers Clock at its Clydebank, Scotland factory which opened in 1885 and closed in 1980. Singer railway station, built to serve the factory, is still in existence to this day.

The famous Singer House, designed by architect Pavel Suzor, was built in 1902-1904 at Nevsky Prospekt in Saint Petersburg for headquarters of the Russian branch of the company. This modern style building (situated just opposite to the Kazan Cathedral) is officially recognized as an object of Russian historical-cultural heritage.

WHITE SEWING MACHINE COMPANY

The White Sewing Machine Company was a sewing machine company founded in 1858 in Templeton, Massachusetts by Thomas H. White and based in Cleveland, Ohio since 1866. Founded as the White Manufacturing Company it took the White Sewing Machine Company name when it was incorporated in 1876. The company took the name White Consolidated Industries in 1964. It was acquired by Electrolux in 1986.

In 2006 Electrolux sold off a number of its lines under the Husqvarna (including variations of the Husqvarna) brand name. The White line of sewing machines was consolidated into the lower end Husqvarna Viking brand after this sale-off. The White-Westinghouse brand name remained with Electrolux and that name is the only remnant of the "White Sewing Machine Company" name, however the brand line included no sewing machines.

New "White" branded sewing machines models have not been manufactured since the Husqvarna spin off in 2006. Husqvarna Viking, Singer, and Pfaff brands are now all owned by SVP Worldwide. Though not advertised, SVP provides all of their user manuals for all of their brands including the "White" brand through the Singer brand web site. The White Motor Company was founded by the same family as a separate company.

11

Yarn Dyeing Wastewater Treatment

INTRODUCTION

Yarn industry can be classified into three categories *viz.*, cotton, woolen, and synthetic fibres depending upon the used raw materials. The cotton yarn industry is one of the oldest industries in China. The yarn dyeing industry consumes large quantities of water and produces large volumes of wastewater from different steps in the dyeing and finishing processes. Wastewater from printing and dyeing units is often rich in colour, containing residues of reactive dyes and chemicals, such as complex components,many aerosols,high chroma,high COD and BOD concentration as well as much more hard-degradation materials. The toxic effects of dyestuffs and other organic compounds, as well as acidic and alkaline contaminants, from industrial establishments on the general public are widely accepted. At present, the dyes are mainly aromatic and heterocyclic compounds, with colour-display groups and polar groups. The structure is more complicated and stable, resulting in greater difficulty to degrade the printing and dyeing wastewater (Shaolan Ding et al.,2010). According to recent statistics, China's annual sewage has already reached 390 million tons, including 51 per cent of industrial sewage, and it has been increasing with the rate of 1 per cent every year. Each year about 70 billion tons of wastewater from yarn and dyeing industry are produced and requires proper treatment before being released into the environment (State Environmental Protection Administration,1994). Therefore, understanding and developing effective printing-dye industrial wastewater treatment technology is environmentally important.

YARN PRINTING AND DYEING PROCESS

Yarn Printing and dyeing processes include pretreatment, dyeing/ printing, finishing and other technologies. Pre-treatment includes desizing, scouring, washing, and other processes. Dyeing mainly aims at dissolving the dye in water, which will be transferred to the fabric to produce colored fabric under certain conditions. Printing is a branch of dyeing which generally is defined as 'localized dyeing' *i.e.* dyeing that is confirmed to a certain portion of the fabric that constitutes the design. It is really a form of dyeing in which the essential reactions involved are the same as those in dyeing. In dyeing, colour is applied in the form of solutions, whereas colour is applied in the form of a thick paste of the dye in printing. Both natural and synthetic yarns are subjected to a variety of finishing processes. This is done to improve specific properties in the finished fabric and involves the use of a large number of finishing agents for softening, crosslinking, and waterproofing. All of the finishing processes contribute to water pollution. In addition, in different circumstances, the singeing, mercerized, base reduction, and other processes should have been done before dyeing/ printing.

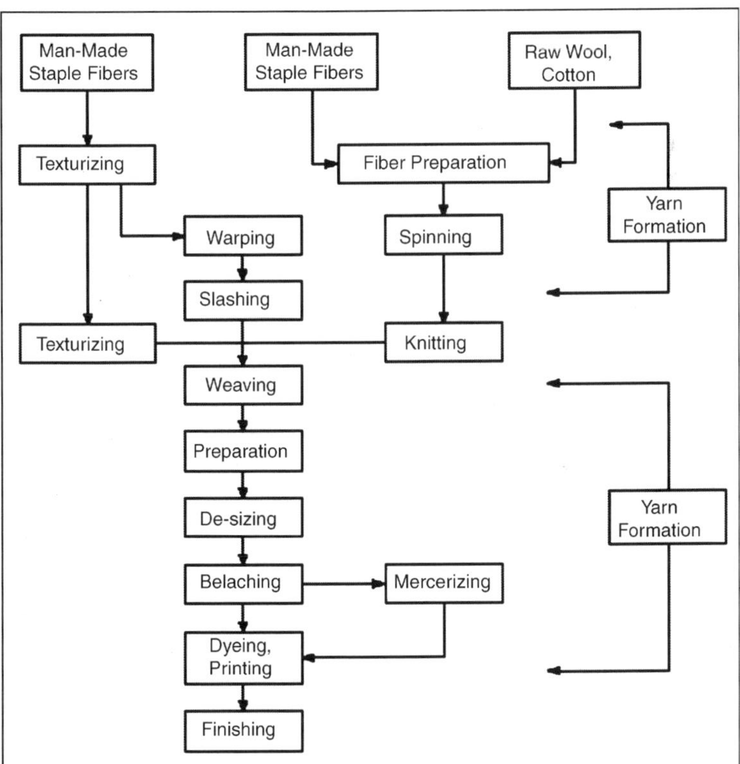

Fig. Various Steps Involved in Processing Yarn in a Cotton Mill (B. Ramesh Babu et al.,2007).

In the yarn dyeing industry, bleaching is an important process. It has three technologies: sodium hypochlorite bleaching; hydrogen peroxide bleaching and sodium chlorite bleaching. Sodium hypochlorite bleaching and sodium chlorite bleaching are the most commonly used processes. Normal concentration of chlorine dioxide in bleaching effluent is 10-200 mg/L. As chlorine dioxide is a strong oxidant, it is very corrosive and toxic as well. The typical printing and dyeing process is shown in Fig. and the main used fibre dyes at present have been shown in Table.

Table. The Varieties of Common Used Fibre

The Variety of Fibre	The Commonly Used Dyes
Cellulose fibre	Direct dyes, Reactive dyes,Vat dyes, Sulfide dyes, Azo dyes
Wool	Acid dyes
Silk	Direct dyes,Acid dyes
Polyester	Azo dyes,Disperse dyes
Polyester-cotton	Disperse / Vat dyes,Disperse / Insoluble dye
polyacrylonitrile fibre	Cationic dyes, Disperse dyes
polyacrylonitrile fibre -wool	Cationic dyes,Acid dyes
vinylon	Direct dyes,Vat dyes,Sulfur dyes,Aciddyes,

PRODUCTION OF YARN INDUSTRY POLLUTION

Yarn Printing and dyeing processes include pre-treatment, dyeing and printing, finishing. The main pollutants are organic matters which come from the pre-treatment process of pulp, cotton gum, cellulose, hemicellulose and alkali, as well as additives and dyes using in dyeing and printing processes. Pre-treatment wastewater accounts for about 45 per cent of the total, and dyeing/printing process wastewater accounts for about 50 per cent~55 per cent, while finishing process produces little. In China, chemical fibre accounts for about 69 per cent of total in which polyester fibres accounts for more than 80 per cent. Cotton accounts for 80 per cent of the natural fibre production. Therefore, the dyeing wastewater analysis of production and pollution is based on these two fibres. Pre-treatment of cotton includes desizing and scouring. The main pollutants are the impurities in the cotton, cotton gum, hemicellulose and the slurry, alkali in weaving process. The current average COD concentration in the pre-treatment is 3000 mg/L.

The main pollutants in dyeing/printing are auxiliaries and the residual dyes. The average concentration of COD is 1000 mg/L and the total average concentration is 2000 mg/L after mixing. Pre-treatment of polyester fibres mainly involves in the reduction with alkali. The so-called reduction is treating the polyester fabric with 8 per cent of sodium hydroxide at 90 °C for about 45 minutes. Some polyester fabrics will peel off and decompose into terephthalic acid and ethylene glycol so that a thin polyester fabric will have the feel of silk. This process can be divided into continuous and batch type. Taking the batch type as an example, the concentration of COD is up to 20000 mg/L-60000 mg/L. The wastewater from reduction process may account for only 5 per cent of the volume of wastewater, while COD accounts for 60 per cent or more in the conventional dyeing and finishing business. The chroma is one pollutant of the wastewater which causes a lot of concerns. In the dyeing process, the average dyeing rate is more than 90 per cent. It means that the residual dyeing rate in finishing wastewater is about 10 per cent, which is the main reason of contamination. According to the different dyes and process, the chroma is 200 to 500 times higher than before. pH is another factor of the dyeing wastewater.

Before the printing and the dyeing process, pH is another factor,the pH of dyeing wastewater remains between 10 to 11 when treated by alkali at high temperature around 90°C in the process of desizing, scouring and mercerization. Polyester base reduction process mainly uses sodium hydroxide, and the total pH is also 10 to 11. Therefore, most dyeing water is alkaline and the first process is to adjust the pH value of the yarn dyeing wastewater. The total nitrogen and ammonia nitrogen come from dyes and raw materials, which is not very high, about 10 mg/L. But the urea is needed while using batik techniques. Its total nitrogen is 300 mg/L, which is hard to treat. The phosphorus in the wastewater comes from the phosphor detergents. Considering the serious eutrophication of surface water, it needs to be controlled. Some enterprises use trisodium phosphate so that the concentration of phosphorus will reach 10 mg/L. So, this phosphorus must be removed in the pre-treatment. In the production process, suspended substance comes from fibre scrap and undissolved raw materials. It will be removed through the grille, grid, etc.

The suspended solids (SS) in the outflow mainly come from the secondary sedimentation tank, whose sludge has not been separated completely which will reach 10-100 mg/L as usual. Sulfide mainly comes from the sulfur, which is a kind of cheap and qualified dye. Due to its toxicity, it has been forbidden in developed countries. However, in China, some enterprises are still using it, so it has been included in the wastewater standards. The sulfide in the wastewater is about 10 mg/L.

Table. Specific Pollutants from Yarn and Dyeing Processing Operations

Process	Compounds
Desizing	Sizes, enzymes, starch, waxes, ammonia. Disinfectants and insecticides residues, NaOH, surfactants,
Scouring	soaps, fats, waxes, pectin, oils, sizes, anti-static agents, spent solvents, enzymes.
Bleaching	H_2O_2, AOX, sodium silicate or organic stabiliser, high pH.
Mercerizing	High pH, NaOH
Dyeing	Colour, metals, salts, surfactants, organic processing assistants, sulphide, acidity/alkalinity, formaldehyde.
Printing	Urea, solvents, colour, metals.
Finishing	Resins, waxes, chlorinated compounds, acetate, stearate, spent solvents, softeners.

There are two main sources of hexavalent chromium. Cylinder engraving makes the wastewater containing hexavalent chromium. However, this technology has not been used. Another possible source is the use of potassium dichromate additive in hair dyeing process. Aniline mainly comes from the dyes. The colour of the dye comes from the chromophore. Some dyes have a benzene ring, amino, etc., which will be decomposed in the wastewater treatment process. The potential specific pollutants from yarn Printing and dyeing is shown in Table.

YARN DYEING WASTEWATER RISK

Discharged wastewater by some industries under uncontrolled and unsuitable conditions is causing significant environmental problems. The importance of the pollution control and treatment is undoubtedly the key factor in the human future. If a yarn mill discharges the wastewater into the local environment without any treatment, it will has a serious impact on natural water bodies and land in the surrounding area. High values of COD and BOD5, presence of particulate matter and sediments, and oil and grease in the effluent causes depletion of dissolved oxygen, which has an adverse effect on the aquatic ecological system. Effluent from yarn mills also contains chromium, which has a cumulative effect, and higher possibilities for entering into the food chain. Due to usage of dyes and chemicals, effluents are dark in colour, which increases the turbidity of water body. This in turn hampers the photosynthesis process, causing alteration in the habitat (Joseph Egli,2007).

THE YARN INDUSTRY STANDARDS FOR WATER POLLUTANTS

As the wastewater is harmful to the environment and people, there are strict requirements for the emission of the wastewater. However, due to the difference in the raw materials, products, dyes, technology and equipment, the standards of the wastewater emission have too much items. It is developed by the national environmental protection department according to the local conditions and environmental protection requirements which is not fixed. It varies according to the situation in different regions. Therefore, the nature of emission targets is priorities of the points.

Table. "Yarn Industry Standards for Water Pollutants"

Serial Number	Parameters	The Limits of Discharged Concentration for New Factory	The Limits of Discharged Concentration Concentration	The Special Limits of Discharged
1	COD	100mg/L	80 mg/L	60 mg/L
2	BOD	25 mg/L	20 mg/L	15 mg/L
3	pH	6~9	6~9	6~9
4	SS	70 mg/L	60 mg/L	20 mg/L
5	Chrominance	80	60	40
6	TN	20 mg/L	15 mg/L	12 mg/L
7	NH3-N	15 mg/L	12 mg/L	10 mg/L

8	TP	1.0 mg/L	0.5 mg/L	0.5 mg/L
9	S	1.0 mg/L	Can not be detected	Can not be detected
10	ClO2	0.5 mg/L	0.5 mg/L	0.5 mg/L
11	Cr6+	0.5 mg/L	Can not be detected	Can not be detected
12	Aniline	1.0 mg/L	Can not be detected	Can not be detected

For printing and dyeing wastewater, the first consideration is the organic pollutants, colour and heavy metal ions. Recently, as the lack of water, the recovery of wastewater should be considered. So the decolorization of the printing and dyeing wastewater increased heavily. The standards of printing and dyeing are different in different countries.Through access to the relevant information, the yarn industry standards for water pollutants in China, Germany, U.S have been found.

Yarn Industry Standards for Water Pollutants in China

For the emission standards of the yarn dyeing wastewater in China, it is the very stringentstandards in the world. The emission standards for different indicators in yarn industry standards for water pollutants in China have been shown in table ("Discharge standard of water pollutants for dyeing and finishing of yarn industry").

Yarn Industry Standards for Water Pollutants in Germany

The emission standards for different indicators in yarn industry standards for water pollutants in Germany have been shown in table.

Table. Yarn Industry Standards for Water Pollutants

Serial Number	Parameters	The Limits of Discharged Concentration
1	COD	160mg/L
2	BOD	25 mg/L
3	TP	2.0 mg/L
4	TN	20 mg/L
5	NH_3-N	10 mg/L
6	Nitrite	1.0 mg/L

The requirements of ammonia and total nitrogen are adjusted for the biochemical outflow at 12°C or above. Besides, the standard has also made the following emission requirements for the wastewater at the production stain.

There must not be in the wastewater:

- Organic chlorine carriers (dyed acceleration)
- Separation of chlorine bleach materials, except the sodium chlorite from the bleached synthetic fibres

- The free chlorine after using sodium chlorite
- Arsenic, mercury and their mixtures
- Alkyl phenol as a bleaching agent (APEO)
- Cr6 + compounds in the oxidizing of sulfur dyes and vat dyes
- EDTA, DTPA, and phosphate in the water treatment softeners
- Accumulation of chemicals, dyes and yarn auxiliaries

Yarn Industry Standards for Water Pollutants in U.S.

Printing and Dyeing Wastewater

It is the order for the printing and dyeing, including rinsing, dyeing, bleaching, washing, drying and other similar processes.

Fabric Printing and Dyeing Wastewater

It is adjusted for the fabric printing and dyeing wastewater, including bleaching, mercerization, dyeing, resin processing, washing, drying and so on. The requirements using BPT (best practical control tech.) to treat the fabric printing and dyeing wastewater has been shown in Table ("Discharge standard of water pollutants for dyeing and finishing of yarn industry").

Table. Emission Standards for Gross Printing and Dyeing Wastewater

Serial Number	Parameters	BPT Maximum	Average of 30 Days Kg/t(Fabric)
1	BOD_5	22.4	11.2
2	COD	163.0	81.5
3	TSS	35.2	17.6
4	S	0.28	0.14
5	Phenol	0.14	0.07
6	Cr	0.14	0.07
7	pH	6.0~9.0	6.0~9.0

Yarn Printing and Dyeing Wastewater

Table. Emission Standards for Fabric Printing and Dyeing Wastewater

Serial Number	Parameters	BPT Maximum	Average of 30 Days Kg/t(Fabric)
1	BOD_5	5.0	2.5
2	COD	60	30
3	TSS	21.8	10.9
4	S	0.20	0.10

5	Phenol	0.10	0.05
6	Cr	0.10	0.05
7	pH	6.0~9.0	6.0~9.0

It is adjusted for the Yarn printing and dyeing wastewater, including washing, mercerizati on, resin processing, dyeing and special finishing. The requirements using BPT (best practical control tech.) to treat the yarn printing and dyeing wastewater has been shown in Table ("Discharge standard of water pollutants for dyeing and finishing of yarn industry").

Table. Emission Standards for Yarn Printing and Dyeing Wastewater

Serial Number	Parameters	BPT	
		Maximum	Average of 30 Days Kg/t(Fabric)
1	BOD_5	6.8	3.4
2	COD	84.6	42.3
3	TSS	17.4	8.7
4	S	0.24	0.12
5	Phenol	0.12	0.06
6	Cr	0.12	0.06
7	pH	6.0~9.0	6.0~9.0

YARN DYEING WASTEWATER TREATMENT PROCESSES

The yarn dyeing wastewater has a large amount of complex components with high concentrations of organic, high-colour and changing greatly characteristics. Owing to their high BOD/COD, their coloration and their salt load, the wastewater resulting from dyeing cotton with reactive dyes are seriously polluted. As aquatic organisms need light in order to develop, any deficit in this respect caused by colored water leads to an imbalance of the ecosystem. Moreover, the water of rivers that are used for drinking water must not be colored, as otherwise the treatment costs will be increased. Obviously, when legal limits exist (not in all the countries) these should be taken as justification. Studies concerning the feasibility of treating dyeing wastewater are very important (C. All'egre et al.,2006). In the past several decades, many techniques have been developed to find an economic and efficient way to treat the yarn dyeing wastewater, including physicochemical, biochemical, combined treatment processes and other technologies. These technologies are usually highly efficient for the yarn dyeing wastewater.

PHYSICOCHEMICAL WASTEWATER TREATMENT

Wastewater treatment is a mixture of unit processes, some physical, others chemical or biological in their action. A conventional treatment process is comprised of a series of individual unit processes, with the output (or effluent) of one process becoming the input (influent) of the next process. The first stage will usually be made up of physical processes. Physicochemical wastewater treatment has been widely used in the sewage treatment plant which has a high removal of chroma and suspended substances, while it has a low removal of COD. The common physicochemical methods are shown as followed.

Equalization and Homogenization

Because of water quality highly polluted and quantity fluctuations, complex components, yarn dyeing wastewater is generally required pretreatment to ensure the treatment effect and stable operation. In general, the regulating tank is set to treat the wastewater. Meantime'to prevent the lint, cotton seed shell, and the slurry Settle to the bottom of the tank, it's usually mixed the wastewater with air or mechanical mixing equipment in the tank. The hydraulic retention time is generally about 8 h.

Floatation

The floatation produces a large number of micro-bubbles in order to form the three-phase substances of water, gas, and solid. Dissolved air under pressure may be added to cause the formation of tiny bubbles which will attach to particles. Under the effect of interfacial tension, buoyancy of bubble rising, hydrostatic pressure and variety of other forces, the microbubble adheres to the tiny fibres. Due to its low density, the mixtures float to the surface so that the oil particles are separated from the water. So, this method can effectively remove the fibres in wastewater.

Coagulation Flocculation Sedimentation

Coagulation flocculation sedimentation is one of the most used methods, especially in the conventional treatment process. Active on suspended matter, colloidal type of very small size, their electrical charge give repulsion and prevent their aggregation. Adding in water electrolytic products such as aluminum sulphate, ferric sulphate, ferric chloride, giving hydrolysable metallic ions or organic hydrolysable polymers (polyelectrolyte) can eliminate the surface electrical charges of the colloids. This effect is named coagulation. Normally the colloids bring negative charges'so the coagulants are usually inorganic or organic cationic coagulants (with positive charge in water). The metallic hydroxides and the organic polymers, besides giving the coagulation, can help the particle aggregation into flocks, thereby increasing the sedimentation. The combined action of coagulation, flocculation and settling is named clariflocculation. Settling needs stillness and flow velocity, so these three processes need different reactions tanks. This processes use mechanical separation among heterogeneous matters, while the dissolved matter is not well removed (clariflocculation can eliminate a part of it by absorption into the flocks). The dissolved matter can be better removed by biological or by other physical chemical processes (Sheng.H et al.,1997). But additional chemical load on the effluent (which normally increases salt concentration) increases the sludge production and leads to the uncompleted dye removal.

Chemical Oxidation

Chemical operations, as the name suggests, are those in which strictly chemical reactions occur, such as precipitation. Chemical treatment relies upon the chemical interactions of the contaminants we wish to remove from water, and the application of chemicals that either aid in the separation of contaminants from water, or assist in the destruction or neutralization of harmful effects associated with contaminants. Chemical treatment methods are applied both as stand-alone technologies and as an integral part of the treatment process with physical methods (K.Ranganathan et al.,2007).Chemical operations can oxidize the pigment in the printing and dyeing wastewater as well as bleaching the effluent. Currently, Fenton oxidation and ozone oxidation are often used in the wastewater treatment.

Fenton Reaction

Oxidative processes represent a widely used chemical method for the treatment of yarn effluent, where decolourisation is the main concern. Among the oxidizing agents, the main chemical is hydrogen peroxide (H_2O_2), variously activated to form hydroxyl radicals, which are among the strongest existing oxidizing agents and are able to decolourise a wide range of dyes. A first method to activate hydroxyl radical formation from H_2O_2 is the so called Fenton reaction, where hydrogen peroxide is added to an acidic solution (pH=2-3) containing Fe^{2+} ions. Fenton reaction is mainly used as a pre-treatment for wastewater resistant to biological treatment or/and toxic to biomass. The reaction is exothermic and should take place at temperature higher than ambient. In large scale plants, however, the reaction is commonly carried out at ambient temperature using a large excess of iron as well as hydrogen peroxide. In such conditions ions do not act as catalyst and the great amount of total COD removed has to be mainly ascribed to the $Fe(OH)_3$ co-precipitation. The main drawbacks of the method are the significant addition of acid and alkali to reach the required pH, the necessity to abate the residual iron concentration, too high for discharge in final effluent, and the related high sludge production (Sheng.H et al.,1997).

Ozone Oxidation

It is a very effective and fast decolourising treatment, which can easily break the double bonds present in most of the dyes. Ozonation can also inhibit or destroy the foaming properties of residual surfactants and it can oxidize a significant portion of COD. Moreover, it can improve the biodegradability of those effluents which contain a high fraction of non-biodegradable and toxic components through the conversion (by a limited oxidation) of recalcitrant pollutants into more easily biodegradable intermediates. As a further advantage, the treatment does increase neither the volume of wastewater nor the sludge mass. Full scale applications are growing in number, mainly as final polishing treatment, generally requiring up-stream treatments such as at least filtration to reduce the suspended solids contents and improve the efficiency of decolourisation. Sodium hypochlorite has been widely used in the past as oxidizing agent. In yarn effluent it initiates and accelerates azo bond cleavage. The negative effect is the release of carcinogenic aromatic amines and otherwise toxic molecules and, therefore, it should not be used (Sheng.H et al.,1997).

Adsorption

Adsorption is the most used method in physicochemical wastewater treatment, which can mix the wastewater and the porous material powder or granules, such as activated carbon and clay, or let the wastewater through its filter bed composed of granular materials. Through this method, pollutants in the wastewater are adsorbed and removed on the surface of the porous material or filter. Commonly used adsorbents are activated carbon, silicon polymers and kaolin. Different adsorbents have selective adsorption of dyes. But, so far, activated carbon is still the best adsorbent of dye wastewater. The chroma can be removed 92.17 per cent and COD can be reduced 91.15 per cent in series adsorption reactors, which meet the wastewater standard in the yarn industry and can be reused as the washing water. Because activated carbon has selection to adsorb dyes, it can effectively remove the water-soluble dyes in wastewater, such as reactive dyes, basic dyes and azo dyes, but it can't adsorb the suspended solids and insoluble dyes. Moreover, the activated carbon can not be directly used in the original yarn dyeing wastewater treatment, while generally used in lower concentration of dye wastewater treatment or advanced treatment because of the high cost of regeneration.

Membrane Separation Process

Membrane separation process is the method that uses the membrane's micropores to filter and makes use of membrane's selective permeability to separate certain substances in wastewater. Currently, the membrane separation process is often used for treatment of dyeing wastewater mainly based on membrane pressure, such as reverse osmosis, ultrafiltration, nanofiltration and microfiltration. Membrane separation process is a new separation technology, with high separation efficiency, low energy consumption, easy operation, no pollution and so on. However, this technology is still not large-scale promoted because it has the limitation of requiring special equipment, and having high investment and the membrane fouling and so on (K.Ranganathan et al.,2007).

Reverse Osmosis

Reverse osmosis membranes have a retention rate of 90 per cent or more for most types of ionic compounds and produce a high quality of permeate. Decolorization and elimination of chemical auxiliaries in dye house wastewater can be carried out in a single step by reverse osmosis. Reverse osmosis permits the removal of all mineral salts, hydrolyzed reactive dyes, and chemical auxiliaries. It must be noted that higher the concentration of dissolved salt, the more important the osmotic pressure becomes; therefore, the greater the energy required for the separation process (B. Ramesh Babu et al.,2007).

Nanofiltration

Nanofiltration has been applied for the treatment of colored effluents from the yarn industry. Its aperture is only about several nanometers, the retention molecular weaght by which is about 80-1000da, A combination of adsorption and nanofiltration can be adopted for the treatment of yarn dye effluents. The adsorption step precedes nanofiltration, because this sequence decreases concentration polarization during the filtration process, which increases the process output. Nanofiltration membranes retain low molecular weight organic compounds, divalent ions, large monovalent ions, hydrolyzed reactive dyes, and dyeing auxiliaries. Harmful effects of high concentrations of dye and salts in dye house effluents have frequently been reported. In most published studies concerning dye house effluents, the concentration of mineral salts does not exceed 20 g/L, and the concentration of dyestuff does not exceed 1.5 g/L. Generally, the effluents are reconstituted with only one dye, and the volume studied is also low. The treatment of dyeing wastewater by nanofiltration represents one of the rare applications possible for the treatment of solutions with highly c oncentrated and complex solutions (B. Ramesh Babu et al.,2007). A major problem is the accumulation of dissolved solids, which makes discharging the treated effluents into water streams impossible. Various research groups have tried to develop economically feasible technologies for effective treatment of dye effluents. Nanofiltration treatment as an alternative has been found to be fairly satisfactory. The technique is also favourable in terms of environmental regulating.

Ultrafiltration

Ultrafiltration whose aperture is only about 1nm-0.05μm, enables elimination of macromolecules and particles, but the elimination of polluting substances, such as dyes, is never complete. Even in the best of cases, the quality of the treated wastewater does not permit its reuse for sensitive processes, such as dyeing of yarn. So the retention molecular weaght is range from 1000-300000da. Rott and Minke (1999) emphasize that 40 per cent of the water treated by ultrafiltration can be recycled to feed processes termed "minor" in the yarn industry (rinsing, washing) in which salinity is not a problem. Ultrafiltration can only be used as a pretreatment for reverse osmosis or in combination with a biological reactor (B. Ramesh Babu et al.,2007).

Microfiltration

Microfiltration whose aperture is about 0.1-1μm is suitable for treating dye baths containing pigment dyes, as well as for subsequent rinsing baths. The chemicals used in dye bath, which are not filtered by microfiltration, will remain in the bath. Microfiltration can also be used as a pretreatment for nanofiltration or reverse osmosis (B. Ramesh Babu et al.,2007). Yarn wastewater contains large amounts of difficult biodegradable organic matter and inorganic. At present'many factories have adopted physicochemical treatment process. Some typical physicochemical treatment process is shown in Table and Fig.

BIOLOGICAL WASTEWATER TREATMENT METHOD

The biological process removes dissolved matter in a way similar to the self depuration but in a further and more efficient way than clariflocculation. The removal efficiency depends upon the ratio between organic load and the bio mass present in the oxidation tank, its temperature, and oxygen concentration. The bio mass concentration can increase, by aeration the suspension effect but it is important not to reach a mixing energy that can destroy the flocks, because it can inhibit the following settling. Normally, the biomass concentration ranges between 2500-4500 mg/l, oxygen about 2 mg/l. With aeration time till 24 hours the oxygen demand can be reduced till 99 per cent. According to the different oxygen demand, biological treatment methods can be divided into aerobic and anaerobic treatment. Because of high efficiency and wide application of the aerobic biological treatment, it naturally becomes the mainstream of biological treatment.

Aerobic Biological Treatment

According to the oxygen requirements of the different bacteria, the bacteria can

Name	Dyes and Additives in Sewage	Water Quantity (t/d)	Main Process	Amount of Coagulant (mg/L)	Water Quality		Treatment Efficiency	
					Colour (times)	COD (mg/L)	Colour (%)	COD (%)
A Knitting Mill in Kunming Yunan	Naphthol, Direct dye, Acidic dye, Reactive dye	1000	Wastewater → Pump → (PAC) Mixed reaction tank sedimentation tank → Effluent	60-80	70-120	267	>90	60
A Printing and Dyeing Mill in Shanghai	Vat dye, Naphthol, Dope	Pilot Scale test	Wastewater → Regulating tank → Pump → Dissoved Vassel → Floatation tank → Effluent	400	400	600-800	80-90	60
A Printing and Dyeing Mill in Beijing	Acidic dye, Disperse dye, Reactive dye, Sulfide dye	120	Wastewater → Regulating tank → Coaulatio Tube Sedimentation tank → Floatation tank → Sand Filter tank → Effluent	–	174-347	228-352	97.41	74.69
A Silk and Dyeing Mill in Shaoxing, Zhejiang	Disperse dye, Reactive dye, Direct dye, Sulfide dye, Acidic dye	500	Wastewater → Regulating tank → Reaction tank → Coaulation tank → Tube sedimentation tank → Effluent tank	$FeSO_4$ 0.7kg/t sewage, Lime 0.38kg/t sewage	720-830	1114-1153	92	55-59

Fig. Physicochemical Treatment Instance of Yarn Dyeing Wastewater

be divided into aerobic bacteria, anaerobic bacteria and facultative bacteria. Aerobic biological treatment can purify the water with the help of aerobic bacteria and facultative bacteria in the aerobic environment. Aerobic biological treatment can be divided into two major categories: activated sludge process and biofilm process.

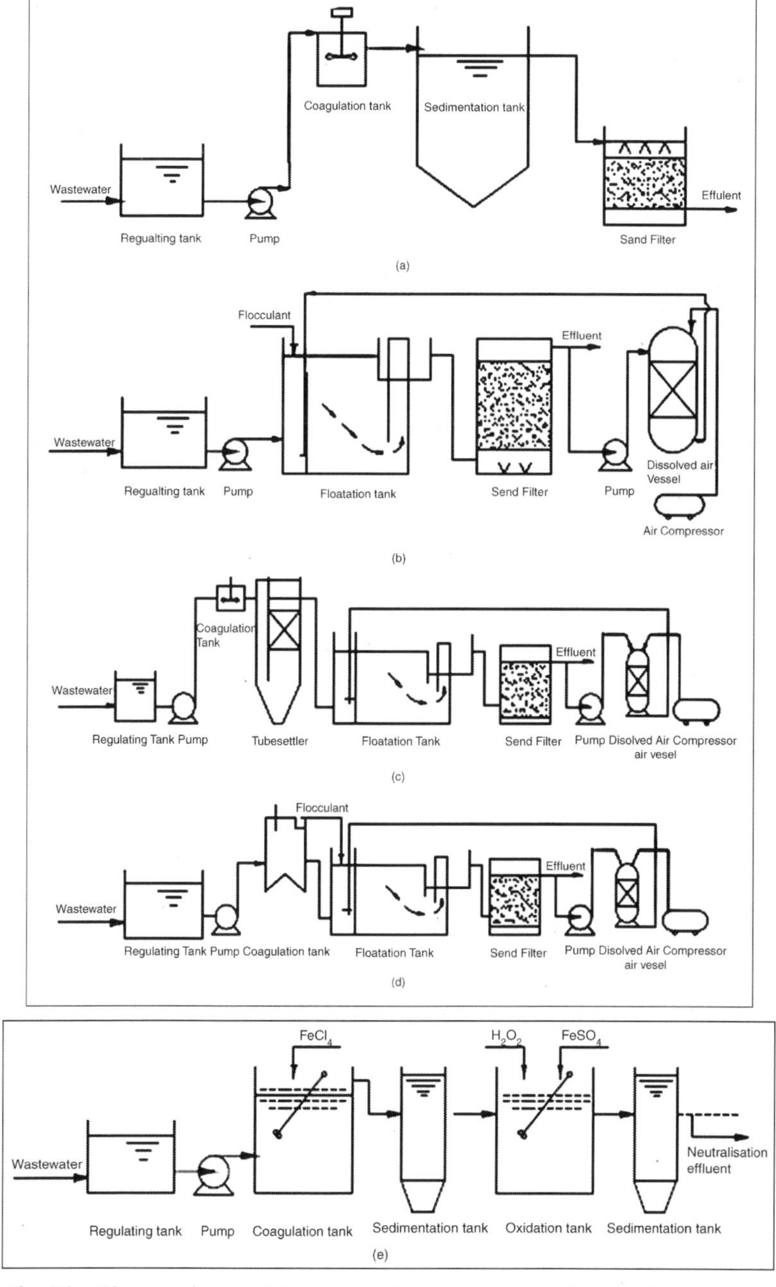

Fig. The Physicochemical Treatment Process in Yarn Dyeing Wastewater

Activated Sludge Process

Activated sludge is a kind of floc which is mainly comprised of many microorganisms, which has strong decomposition and adsorption of the organics, so it is called "activated sludge". The wastewater can be clarified and purified after the separation of activated sludge. Activated sludge process is based on the activated sludge whose main structure is the aeration tank. Activated sludge Process is an effective method. As long as according to the scientific laws, after getting some experiences, higher removal efficiency can be got. In present treatment, the oxidation ditch and SBR process are the commonly used activated sludge process.

Oxidation Ditch Process

Oxidation ditch is a kind of biological wastewater treatment technology, which developed by the netherlands health engineering research institute in the 50's of last century. It is a variant of activated sludge, which is a special form of extended aeration. And the biological sewage treatment process which mainly composed by the oxidation ditch.

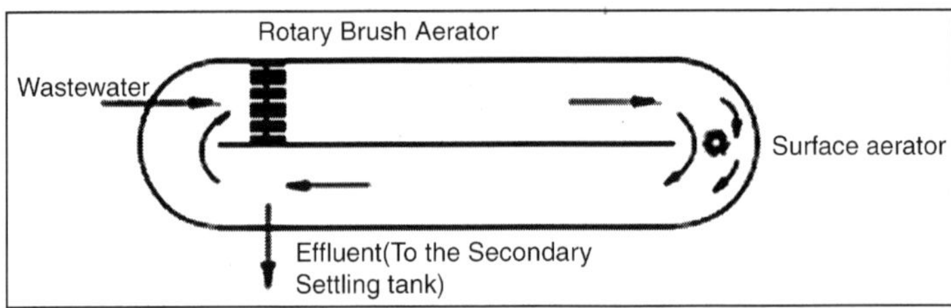

Fig. Oxidation Ditch Plan

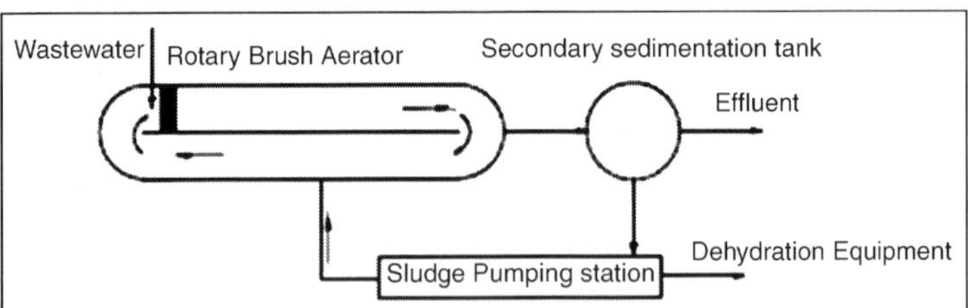

Fig. The Oxidation Ditch for the Biological Treatment Process

The oxidation ditch is generally consisted of the ditch body, aeration equipment, equipment of the water in or out, diversion and mixing equipment. The shape of the ditch body is usually ring; it also can be rectangular, L-shaped, round or other shape. The side shape of the oxidation ditch generally is rectangular or trapezoidal. Wastewater, activated sludge, and various microorganisms are mixed in a continuous loop ditch in order to complete nitrification and denitrification. Oxidation ditch has the characteristics of completely mixed, plug-flow and oxidation tank. Since oxidation ditch has long hydraulic retention time(HRT), low organic loading and long sludge age, compared to conventional activated sludge process, the equalization tank, primary sedimentation tank, sludge digestion tank can be omitted. The secondary sedimentation tank also can be omitted in some processes. It has many characteristics, such as high degree of purification, impact resistance, stable, reliable, simple, easy operation and management, easy maintenance, low investment and energy consumption. Oxidation ditch forms aerobic zone, anoxic zone and the anaerobic zone in space, which has a good function of denitrification.

Sequencing Batch Reactor Activated Sludge Process

Sequencing Batch Reactor Activated Sludge Process (SBR Process) is a reform process from activated sludge, which is a new operating mode.

Its operation is mainly composed of five processes:

1. Inflow;
2. Reaction;
3. Sedimentation;
4. Outflow;
5. Standby.

The reaction process plan has been showed in Fig. SBR treatment process not only has a high removal rate of COD, but also has a high removal efficiency of colour. Compared to the traditional methods, using SBR process has the following advantages (Li-yan Fu et al.,2001):

- It has great resistance to shock loading. The reserved water can effectively resist the impact of water and organic matter.

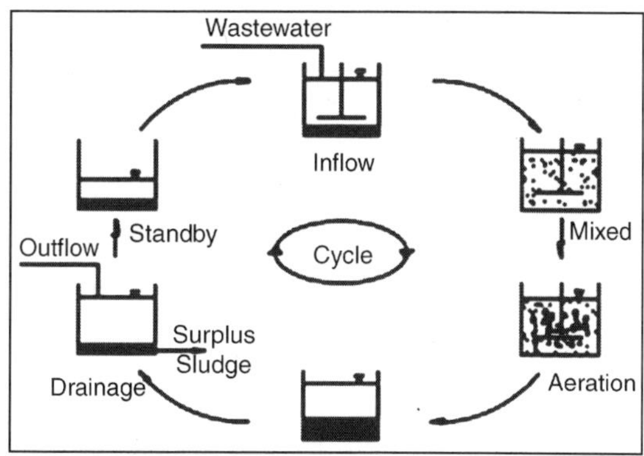

Fig. Reaction Process of SBR

- Flexible control. The run time, the total residence time and gas supply of each stage can be adjusted according to the quantity of inflow or outflow.
- The activated sludge has good traits and low sludge production rate. Since the original water contains many organic matters, which is

suitable for the bacteria to grow, the sludge is in good characters. In the standby period, the sludge is in the endogenous respiration phase, so the sludge's yield is low.

- It has less processing equipment, simple structure'and it is easy to be operated.

Table. The Removal Efficiency of Various Indicators in This Dyeing Factory

Items	wastewater	Effluent	Removal Rate
CODcr(mg/L)	689	61.2	91.2
Colour(times)	800	50	93.8
SS(mg/L)	384.3	32.6	91.5

There is a dyeing factory in Jiangsu, which produces 60,000 tons wastewater each year. The wastewater mainly comes from dyeing and rinse. The major pollutant in wastewater is seriously exceeded, such as COD, colour and SS. The wastewater treatment process which is used in this factory has been shown in Fig.

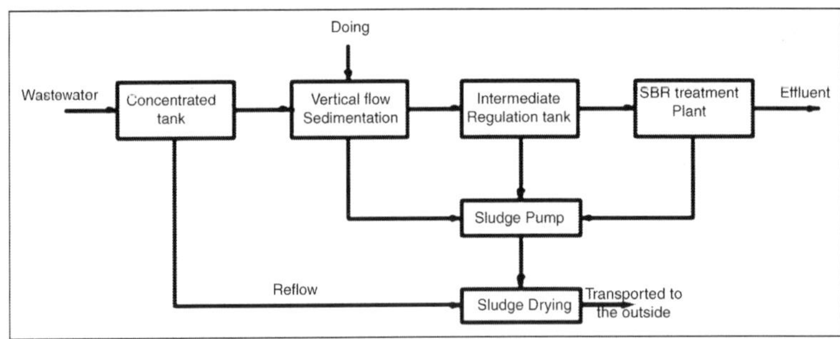

Fig. The Wastewater Treatment Process Based on SBR

Biofilm

The biofilm process is a kind of biological treatment that making the numerous microorganisms to attach to some fixed object surface, while letting the wastewater flow on its surface to purify it by contact. The main types of the biofilm process are biological contact oxidation, rotating biological contactors and biological fluidized bed.

Biological Contact Oxidation

Biological contact oxidation is widely used in the dyeing wastewater treatment. The main feature of the process is to set fillers in the aeration tanks, so that it has the characteristics of activated sludge and biofilm. The wastewater in oxidation tank contains a certain amount of activated sludge, while the fillers are covered with a large number of biofilm. When the wastewater contact with the fillers, it can be purified under the function of aerobic microorganisms.

- The main features of the biological contact oxidation tank are as following:
  - It has a efficient purification, a short processing time as well as a good and stable water quality.
  - It has a higher ability to adapt the impact load. Under the intermittent operating condition, it is still able to maintain good treatment effect. To uneven drainage enterprises, it has more actual significance.
  - Its operation is simple, convenient and easy maintenance. It don't have sludge return, sludge bulking phenomenon or any filter flies.
  - It produce less sludge and the sludge particles is larger,which is easier to be sedimentated
- The form of biological contact oxidation According to the different direction of the influent and aeration, the design form of the biological contact oxidation can be divided into two types: one is the same flow contact process which makes the wastewater and air flow into the contact tank bottom. This process can guarantee the volume load about 4.5 kgBOD/m^3. Another is the reverse flow contact process which let the air flow into the bottom, and let the wastewater into the top. This process can guarantee the volume load about 8 kgBOD/m^3.
- The choice of filler The filler is not only related to the treatment effect, but also affects the project investment. The surface area, bio-adhesion and whether easily blocked of the filler are undoubtedly the most important conditions, but the economy is also an important factors. The filler in the oxidation tank is the proportion of investment accounted for relatively large, so the price is often the first consideration. For example, the technical performance of some filler is slightly worse, but the price is cheap. The contact time can be increased appropriately. Commonly used fillers are honeycomb packing, corrugated board packing equipment, soft and semi-soft filler.

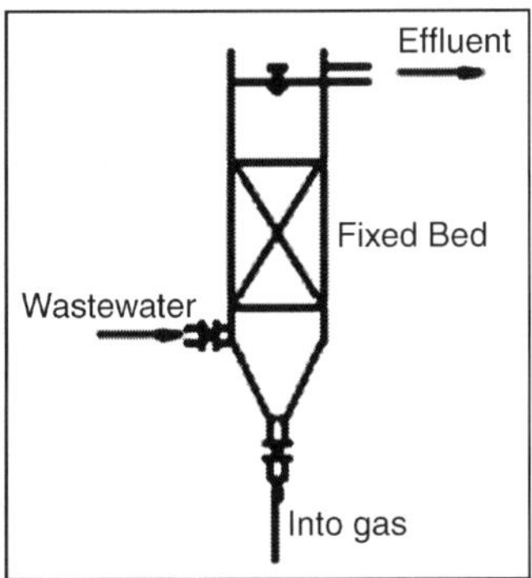

Fig. The Schematic Diagram of the Same Flow Contact

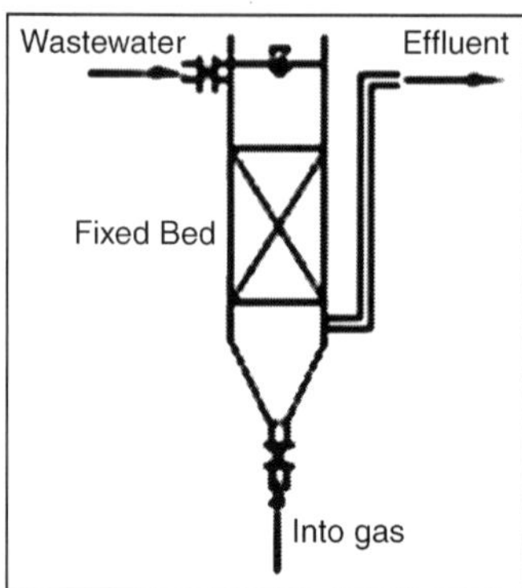

Fig. The Schematic Diagram of the Reverse Flow Contact

Rotating Biological Contactor

Rotating Biological Contactor is an efficient sewage treatment plant developed on the basis of the original biological filter. It is constituted by a series of closed disks which are fixed on a horizontal axis. The disks are made of lightweight materials, such as hard plastic plate, glass plate etc.The disk diameter is about 1-3m generally. Nearly half of the disk area is submerged in the sewage of the oxidation tank, but the upper half is exposed to the air. The rotating horizontal axis which is driven by the Rotating device makes the disk rotating slowly.

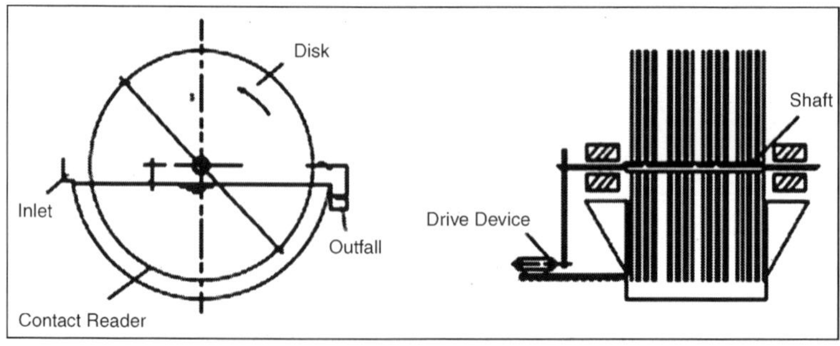

Fig. The Schematic Diagram of Biological Rotating Disk

Due to the disk's rotating, the sewage in the oxidation tank is completely mixed. There is a layer of biofilm on the disk surface, when the rotating disk immersed in the sewage inside of the oxidation tank, the organic matter in the sewage would be adsorbed by the biofilm on the disk. When the rotating disk rotate to the air, the water film which is brought up by the disk will drip down along the biofilm surface, at the same time, the oxygen in the air will dissolve into the water film constantly. Under the catalysis of enzyme, though absorbing the dissolved oxygen in the water film, microorganism can oxidate and decompose the organic matter in the sewage and excrete the metabolite at the same time. In the process of disk rotation, the biofilm on the disk get in touch with the sewage and the air constantly alternating, completing the process of adsorption-oxidation-oxidative decomposition continuously to purify the sewage. The advantages of the Biological Rotating Contactor are power saving, large shock load capability, no sludge return, little sludge generated, little noise, easy maintenance and management and so on. On the contrary, the shortcomings are the large area and needing the maintain buildings.

The main parameters of affecting the process performance are rotate speed, sewage residence time, reactive tank stage, disk submergence and temperature. Towards the low consistency sewage whose BOD consistency is under 300 mg/L, when the rotate speed is under 18 m/min, the disposal efficiency would enhance with the rotate speed increase, otherwise, it would not be any improving. Towards high-BOD consistency sewage, increasing the rotate speed is equivalent to increase the contact, oxidation and stirring. So, it improves the efficiency. On the other hand, increasing the rotate speed increase the energy consumption shapely, so we must do full economic comparison between increasing energy consumption and increasing land acreage.

Biological Fluidized Bed

Fluidized bed process is also called suspended carrier biofilm process, which is a new efficient sewage treatment process. The method which adopts the solid particles fluidization technology can keep the whole system at a fluidized state to enhance the contact of solid particles and fluid and achieve the purpose of sewage purification. The substance of this method is that using activated carbon, sand, anthracite or other particles whose diameter is less than 1mm as the carrier and filling in the container. Through pulse water, the carrier is made into fluidization and covered by the biofilm on its surface. Due to the small size of the carrier, it has a great surface area in unit volume (the carrier surface area can reach 2000-3000 m^2/m^3), and can maintain a high level of microbial biomass. The treatment efficiency is about 10-20 times higher than conventional activated sludge process. It can remove much organic matter in short time, but the land acreage is only about 5 per cent of the common activated sludge process. Therefore, in the treatment of high consistency organic wastewater in the yarn dyeing wastewater, we can consider adopting this suspended carrier fluidized bed process.

Anaerobic Biological Treatment

Anaerobic biological treatment process is a method that make use of the anaerobic bacteria decompose organic matter in anaerobic conditions. This method was first used for sludge digestion. In recent years it was gradually used in high concentration and low concentration organic wastewater treatment. In yarn industry, there are many types of high concentration organic wastewater, such as wool washing sewage, yarn printing and dyeing wastewater etc., which the organic matter content of it is as high as 1000 mg/L or more, the anaerobic wastewater treatment process can achieve good results. The anaerobicaerobic treatment process is usually adopted in actual project that is using anaerobic treatment to treat high concentration wastewater, and using aerobic treatment to treat low concentration wastewater. Currently, the hydrolysis acidification process is the main anaerobic treatment process, which can increases the biodegradability of the sewage to facilitate the following biological treatment process.

The hydrolysis acidification process is the first two stages of the anaerobic treatment. Through making use of the anaerobic bacteria and facultative bacteria, the macromolecule, heterocyclic organic matter and other difficult biodegradable organic matter would be decomposed into small molecular organic matter, thereby enhancing the biodegradability of the wastewater and destructing the colored groups of dye molecules to remove part of the colour in wastewater. More importantly, due to the molecular structure of the organic matter and colored material or the chromophore has been changed by the anaerobic bacteria, it's easy to decompose and decolor under the aerobic conditions, which improve the decolorization effect of the sewage. Operating data shows that the pH value of the effluent from hydrolysis tank usually decrease 1.5 units. The organic acid which is produced in hydrolysis can effectively neutralize some of the alkalinity in wastewater, which can make the pH value of sewage drop to about 8 to provide a good neutral environment for following aerobic treatment. Currently, the anaerobic digestion process is a essential measure in the biological treatment of yarn dyeing wastewater. In addition, there are many other processes used in yarn dyeing wastewater treatment currently, such as upflow anaerobic sludge bed(UASB), upflow anaerobic fluidized bed (UABF), anaerobic baffled reactor (ABR) and anaerobic biological filter and so on.

BIOCHEMICAL AND PHYSICOCHEMICAL COMBINATION PROCESSES

In recent years, as the application of new technologies in yarn and dyeing industry, a large number of difficult biodegradable organic matter such as PVA slurry, surface active agents and new additives enter into the dyeing wastewater, which result in the high concentration of the organic matter, complex and changeable composition and the obvious reduction of the biodegradability. The COD_{Cr} removal rate of the simple aerobic activated sludge process which was used to treat the yarn dyeing wastewater has decreased from 70 per cent to 50 per cent, and the effluent can not meet the discharge standards. More seriously, quite a number of sewage treatment facilities can't normally operate even stop running. Therefore, the biochemical and physicochemical combination processes has been gradually developed. And its application is increasingly widespread (Sheng-Jie You et al.,2008). The types of the combination process are various, and the main adoptions currently are as following:

Hydrolytic Acidification-contact Oxidation-air Floatation Process

This combination process is a typical treatment process of the yarn dyeing wastewater, which is widely used. The wastewater firstly flows through the bar screen, in order to remove a part of the larger fibres and particles, and Then flows into the regulating tank. After well-distributed through a certain amount of time, the sewage flows into the hydrolysis acidification tank to carry out the anaerobic hydrolysis reaction. The reaction mechanism is making use of the anaerobic hydrolysis and acidification reaction of the anaerobic fermentation to degrade the insoluble organic matter into the soluble organic matter by controlling the hydraulic retention time. At the same time, through cooperating with the acid bacteria, the macromolecules and difficult biodegradable organic matter would be turned into biodegradable small molecules, which provide a good condition for the subsequent biological treatment. Next, the sewage enters into the biological contact oxidation tank. After the biochemical treatment, the wastewater directly enters into the flotation tank for flotation treatment, which is adopted the pressurized fulldissolved air flotation process. The polymer flocculants added in flotation tank react with the hazardous substances, which can condense the hazardous substances into tiny particles. Meanwhile, sufficient air is dissolved in the wastewater. And then the pressure suddenly is released to produce uniformly fine bubbles, which would adhere to the small particles. The density of the formation is less than 1kg/m^3, which can make the formation float and achieve the separation of the solid and liquid.

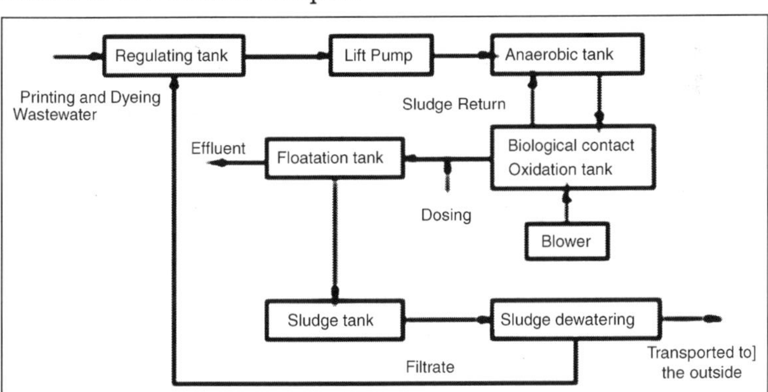

Fig. The Process Flow Diagram

The anaerobic hydrolysis acidification tank equipped with semi-soft padding and the biological contact oxidation tank equipped with the new SNP-based filler. The following physicochemical treatment uses the dosing flotation tank, which has four characteristics. Firstly, the deciduous biofilm and suspended solids removal rate can reach 80 per cent to 90 per cent. Secondly, the colour removal rate can reach 95 per cent. Thirdly, the hydraulic retention time in the flotation tank is short, which is only about 30 min, while the precipitation tank is about 1.15 h to 2 h, so the volume and area of the flotation tank is small. Finally, the sludge moisture content is low, only about 97 per cent to 98 per cent, which can be directly dewatered. But the flotation treatment need an additional air compressor, pressure dissolved gas cylinders, pumps and other auxiliary system. The operation and management is also relatively complicated. After the treatment of this process, the COD_{Cr} removal rate can be up to 95 per cent or more. The actual effluent quality is about: pH=6~9, colour<100times, SS<100mg/L, BOD_5<50mg/L, COD_{Cr}<150mg/L (Honglian Li, 2006).

Anaerobic-aerobic-biological Carbon Contacts

The treatment process is a mature and widely used process in wastewater treatment in recent years. The anaerobic treatment here is not the traditional anaerobic nitrification, but the hydrolysis and acidification. The purpose is aiming at degrading some poorly biodegradable polymer materials and insoluble material in yarn dyeing wastewater to small molecules and soluble substances by hydrolysis and acidification, meanwhile, improving the biodegradability and BOD_5/COD_{Cr} value of the wastewater in order to create a good condition for the subsequent aerobic biological treatment. At the same time, all sludge generated in the aerobic biological treatment return into the anaerobic biological stage through the sedimentation tank. Because of the sludge in the anaerobic biochemical stage has sufficient hydraulic retention time (8h~10h) to carry out anaerobic digest thoroughly, The whole system would not discharge sludge, that is the sludge achieve its own balance (Note: only a small amount of inorganic sludge accumulate in the anaerobic stage, but do not have to set up a special sludge treatment plant).

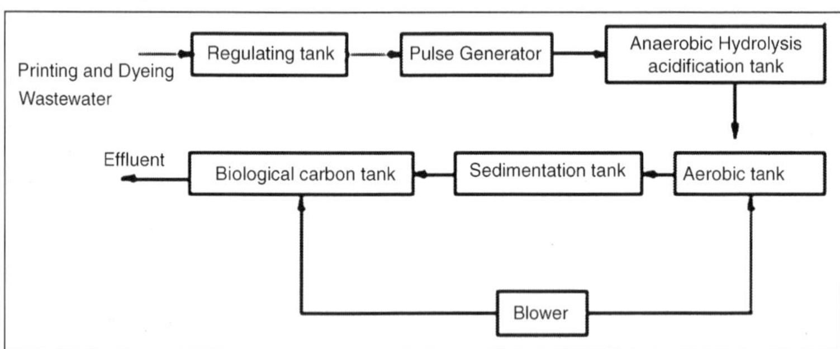

Fig. The Process Flow Diagram

Anaerobic tank and aerobic tank are both installed media, which is a biofilm process. Biological carbon tank is filled with activated carbon and provided oxygen, which has the characteristics both of suspended growth method and fixation growth method. The function of pulse water is mixing in the anaerobic tank. The hydraulic retention time of various parts is about:

Regulating tank: 8h~12h; anaerobic biochemical tank: 8h~10h
Aerobic biochemical tank: 6 ~8h; biological carbon tank: 1h~2h
Pulse generator interval: 5min~10min.

According to the yarn dyeing wastewater standard (COD_{Cr}≤1000mg/L), the effluent can achieve the national emission standards, which can be reused through further advanced treatment. For the five years operation project, the results show that the operation is normal, the treatment effect is steady, there is no efflux of sludge and the sludge was not found excessive growth in the anaerobic biochemical tank.

Coagulation-ABR-oxidation Ditch Process

The treatment has been adopted widely currently, such as a yarn dyeing wastewater treatment plant in Jiangmen of Guangdong Province. The characteristics of the yarn dyeing plant effluent are the variation in water, the higher of the alkaline, colour and organic matter concentration, and the difficulty of the degradation (BOD_5/COD_{Cr} value is about 0.25). The workshop wastewater enter into the regulating tank by pipe network to balance the quantity and quality, after wiping off the large debris by the bar screen before the regulating tank. The adjusted wastewater flow into the coagulation reaction tank, at the same time, the $FeSO_4$ solution was added into it to carry out chemical reaction. Finally, the effluent flows into the primary sedimentation tank for spate separation, meanwhile enhancing the BOD_5/COD_{Cr} ratio.

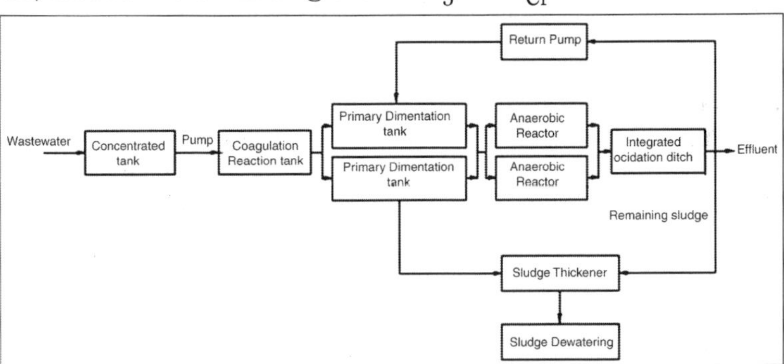

Fig. The Process Flow Diagram

The effluent of primary sedimentation tank flows into the anaerobic baffled reactor (ABR) by gravity. After the anaerobic hydrolysis reaction, it enters into the integrated oxidation ditch for aerobic treatment, and then goes into the secondary sedimentation tank for spate separation. The upper liquid was discharged after meeting the standards, but the settled sludge was returned to the return sludge tank, most of which was returned to ABR anaerobic tank by pump. The remaining sludge was pumped to the sludge thickening tank for concentration. Since the using of this sewage treatment process, the treatment effect is stable. The removal rate of each index is high.

Table. Average Quality of Each Process Effluent

Process	PH	CODcr (mg/L)	BOD_5 (mg/L)	SS (mg/L)	Colour (times)
Wastewater	9-13	1800-2000	400-500	250-350	500
Coagulation effluent	6-9	1327	344	157	102
ABR effluent	6-9	532	292	94	48
Aerobic secondly sedimentation effluent	6-9	80	15	14	30

UASB-aerobic-physicochemical Treatment Process

For the high pH value dyeing wastewater treatment systems, we should adjust the pH value to the range from 6 to 9before entering the treatment system. While towards the difficultly biodegradable yarn dyeing wastewater, effective measures should be taken to increase the biodegradability. At the same time, the volume of the regulating tank must be increased (this is repeatedly demonstrated in engineering practice) to guarantee the water quantity, quality and colour achieving a relatively uniform towards changeable dyeing wastewater. Good inflow condition can be created for following process though this method. Therefore, we put forward the "UASB-aerobic-physicochemical method" treatment process in the actual project, and the technological process is shown in Fig.

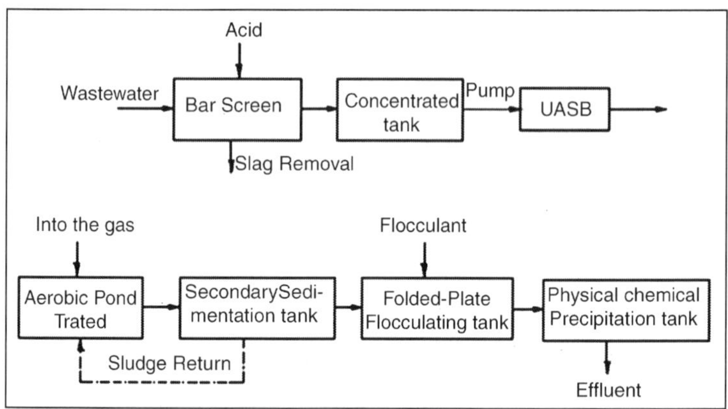

Fig. The Process Flow Diagram

At present, the treatment process has been applied in a number of yarn dyeing factory, such as a Jiangyin yarn dyeing factory, which sewage component is complex. Before the sewage enters into the regulating tank, the suspended particles in which must be removed by bar screen, at the same time, an appropriate amount of acid was added to adjust the pH value of the sewage. Then adjusting the water quantity and quality to make it uniformed. After pretreatment, the yarn dyeing wastewater carries out anaerobic reaction firstly in UASB reactor to improve the biodegradability of wastewater and the decolorization rate, and then flows into the aerobic tank. In the aerobic tank, the organic matter in sewage is removed. Finally, in order to ensure the effective removal of the suspended particles particular the activated sludge, some flocculants was added in the sedimentation tank to improve its effect.

Table. The Removal of the Processing Units

Items	(Raw Water) Regulating Tank	Biochemical Treatment System Effluent	Removal Rate	Physicochemical Treatment System Effluent	Removal Rate
PH	8-12	7-8		6-9	
CODcr(mg/L)	1000-2000	100-200	90	≤100	50
BOD_5(mg/L)	300-600	15-30	95	≤30	
Colour(times)	100-600	60	80	≤40	35

CUTTING-EDGE TREATMENT PROCESS

Photochemical Oxidation

Photochemical oxidation has many advantages of the mild reaction conditions (ambient temperature and pressure), powerful oxidation ability and fast speed, etc. It can be divided into 4 kinds, which are light decomposition, photoactivate oxidation, optical excitation oxidation and photocatalysis oxidation. Among them, the photocatalysis oxidation has been more researched and applied currently. This technology can effectively destroy a lot of organic pollutants whose structure is stable and difficult to biologically degrade. Compared with the physical treatment in traditional wastewater treatment process, the most obvious advantages of this technology are significant energy efficiency, completely pollutants degradation and so on. Almost all of the organic matter can be completely oxidized to CO_2, H_2O and other simple inorganic substances under the light catalyst. However, towards high concentration wastewater, the effect of the photocatalysis oxidation process is not ideal. The research about photocatalysis oxidation degradation of dye mainly focused on the study of photocatalyst.

Electrochemical Oxidation

The mechanism of the electrochemical process treating dyeing wastewater is making use of electrolytic oxidation, electrolytic reduction, electrocoagulation or electrolytic floating destruct the structure or the existence state to make it bleached. It has the advantages of small devices, small area covering, operation and management easily, higher COD_{Cr} removal rate and good bleaching effect, but the precipitation and the consumption of electrode material is great, and the operating cost is high. The traditional electrochemical methods can be divided into power flocculation, electrical float, electro-oxidation, micro-electrolysis and the electrolysis method. With the development of electrochemical technologies and the appearing of a variety of highefficiency reactor, the cost of treatment will decrease largely. Electro-catalytic advanced oxidation process (AEOP) is a new advanced oxidation technology developed recently. Because of its high efficiency, easy operation, and environmental friendliness, it has attracted the attention of researchers. Under normal temperature and pressure, it can produce hydroxyl radicals directly or indirectly through the reactions in the catalytic activity electrode, thus the degradation of the difficultly biodegradable pollutants is effective. It is one of the main directions in future research.

Ultrasonic Technology

Using ultrasonic technology can degrade chemical pollutants, especially the refractory organic pollutants in water. It combines the characteristics of advanced oxidation technology, incineration, supercritical water oxidation and other wastewater treatment technologies. Besides the degradation conditions are mild, degradation speed is fast and application widely, it can also use individually or combined with other water treatment technologies. The principle of this method is that the sewage enters into the air vibration chamber after being added the selected flocculants in regulating tank. Under the intense oscillations in nominal oscillation frequency, a part of organic matter in wastewater is changed into small organic molecule by destructing its chemical bonds. The flocculants flocculation rapidly companied with the colour, COD_{Cr} and the aniline concentration was fall under the accelerating thermal motion of water molecules, which play the role of reducing organic matter concentration in wastewater. At present, the ultrasonic technology in the research of water treatment has achieved great achievements, but most of them are still confined to laboratory research level.

High Energy Physical Process

High energy physical process is a new wastewater treatment technology. When the highenergy particle beam bombard aqueous solution, the water molecules would come up with excitation and ionization, produce ions, excited molecules, secondary electrons. Those products would interact with each other before spreading to the surrounding medium. It would produce highly reactive HO• radicals and H atoms, which would react with organic matter to degrade it. The advantages of using high-energy physics process treat dyeing wastewater are the small size of the equipment, high removal rate and simple operation. However, the device used to generate high energy particle is expensive, technically demanding is high, energy consumption is big, and the energy efficiency is low and so on. Therefore, it needs a lot of research work before put into actual project.

CONCLUSIONS AND RECOMMENDATIONS

Current, the yarn dyeing wastewater is one of the most important source of pollution. The type of this wastewater has the characteristics of higher value of colour, BOD and COD, Complex composition, large emission, widely distributed and difficult degradation. If being directly discharged without being treated, it will bring serious harm to the ecological environment. Because of the dangers of dyeing wastewater, many countries have enacted strict emissions standards, but There is no uniform standard currently. Waste minimization is of great importance in decreasing pollution load and production costs. This book has shown that various methods can be applied to treat cotton yarn effluents and to minimize pollution load.

Traditional technologies to treat yarn wastewater include various combinations of biological, physical, and chemical methods, but these methods require high capital and operating costs. Technologies based on membrane systems are among the best alternative methods that can be adopted for large-scale ecologically friendly treatment processes. A combination methods involving adsorption followed by nanofiltration has also been advocated, although a major drawback in direct nanofiltration is a substantial reduction in pollutants, which causes permeation through flux. It appears that an ideal treatment process for satisfactory recycling and reuse of yarn effluent water should involve the following steps. Initially, refractory organic compounds and dyes may be electrochemically oxidized to biodegradable constituents before the wastewater is subjected to biological treatment under aerobic conditions. Colour and Odour removal may be accomplished by a second electrooxidation process. Microbial life, if any, may be destroyed by a photochemical treatment. The treated water at this stage may be used for rinsing and washing purposes; however, an ion-exchange step may be introduced if the water is desired to be used for industrial processing.

As the improvement of the environmental protection laws and the raise of the awareness of environmental protection, the pollution of printing and dyeing enterprises has caught a lot of attention and the treatment of dyeing wastewater has become a focus. Recently, except the oxidation, filtration and other single method research, it has been introduced a large number of electric, magnetic, optical and thermal method to treat the refractory materials. Several of other techniques have also been carried out to treat the wastewater in order to develop a variety of ways. The interdisciplinary study based on the traditional biological approach will be the direction of the wastewater treatment. The writer expected that the technology which is efficient, clean and reasonable will come out soon. On the other hand, clean production is also an important research, which can shift the focus from end of the treatment to the prevention of pollution and conduct more in-depth research on the printing and dyeing production technology and process management. Moreover, the strategic, comprehensive, preventive measures and advanced production technology can be used to improve the material and energy utilization. Also, we can reduce and eliminate the generation and emissions of wastes as well as the production of excessive use of resources and the risks to humans and the environment. Prevention and treatment of dyeing wastewater pollution are complementary. We can both use preventive measures as well as a variety of methods to control the wastes and make use of treated water. This will not only reduce water consumption, but also effectively reduce the pollution of the printing and dyeing wastewater and achieve sustainable development of society.

12

Photochemical Treatments of Textile Industries Wastewater

INTRODUCTION

Textile industry is one of the most water and chemical intensive industries worldwide due to the fact that 200-400 liters of water are needed to produce 1 kg of textile fabric in textile factories (Correia et al., 1994 and Orhon et al., 2003). The water used in this industry is almost entirely discharged as waste. Moreover, the loss of dye in the effluents of textile industry can reach up to 75 per cent (Couto and Toca-Herrera, 2006). The effluents are considered very complex since they contain salt, surfactants, ionic metals and their metal complexes, toxic organic chemicals, biocides and toxic anions. Azo dyes are regarded as the largest class of synthetic. Approximately, 50–70 per cent of the available dyes for commercial applications are azo dyes followed by the anthraquinone group (Konstantinou and Albanis, 2004). Azo dyes are classified according to the presence of azo bonds (–N=N–) in the molecule *i.e.*, monoazo, diazo, triazo etc. and also sub-classified according to the structure and method of applications such as acid, basic, direct, disperse, azoic and pigments (Bhutani, 2008). Some azo dyes and their dye precursors are well-known of high toxicity and suspected to be human carcinogens as they form toxic aromatic amines (Gomes et al., 2003 and Stylidi et al., 2003).

Different physical, chemical and biological as well as the various combinations of pretreatment and post-treatment techniques have been developed over the last two decades for industrial wastewaters treatment in order to meet the ever-increasing requirements of human beings for water. Though there are numerous studies published in this field, most of the techniques adopted by these researchers are uneconomical, ineffective or impractical uses (Cooper, 1995 and Stephen, 1995). Recent studies have demonstrated that heterogeneous photocatalysis is the most efficient technique in the degradation of colored chemicals(Li et al., 2003; Vione et al., 2003; Antharjanam et al., 2003; Fernandez-Ibanez et al., 2003; Liu et al., 2003; Ohno,

2004; Chen et al., 2004; Alkhateeb et al., 2005 and Attia et al., 2008). These studies used titanium dioxide and/ or zinc oxide in the photolysis processes. The large bang gap of titanium dioxide and zinc oxide (~ 3.2 eV) put a limitation of using these semiconductors in photocatalytic degradation under natural weathering conditions. Only a small part of the overall solar intensity could be useful in such photodegradation processes. However, the existence of dye on the surface of catalyst reduces the energy required for excitation and then increases the efficiency of the excitation process by extending its absorption in the visible region of the spectrum.

METHODS USED IN TREATMENT OF TEXTILE WASTEWATER

There are several factors to choose the appropriate textile wastewater treatment method such as, economic efficiency, treatment efficiency, type of dye, concentration of dye and environmental fate. There is no general method for the treatment of textile industrial wastewater. Wastewaters from textile industry contain various pollutants resulting from various stages of production, such as, fibres preparation, yarn, thread, webbing, dyeing and finishing. Mainly three methods are used for the treatment of textile industrial wastewater.

These are:

1. Physico-chemical methods.
2. Advanced oxidation methods.
3. Biological sludge methods.

The main operational methods used for the treatment of textile industrial water involve physical and chemical processes (Shaw et al., 2002 and Liu et al., 2005). However, these techniques have many disadvantages. These disadvantages include Sludge generation, high cost, formation of bi – products, releasing of toxic molecules, requiring a lot of dissolved oxygen, limitation of activity for specific dyes and requiring of long time. In recent years, advanced oxidation processes (AOPs) have gained more attraction as a powerful technique in photocatalytic degradation of textile industrial wastewater since they are able to deal with the problem of dye destruction in aqueous systems (Konstantinou and Albanis, 2004).

PHOTOCATALYSIS

Photocatalysis is defined as the acceleration of a photoreaction in the presence of a catalyst, while photolysis is defined as a chemical reaction in which a chemical compound is broken down by photons. In catalyzed photolysis, light is absorbed by an adsorbed substrate. Photocatalysis on semiconducting oxides relies on the absorption of photons with energy equal to or greater than the band gap of the oxide, so that electrons are promoted from the valence band to the conduction band:

$$\text{Semiconductor} + h\nu \rightarrow h^{+} + e^{-}$$

If the photoholes and photoelectrons produced by this process migrate to the surface, they may interact with adsorbed species in the elementary steps, which collectively constitute photocatalysis. The numerous gas- and liquid-phase reactions photocatalysed by TiO_2 and ZnO have been reviewed. There is a general agreement that adsorption is necessary since the surface species act as traps for both; photogenerated holes and electrons, which otherwise recombine. In addition to participating in conventional surface reaction steps, adsorbed dye molecules assist in the separation of photoholes and photoelectrons, which may otherwise recombine within the semiconductor particles. A major factor affecting the efficiency of photocatalysis process is electron/hole recombination. If the electrons and holes are used in a reaction, a steady state will be reached when the removal of electrons and holes equals the rate of generation by illumination. Recombination and trapping processes are the de-excitation processes which are responsible for the creation of the steady state, if no reaction occurs.

There are three important mechanisms of recombination:

1. Direct recombination.
2. Recombination at recombination centers.
3. Surface recombination.

There are different types of semiconductors, whose band gaps range between 1.4 and 3.9 eV, *i.e.*, it could be excited with a light of 318–886 nm wavelengths. This means that most of the known semiconductors could be excited by using visible light. However, not all these semiconductors could be used in the photocatalytic reactions. Bahnemann et al., (1994) report that the most appropriate photocatalysts should be stable towards chemicals and illumination and devoid of any toxic constituents, especially for those used in environmental studies. The authors also explaine that TiO_2 and ZnO are the most commonly used in photocatalytic reactions due to their efficient absorbtion of long wavelength radiation as well as their stability towards chemicals. Other semiconductors like WO_3, CdS, GaP, CdSe and GaAs absorb a wide range of the solar spectrum and can form chemically activated surface-bound intermediates, but unfortunately these photocatalysts are degraded during the repeated catalytic cycles involved in heterogeneous photocatalysis.

PHOTOSENSITIZATION

The illumination of suspended semiconductor in an aqueous solution of dye with unfiltered light (polychromatic light) leads to the possibility of the existence of two pathways (Hussein et al., 2008):

1. In the first pathway, the part of light with energy equal to or more than the band gap of the illuminated semiconductor will cause a promotion of an electron to conduction band of the semiconductor

and as a result, a positive hole will be created in the valence band. The formed photoholes and photoelectrons can move to the surface of the semiconductor in the presence of light energy. The positive hole will react with adsorbed water molecules on the surface of semiconductor producing $^{\bullet}OH$ radicals and the electron will react with adsorbed oxygen on the surface. Moreover, they can react with deliquescent oxygen and water in suspended liquid and produce perhydroxyl radicals ($HO_2^{\bullet}$) with high chemical activity (Zhao and Zhang, 2008). The processes in this pathway can be summarized by the following equations:

$$h^+ + OH^- \rightarrow \dot{O}H$$

$$h^+ + H_2O \rightarrow H^+ + \dot{O}H$$

$$e^- + O_2 \rightarrow O_2^{\bullet -}$$

$$O_2^{\bullet -} + H^+ \rightarrow HO_2^{\bullet}$$

$$HO_2^{\bullet} + O_2^{\bullet -} \xrightarrow{H^+} H_2O_2 + O_2$$

$$H_2O_2 \rightarrow 2\dot{O}H$$

$$\text{Dye} + \text{Semiconductor } (h^+_{VB}) \rightarrow \text{Dye}\cdot^+ + \text{Semiconductor}$$

$$\text{Dye}\cdot^+ + O\cdot_2^- \rightarrow DO_2 \rightarrow \text{degradation products}$$

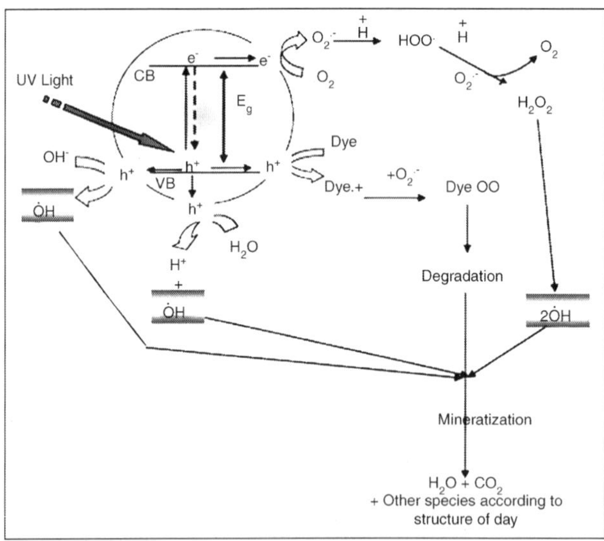

Fig. Dye/Semiconductor/UV Light System

2. In the second pathway, the other part of light with energy which is less than the band gap of the illuminated semiconductor will be absorbed by the adsorbed dye molecules. Dye molecules will be decolorized by a photosensitization process. The photocatalytic

decolorization of dyes, which is described as photosensitization processes, is also characterized by a free radical mechanism. In this process, the adsorbed dyes molecules on the surface of the semiconductor can absorb a radiation in the visible range in addition to the radiation with a short wavelengths (Fernandez-Ibanez et al., 2003; Ohno, 2004 and Alkhateeb et al., 2005). The excited colored dye (dye*) (in the singlet or triplet state) will inject an electron to the conduction band of the semiconductor (Hussein and Alkhateeb, 2007). The processes in this pathway can be summarized by the following equations:

$$\text{Dye} + \text{hv}\left(\text{VIS or UV reagion}\right) \rightarrow {}^{1}\text{Dye}^{*} \text{or } {}^{3}\text{Dye}^{*}$$

$${}^{1}\text{Dye}^{*} \text{or } {}^{3}\text{Dye}^{*} + \text{semiconducto} \rightarrow \text{Dye}^{\cdot+} + e^{-} \left(\text{to the conduction band of semiconductor}\right)$$

$$e^{-} + O_{2} \rightarrow O_{2}^{\bullet -} + \text{Semiconductor}$$

$$\text{Dye}^{\cdot +} + O_{2}^{\cdot -} \rightarrow \text{Dye } O_{2} \rightarrow \textit{degradation} \text{ products}$$

$$\text{Dye}^{\cdot +} + HO_{2}^{\cdot} \text{ (orHO)} \rightarrow \textit{degradation} \text{ products}$$

$$\text{Dye} + 2\dot{O}H \rightarrow H_{2}O + \text{oxidation products}$$

$$\text{Dye}^{\cdot +} + \bar{O}H \rightarrow \text{Dye} + \dot{O}H$$

$$\text{Dye}^{\cdot +} + H_{2}O \rightarrow \text{Dye} + \dot{O}H + H^{+}$$

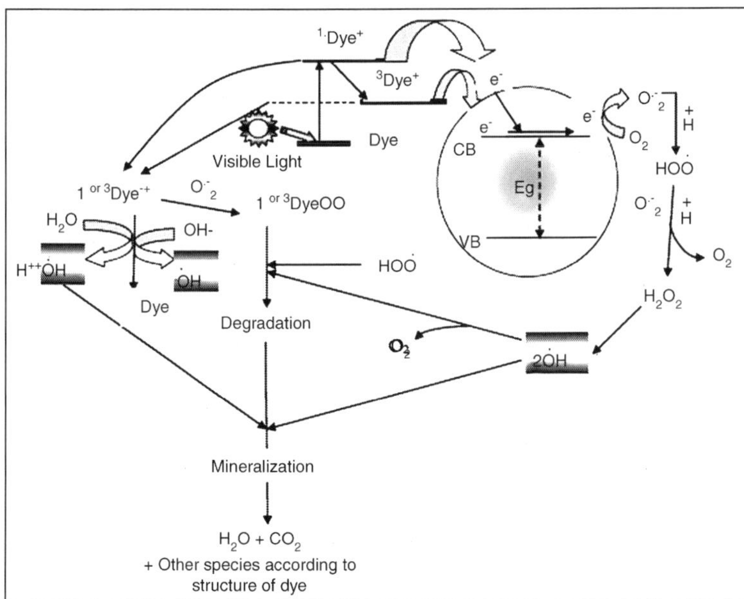

Fig. Non-regenerative Dye/Semiconductor/Visible Light System

The mechanism above is favoured by non-regenerative organic dye where dye/semiconductor/visible light system and the sensitizer itself degrade. However, in regenerative semiconductor system, the following mechanism may be followed:

$$Sen + TiO_2 \leftrightarrow Sen - TiO_2$$

$$Sen - TiO_2 + RX \leftrightarrow Sen - TiO_2......RX_{(ads)}$$

$$Sen - TiO_2 + O_2 \leftrightarrow Sen - TiO_2......O_{2(ads)}$$

$$Sen - TiO_2 \underset{hv}{\overset{hv\ visible\ light}{\rightleftarrows}} Sen^* - TiO_2$$

$$Sen^* - TiO_2......O_{2(ads)} \rightarrow Sen^+ - TiO_2 + O^{\cdot}_2$$

$$Sen^+ - TiO_2 + O_2^- \rightarrow Sen - TiO_2 + O_2$$

$$Sen^* - TiO_2\ Sen^+ - TiO_2\ (e^-_{C.B})$$

$$Sen^+ - TiO_2\ (e^-_{C.B}) \rightarrow Sen - TiO_2$$

$$Sen^+ - TiO_2 \left(e^-_{C.B}\right)......RX \rightarrow Sen^+ - TiO_2 + \dot{R}X + X^-$$

$$Sen^+ - TiO_2 \left(e^-_{C.B}\right)......O_{2(ads)} \rightarrow Sen^+ - TiO_2 + O_2^-$$

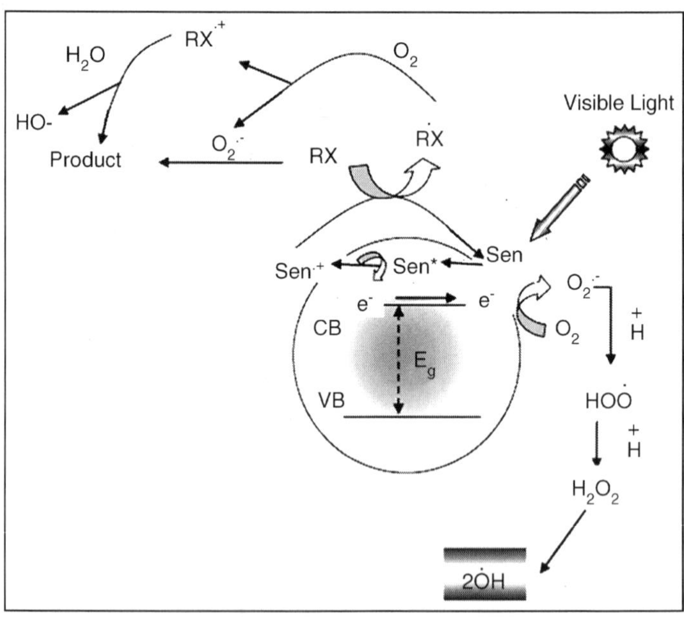

Fig. Regenerative Dye/Semiconductor/Visible Light System

Cho et al., (2001) conclude that there is no direct electron transfer between an excited sensitizer and CCl_4 molecules, in homogeneous solution, and the existence of semiconductor is essential for sensitized photocatalysis. Platinum supported on titanium dioxide acts as an excellent sensitizer and could have practical advantages as a mild and convenient photocatalyst for selective oxidation processes (Hussein et al., 1984). The addition of Rhodamine-B, as sensitizer to TiO_2 dispersion system increases the rate of photooxidation properties (Hussein et al., 1991). The authors explained that due to the fact that more light absorbed by Rhodamine-B between 460-580 nm, then the energy transfer from sensitizer to TiO_2 or to any other active species and hence promote the photocatalytic activity of titanium dioxide.

ADVANCED OXIDATION PROCESSES

Glaze et al., (1987) define Advanced Oxidation Processes (AOPs) for water treatments as the processes that occur near ambient temperature and pressure which involve the generation of highly reactive radicals,especially hydroxyl radicals (•OH), in sufficient quantity for water purification. Advanced oxidation processes can also be easily defined as techniques of destruction of organic pollutants from wastewaters. These processes include chemical oxidation processes using hydrogen peroxide, ozone, combined ozone and hydrogen peroxide, hypochlorite, Fenton's reagent, ultra-violet enhanced oxidation such as UV/O_3, UV/ H_2O_2, UV/air, wet air oxidation and catalytic wet air. Hydroxyl radicals are strong reactive species, which are capable of destroying a wide range of organic pollutants. Table shows hydroxyl radical as the second strongest oxidant.

Table. Standard Reduction Potential of Common Oxidants against Standard Hydrogen Electrode

Oxidant	E° (V)
Fluorine (F_2)	3.03
Hydroxyl radical (•OH)	2.80
Atomic oxygen (O)	2.42
Ozone (O_3)	2.07
Hydrogen peroxide(H_2O_2)	1.78
Hydroperoxyl radical ($O_2H\cdot$)	1.70
Potassium permanganate ($KMnO_4$)	1.67
Hypobromous acid (HBrO)	1.59
Chlorine dioxide (ClO_2)	1.50
Hypochlorous acid (HClO)	1.49
Hypochloric acid	1.45
Chlorine (Cl_2)	1.36
Bromine (Br_2)	1.09
Iodine (I_2)	0.54

The attack of organic pollutants by hydroxyl radicals occurs via the following mechanisms (Buxton et al., 1988; Legrini et al., 1993 and Pignatello et al., 2006):

- Electron transfer from organic pollutants to hydroxyl radicals:

$$\dot{O}H + RX \rightarrow RX^{\cdot+} + OH^-$$

- Hydrogen atom abstraction from the C–H, N–H or O–H bonds of organic pollutants:

$$\dot{O}H + RH \rightarrow R^{\cdot} + H_2O$$

- Addition of hydroxyl radical to one atom of a multiple atom compound:

$$\dot{O}H + Ph \rightarrow HOPh^{\cdot}$$

FUNDAMENTAL PARAMETERS IN PHOTOCATALYSIS

In semiconductor photocatalysis of industrial wastewater treatment, there are different parameters affecting the efficiency of treatment. These parameters include mass of catalyst, dye concentration, pH, light intensity, addition of oxidizing agent, temperature and type of photocatalyst. Other factors, such as, ionic components in water, solvent types, mode of catalyst application and calcinations temperature can also play an important role on the photocatalytic degradation of organic compounds in water environment (Guillard et al., 2005 and Ahmed et al., 2011).

EFFECT OF TYPE OF CATALYST

Haque and Muneer (2007) observe that Degussa P25 is more reactive for degradation of a textile dye derivative, bromothymol blue, in aqueous suspensions than other commercially available photocatalysts types of titanium dioxide, namely Hombikat UV100, PC500 and TTP. They explain the high activity of Degussa P25 is due to composing of small nano-crystallites of rutile dispersed within the anatase matrix. The band gap of rutile is less than that of anatase and as a result electron will transfer from the rutile conduction band to electron traps in anatase and the recombination of electrons and holes will be reduced. Hussein, (2002) reported that anatase has higher photoactivity than rutile due to the difference in surface area. Decolorization percentage of real textile industrial wastewater on rutile, anatase, and zinc oxide shows that the activity of different catalysts falls in the following sequence:

$$ZnO > TiO_2 \text{ (Anatase)} > TiO_2 \text{ (Rutile)}$$

ZnO is more active than TiO_2 due to the absorption of wider spectrum light (Sakthivela et al., 2003). However, the amount of zinc oxide required to reach the optimum activity is two times more than that for titanium dioxide (anatase or rutile) (Hussein and Abass, 2010 a). In another study, Hussein et al., (2008) observed that ZnO is less active than anatase when the same weight of catalysts is used for photocatalytic degradation of textile wastewater. Akyol et al., (2004) reported that ZnO is more active than TiO_2 for the decolorization efficiency of aqueous solution of a commercial textile dye due to the band gap energy, the charge carrier density, and the crystal structure. Decolorization efficiency of real textile industrial wastewater in the presence and absence of catalyst and/or solar radiation was also investigated (Hussein and Abass, 2010 b). The results indicate that the activity of different catalysts fall in the sequence:

ZnO > TiO_2 (Anatase) > TiO_2 (Rutile) > in the absence of catalyst = in the absence of solar radiation or artificial radiation = 0

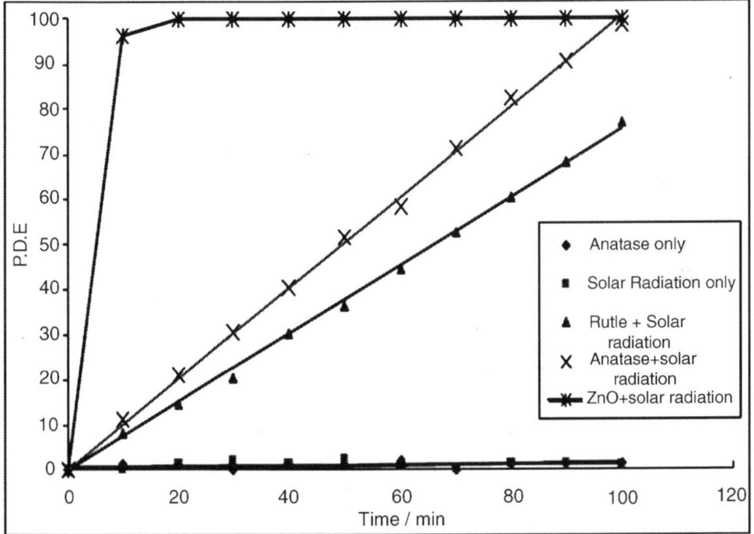

Fig. Photocatalytic Decolorization of Real Textile Industrial Wastewater at Different Conditions.

These results also indicate that there has been no dark reaction. Incubations of colored industrial wastewater without solar radiation and/or without catalyst has been performed to demonstrate that decolorization of the dye is dependent on the presence of both; light and catalyst.

EFFECT OF MASS OF CATALYST

Photocatalysts dosage added to the reaction vessel is a major parameter affecting the photocatalytic degradation efficiency (Dong et al., 2010). Photocatalytic degradation efficiency increases with an increase in catalysts mass. This behaviour may be due to an increase in the amount of active site on surface of photocatalyst particles. As a result, an increasing the number of dye molecule adsorbed on the surface of photocatalyst lead to an increase in the density of particles in the area of illumination (Kim and Lee, 2010). The extrapolation of Hird's data (Hird, 1976) indicates that only 7.5 mg of TiO_2 was sufficient to absorb all incident 366 nm radiation. It follows that the mass effect must be caused by changes in the effective utilization of the absorbed radiation rather than by the increased absorption. Photocatalyst with small particles are more efficient than larger particles.

This behaviour may be due to (Hussein, 1984):

- Photoholes and photoelectrons generated in the bulk would have fewer traps and recombination centers to overcome before reaching the surface.
- A greater proportion of material would be within the space charge arising from depletive oxygen chemisorptions, which favour exciton dissociation and photohole migration to the surface.

Hence, increasing the catalyst's mass will increase the concentration of the efficient small particles within the illuminated region of the reaction vessel. The direct proportionality between photocatalytic degradation efficiency and catalyst loading is real within low concentrations of photocatalyst where there are excess active sites reaching plateau reign. The plateau is reached when this effect can no longer increase the overall efficiency of utilizing incident radiation. Moreover, after the plateau region is achieved, the activity of photocatalytic decolorization decrease with increase of catalyst concentration for all types of catalysts. This behaviour is more likely to emanate from variation in the intensity of radiation entering the reaction vessel and the way the catalyst utilizes that radiation. Light scattering by catalyst particles at higher concentration lead to decrease in the passage of irradiation through the sample leading to poor light utilization (Gaya et al., 2010; Kavitha and Palanisamy, 2011). Deactivation of activated photocatalyst molecules colliding ground state molecules with increasing the load of photocatalyst may be also cause reduction in photocatalyst activity (Kim and Lee, 2010). Photocatalytic decolorization efficiency (PDE) per cent of real textile industrial wastewater has been investigated by employing different masses of TiO_2 (anatase or rutile) or ZnO under natural weathering conditions for 20 minutes of irradiation.

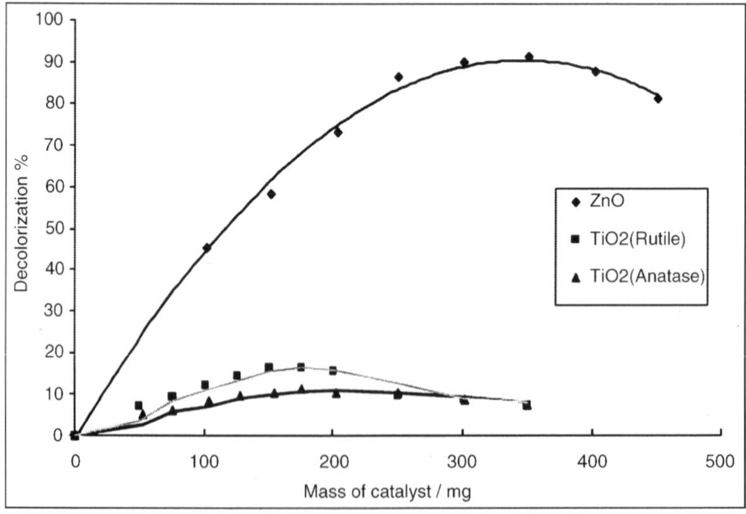

Fig. Effect of Mass on Photocatalytic Decolorization Efficiency of Real Textile Industrial Wastewater

The results in all cases indicate that the decolorization efficiency increases with increase in catalysts mass and then it becomes constant. It is clear from consideration of the catalyst concentrations at which the activity plateau were achieved that the mass effect does not depend upon the type of dye and source of irradiation. Moreover, plateau regions were achieved and then the activity of decolorization decreased with increasing catalyst concentration, for all types of catalysts used in this project.

EFFECT OF PH

Aqueous solution pH is an important variable in the evaluation of aqueous phase mediated photocatalytic decolorization reactions. pH change effects the adsorption quantity of organic pollutants and the ways of adsorption on the surface of photocatalyst (coordination). As a result, the photocatalytic degradation efficiency will greatly be influenced by pH changes. Zero Point Charge (pHzpc), is a concept relating to adsorption phenomenon and defined as the pH at which the surface of an oxide is uncharged. If positive and negative charges are both present in equal amounts, then this is the isoelectric point (iep). However, the zpc is the same as iep when there is no adsorption of other ions than the potential determining H^+/OH^- at the surface. In aqueous solution, at pH higher than pHzpc, the oxide surface is negatively charged and then the adsorption of cations is favoured and as a consequence, oxidation of cationic electron donors and acceptors are favoured. At pH lower than pHzpc, the adsorbent surface is positively charged and then the adsorption of anions is favoured and as a consequence, the acidic water donates more protons than hydroxide groups.

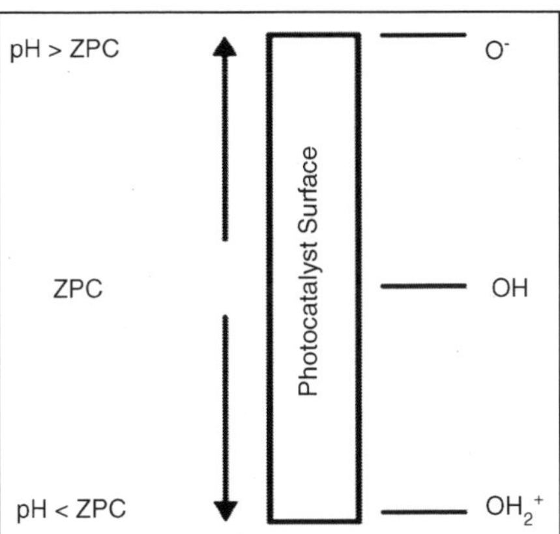

Fig. Effect of pH on ZPC.

Infrared spectroscopy study of Szczepantiewicz et al., (2000) shows that the TiOH sites are the major electron traps when TiO_2 is illuminated. The distribution of other species ($TiOH_2^+$ and TiO^-) with changing pH has been proposed by Kormann and co-workers. Figures show that, at pH below ZPC the surface is mostly positively charged and TiOH sites increase as pH increases and reach maximum value at ZPC of semiconductor. However, $TiOH_2^+$ as pH increases and reaches zero value at ZPC. At pH higher than ZPC the density of TiO- groups on the surface start to form and reached 100 per cent value at pH 14. The importance of pH during the reaction is not less than that of initial state. The formation of intermediate products, sometimes, changes the pH of aqueous solution and as a result, it affects the rate of photodegeradation (Galvez, 2003).

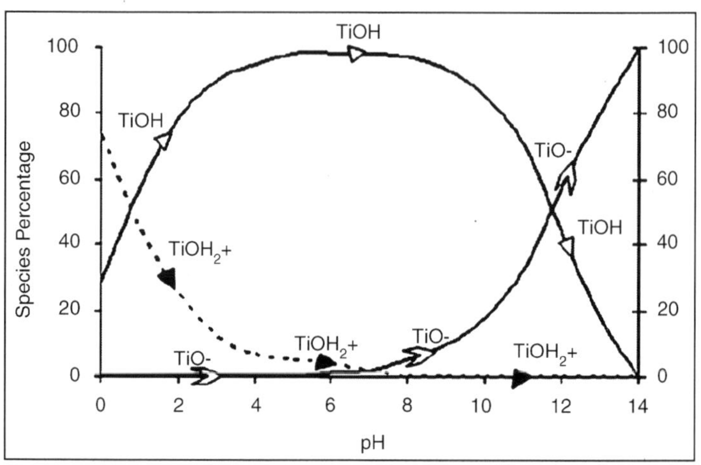

Fig. The Distribution of TiO_2 Surface Species at Different pH

At pH above and below pHzpc, the surface of zinc oxide and titanium dioxide are negatively or positively charged according to the following equations:

$$ZnOH + H^+ \leftrightarrow ZnOH_2^+$$

$$ZnOH + OH^- \leftrightarrow ZnO^- + H_2O$$

$$TiOH + H^+ \leftrightarrow TiOH_2^+$$

$$TiOH + OH^- \leftrightarrow TiO^- + H_2O$$

Effect of pH on Photocatalytic Decolorization of Bismarck Brown R

Under the determined experimental condition with initial dye concentration equal to 10^{-4} M, ZnO dosage 3.75 $gm.L^{-1}$, light intensity equal to 2.93 $mW.cm^{-2}$ and temperature equal to 298.15 K, the effect of change in solution pH on decolorization percentage has been studied in the range 2-12. The decolorization percent has been found to be strongly dependent on pH of solution because the reaction takes place on the surface of semiconductor. The decolorization percentage of Bismarck brown R increases with the increase of pH, exhibiting maximum decolorization at pH 9

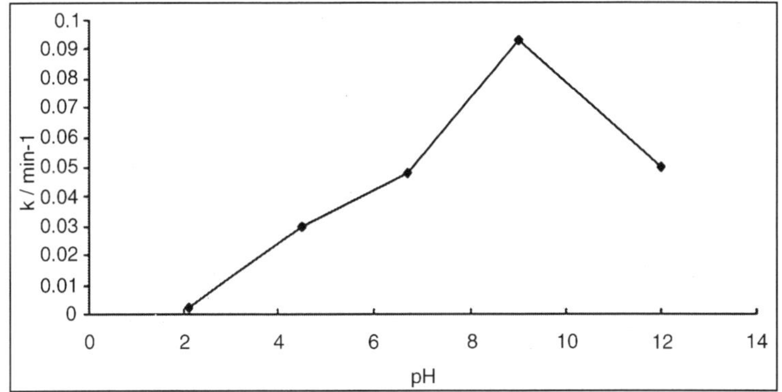

Fig. Effect of pH on Photocatalytic Decolorization Efficiency of Bismarck R Brown on ZnO

Under the determined experimental condition with initial dye concentration equal to 10^{-4}M, TiO_2 dosage 1.75 gm.L^{-1}, light intensity equal to 2.93 mW.cm^{-2} and temperature equal to 298.15 K, the effect of change in solution pH on decolorization percentage has been studied in the range 2-10. TiO_2 (DegussaP25), TiO_2 (HombikatUV100), TiO_2 (MillenniumPC105) and TiO_2 (Koronose2073). It was observed that the decolorization percentage strongly depends on the pH of solution because the reaction takes place on the surface of semiconductor. The decolorization percentage of Bismarck brown R increases with the increase of pH, exhibiting maximum decolorization at pH that is equal to 6.61, 6.54, 6.75, 6.63 for TiO2 (DegussaP25), TiO_2 (HombikatUV100), TiO_2 (MillenniumPC105) and TiO_2 (Koronose2073), respectively.

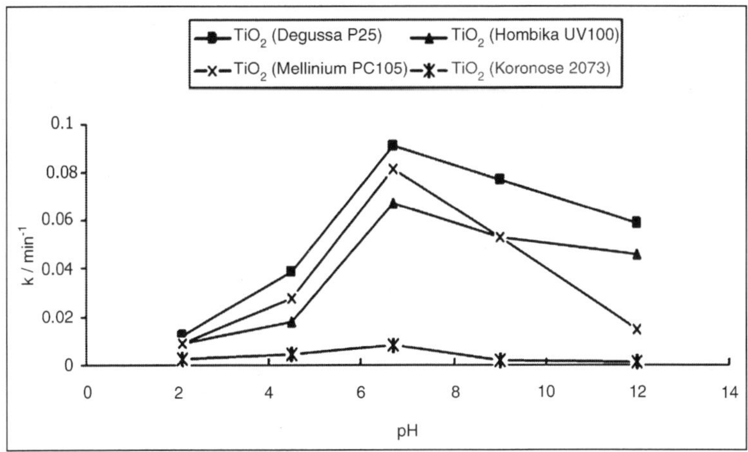

Fig. Effect of pH on Photocatalytic Decolorization Efficiency of Bismarck R Brown on Different Types of TiO_2

This behaviour could be explained on the basis of zero point charge (ZPC). The zero point charge is equal to 9.00 for ZnO and 6.25 for TiO_2 (Degussa P25). With the increase of the pH of solution, the surface of catalyst will be negatively charged by adsorbed hydroxyl ions. The presence of large quantities of adsorbed OH^- ions on the surface of catalyst favour the formation of •OH radical. However, if pH is lower than ZPC, the hydroxyl ions adsorbed on the surface will be decreased and, therefore, hydrogen ions adsorbed on the surface will increase and the surface will become positive charged. Both the acidic and basic media leave an inverse impact on the photodecolorization efficiency because of the decrease of the formation of the hydroxyl radical. The decolorization of Bismarck brown R decreases dramatically at strong acid media (pH = 2.1) for ZnO. This could be explained due to photocorrosion of ZnO as shown in the following equations:

$$ZnO \xrightarrow{hv} e^-_{CB} + h^+_{VB}$$

$$ZnO + 2h^+_{VB} \rightarrow Zn^{2+} + \frac{1}{2}O_2$$

EFFECT OF LIGHT INTENSITY

Egerton and King, (1979) show that square root of light intensity depends on the activity of titanium dioxide for different wavelengths of light. However, this relationship cannot be applied to all range of light intensities. The primary electronic processes which occur in the absorption of photons with energy equal or greater than the band gap of semiconductor are:

$$\text{Semiconductor} + hv \xrightarrow{k_1} (h-e)\text{exciton}$$

$$(h-e) \xrightarrow{k_2} h^+ + e^-$$

$$h^+ + e^- \xrightarrow{k_3} \text{Radiationless Recombination}$$

For photocatalysis processes, it is necessary that the excitons dissociate and the photoholes and photoelectrons reach the catalyst surface where they are trapped by surface species:

$$h^+OH^-_{(S)} \xrightarrow{k_4} \dot{O}H_{(S)}$$

$$e^- + O_2 \xrightarrow{k_5} O^-_{2(ads)}$$

The concentration of excitons, photoholes and photoelectrons may be considered by applying a steady state treatment:

$$\frac{d[h-e]}{dt} = k_1 I_{(ads)} - k_2[(h-e)] = 0$$

So that:

$$[(h-e)] = \frac{k_1}{k_2} I_{(ads)}$$

Similarly:

$$\frac{d[h]}{dt} = k_2[(h-e)] - k_3\left[h^+\right]\left[e^-\right] - k_4\left[h^+\right]\left[OH_{(S)}\right] = 0$$

Since:

$$\left[h^+\right] = \left[e^-\right]$$

Then:

$$\frac{d[h]}{dt} = k_1 I_{(ads)} - k_3\left[h^+\right]^2 - k_4\left[h^+\right]\left[OH^-_{(S)}\right] = 0$$

So that:

$$k_1 I_{(abs)} = k_3\left[h^+\right]^2 + k_4\left[h^+\right]\left[OH^-_{(S)}\right]$$

There are two possibilities concerning the light intensity:

- At high light intensities, where the recombination of photoholes and photoelectrons is predominate, then:

$$k_3\left[h^+\right]^2 \gg k_4\left[h^+\right]\left[OH^-_{(S)}\right]$$

So equation 48 becomes:

$$k_1 I_{(abs)} = k_3\left[h^+\right]^2$$

Then:

$$\left[h^+\right] = \left[\frac{k_1}{k_3}\right]^{\frac{1}{2}} I^{\frac{1}{2}}_{(abs)}$$

If the rate controlling step in the overall photocatalysis processes involves the surface trapping of photoholes at surface OH-, then the rate will be given by equation 40. The reaction rate is given by:

$$\text{Reaction rate} = k_4\left(\frac{k_1}{k_3}\right)^{\frac{1}{2}} I^{\frac{1}{2}}_{(abs)}\left[OH^-_{(S)}\right]$$

Then:

$$\text{Reaction rate}\ \alpha \left[I_{(abs)}\right]^{\frac{1}{2}}$$

If, on the other hand, the rate controlling step involves photoelectron trapping by oxygen, then the rate controlling step will be equation 41. The reaction rate is given by:

$$\text{Reaction rate} = k_5\left(\frac{k_1}{k_3}\right)^{\frac{1}{2}} I_{abs}\left[O_{2(abs)}\right]$$

Then:

$$\text{Reaction rate}\,\alpha\left[I_{(abs)}\right]^{\frac{1}{2}}$$

- At low light intensities, it is expected that recombination of photoholes and photoelectrons will be low, then:

$$k_4\left[h^+\right]\left[OH^-_{(S)}\right]\rangle\rangle k_3\left[h^+\right]^2$$

So equation 47 becomes:

$$k_1 I_{(abs)} = k_4\left[h^+\right]\left[OH^-_{(S)}\right]$$

Then:

$$\left[h^+\right] = \frac{k_1\left[I_{(abs)}\right]}{k_4\left[OH^-_{(S)}\right]}$$

It follows that:

$$\text{Reaction rate} = k_1\left[I_{(abs)}\right]$$

Alternatively, if photoelectron trapping is considered to be rate controlling, then:

$$\text{Reaction rate} = k_5\frac{k_1}{k_4}\left[\left(I_{(abs)}\right)\right]$$

Hence, a linear dependence would be expected at low light intensities. Square-root intensity dependence was observed with rutile I, rutile 11, anatase, uncoated anatase pigment and platinized anatase and also independent on wavelength of incident radiation (Harvey et al., 1983 a; Hussein and Rudham, 1987). Bahnemann et al., (1991) reported that the change in kinetic constant is a function of the square root of the radiation entering at high light intensities, while this change can be linear with light intensity of incident radiation low light intensities (Peterson et al., 1991). Ollis et al., (1991) summarize the effect of light intensity on the kinetics of the photocatalytic degradation of dye as follows:

- At low light intensities (0–20 mW/cm^2), the rate of photocatalytic degradation is proportional directly with light intensity (first order).
- At high light intensities (25 mW/cm^2), the rate of photocatalytic degradation is proportional directly with the square root of the light intensity (half order).
- At high light intensities the rate of photocatalytic degradation is independent of light intensity (zero order).

However, Hussein et al (2011) found that the rate of photocatalytic decolorization of Bismarck brown R on ZnO and different types of titanium dioxide is proportional directly with the light intensity of incident UVA radiation in the range of 0- 2.0 mW/cm^2 and with the square root of light intensity in the range of 2.0- 3.5 mW/cm^2.

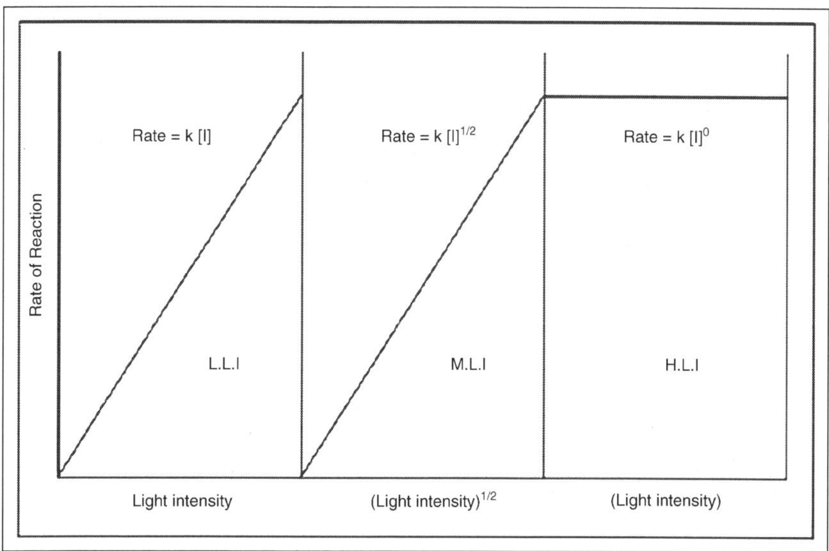

Fig. Effect of Light Intensity on the Kinetics of the Photocatalytic Degradation of Dye

Figures illustrate the impact of initial light intensity on the value of rate constant for photocatalytic decolorization of Bismarck brown R on ZnO and different types of titanium dioxide, respectively.

The results indicate that the photocatalytic decolorization of Bismarck brown R increases with the increase in light intensity, attaining a maximum value at 3.52 mW.cm^{-2}.

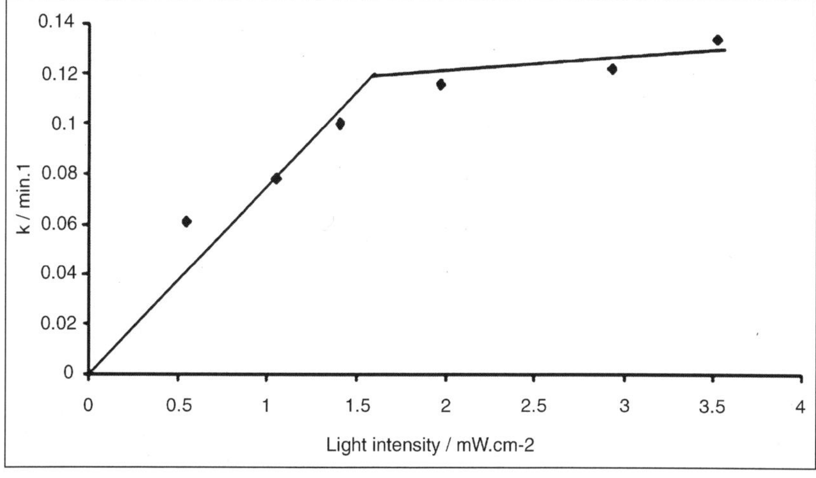

Fig. Effect of Initial Light Intensity on Rate Constant of Photocatalytic Decolorization of Bismarck Brown R on ZnO

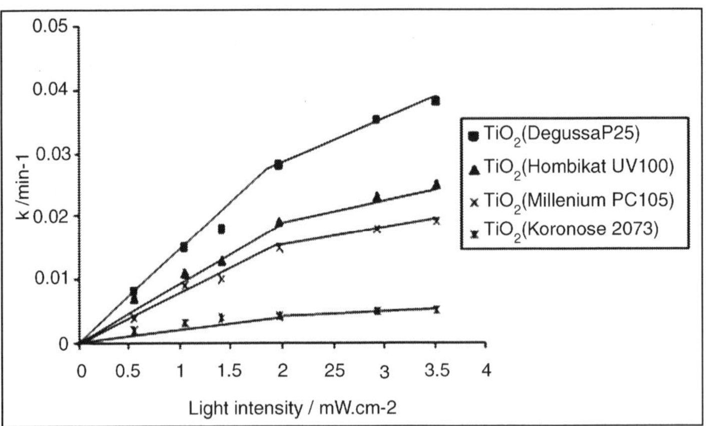

Fig. Effect of Initial Light Intensity on Rate Constant of Photocatalytic Decolorization of Bismarck Brown R using Different Types of TiO_2

EFFECT OF TEMPERATURE

One of the advantages of photoreaction is that it is not affected or slightly affected by temperature change. Temperature dependent steps in photocatalytic reaction are adsorption and desorption of reactants and products on the surface of photocatalyst. None of these steps appears to be rate determining. The impact of temperature is explained as the variable with the least effect on photocatalytic degradation of aqueous solution of azo dyes (Obies, 2011). Attia et al., (2008) have found that the activation energy of photodegeradation of real textile industrial wastewater is equal to 21 ± 1 kJ mol^{-1} on titanium dioxide and 24 ± 1 kJ mol^{-1} on zinc oxide.

The activation energy for the photocatalytic degradation of textile industrial wastewater on titanium dioxide is similar to previous findings for photocatalytic oxidation of different types of alcohols on titanium dioxide and metalized titanium dioxide (Al-zahra et al., 2007; Hussein and Rudham, 1984, 1987). The single value of activation energy (21 ± 1 kJ mol^{-1}) that can be related to the calculated activation energy of photooxidation of different species of titanium oxide is associated with the transport of photoelectron through the catalyst to the adsorbed oxygen on the surface (Harvey et al., 1983 a & b). Kim and Lee, (2010) explained that the very small activation energy in photocatalytic reactions is the apparent activation energy E_a, whereas the true activation energy E_t is nil. These types of reactions are operating at room temperature. Palmer et al., (2002) observed that the effect of temperature on the photocatalytic degradation is insignificant in the range of 10-68°C. High temperatures may have a negative impact on the concentration of dissolved oxygen in the solution and consequently, the recombination of holes and electrons increases at the surface of photocatalyst. However, Trillas et al., (1995); Chen and Ray, (1998) reported that raising the temperature of reaction enhances the rate of photocatalytic degradation significantly. Hussein and Abbas (2010 b) reported that the decolorization efficiency of real textile industrial wastewater increases with increasing of temperature as shown in fig.

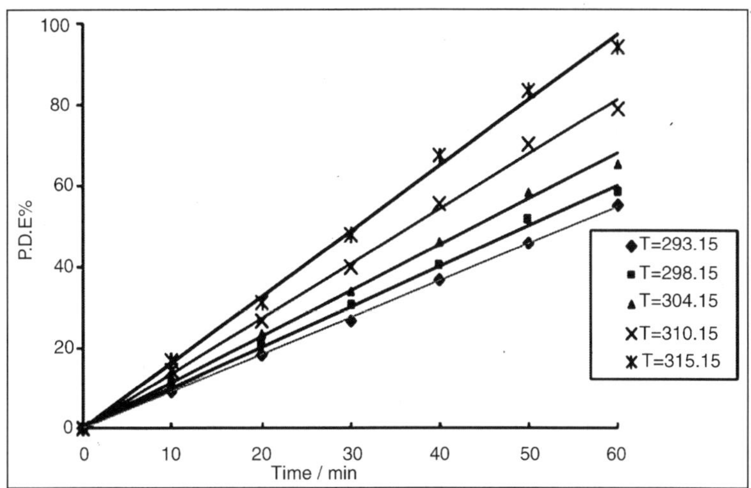

Fig. Effect of Temperature on P.D.E. of Real Textile Industrial Wastewater on Anatase under Solar Radiation

Hussein et al., (2011) have found that the rate of decolorization of Bismarck brown R on ZnO and different types of TiO_2 increases slightly with the increase of the temperature and the activation energy 24 ± 1 kJ.mol^{-1} for ZnO and 14 ± 1, 16 ± 1, 21 ± 1 and 22 ± 1 kJ.mol^{-1} for TiO_2 (Degussa P25), TiO_2 (Hombikat UV100), TiO_2 (Millennium PC105), and TiO_2 (Koronose 2073), respectively. Figure shows the impact of temperature on photodecolorization of Bismarck brown R by using TiO_2 (Hombikat UV100).

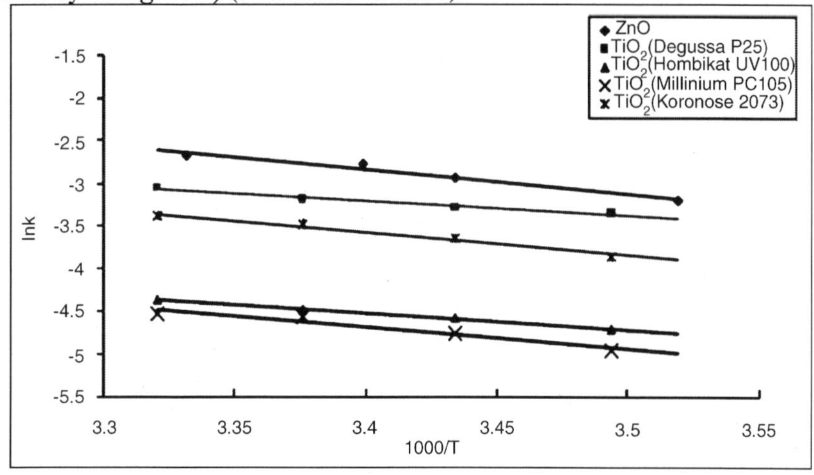

Fig. Arrhenius Plot by Different Types of Catalyst with Bismarck Brown R

EFFECT OF ADDITION OF OXIDANTS

It is well known that the addition of oxidants increases the rate of photocatalytic degradation of dyes by the formation of hydroxyl radicals (Salvador and Decker, 1984; Jenny and Pichat, 1991). However, this is not general for all types of dyes (Hachem et al., 2001).

Production of additional hydroxyl radicals occurs when hydrogen peroxide is added through the following mechanisms (Galvez, 2003 and Dong et al., 2010):

- Trapping of photogenerated electrons.

$$H_2O_2 + 2e^- \rightarrow 2OH^-$$

- Self-decomposition by photolysis

$$H_2O_2 + hv \rightarrow 2\dot{O}H$$

- Reaction with superoxide radical anion $O2^{\bullet -}$

$$H_2O_2 + O_2^- \rightarrow \dot{O}H + OH^- + O_2$$

The addition of persulphate leads to form sulphate radical anion by trapping the photogenerated electrons (Konstantinou and Albanis, 2004)

$$S_2O_8^{2-} + e^- \rightarrow SO_4^{2-} + SO_4^-$$

The formed sulphate radical anion is a strong oxidant and reacts with organic molecules pollutants as follows (Galvez, 2003):

- Abstracting a hydrogen atom from saturated carbon.
- Adding hydrogen to unsaturated or aromatic carbon.
- Removing one electron from carboxylate anions and from certain neutral molecules.

Sulphate radical anion can also react with water molecule to produce hydroxyl radical (Konstantinou and Albanis, 2004):

$$SO_4^- + H_2O \rightarrow SO_4^{2-} + \dot{O}H + H^+$$

Other oxidants such as iodate and bromate can also increase the reaction rate because they are also electron scavengers, while chlorate has been proven insufficient to improve effectiveness (Galvez, 2003). However, these additives are too expensive to be compared to hydrogen peroxide and peroxydisulphate. Moreover, they do not dissociate into harmless products.

The addition of oxidant to reaction mixture serves the rate of photocatalytic degradation by:

- Generation of additional $\bullet OH$ and other oxidizing species.
- Increasing the number of trapped photoelectrons.
- Increasing the oxidation rate of intermediate compounds.
- Replacement of oxygen role in the case of the absence of oxygen in the reaction mixture.

Table shows the effect of addition of hydrogen peroxide on the rate of photocatalytic degradation of red disperse dye on ZnO. The results indicate that the apparent rate constant increases with the increase in H_2O_2 concentration to a certain level and a further increase in H_2O_2 concentration leads to decrease in the degradation rate of the red disperse dye. The presumed reason is that the addition of H_2O_2 to a certain level increases the production of hydroxyl radicals, but the additional amount leads to reduce the amounts of photoholes and hydroxyl radicals (Legrini et al., 1993; Malato, 1998; Daneshvar et al., 2003; Konstantinou and Albanis, 2004):

$$H_2O_{2+}2h^+ \rightarrow O_2 + 2H^+$$

$$H_2O_2 + \dot{O}H \rightarrow H_2O + HO_2^{\cdot}$$

$$HO_2^{\cdot} + \dot{O}H \rightarrow H_2O + O_2$$

Table. Effect of Addition of H_2O_2 on the Apparent Rate Constant of Photocatalytic Decolorization of Disperse Dye.

Conc. of H_2O_2 (mmol.L^{-1})	K_{app} (min^{-1})
0	0.0893
1.5	0.1408
3.0	0.1499
4.5	0.1680
6.0	0.1404
7.5	0.1290

This behaviour relates to the competition between the adsorption of organic pollutants and H_2O_2 on the surface of photocatalyst. The required amount of H_2O_2 reaches the highest level of enhancement for the rate of photodegradation is related to the ratio of the concentration of organic pollutants and H_2O_2 (Galvez, 2003). When the pollutant concentration is low compared with the concentration of H_2O_2, the adsorption of organic pollutants decreases due to the increase of adsorption of hydrogen peroxide and, as a result, the additional hydroxyl radicals generated by H_2O_2 do not react efficiently.

COMPARISON BETWEEN MINERALIZATION AND PHOTOCATALYTIC DECOLORIZATION

Mineralization of dyes is a process in which dyes are converted completely into its inorganic chemical components (minerals), such as carbon dioxide, water and other species according to the structure of dye.

Mineralization of Bismarck brown R was evaluated by analyzing total organic carbon (TOC) (Hussein et al. 2011). The results shown in Fig. indicate that photocatalytic decolorization of Bismarck R brown was faster than the decrease of TOC. The results indicate that per cent TOC reduction was about 73 per cent after 60 minutes of irradiation while the per cent of decolorization achieved 88 per cent for the same period of irradiation. These findings in good agreement with those reported before (He et al. 2007 and Chen 2009). This could be explained to the formation of some by product, which resist the photocatalytic degradation. Furthermore, the formed by products need more time to destruct.

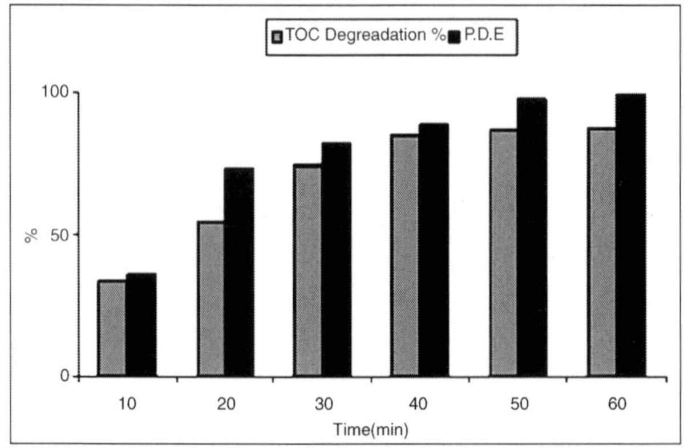

Fig. Comparison between Mineralization and Photocatalytic Decolorization of Bismarck Brown R on ZnO

EFFECT OF IRRADIATION SOURCES

Table summarized the obtained results from the different techniques used for the treatments of textile industrial wastewater and different types of industrial dyes (Hussein, 2010). The results show that the decolorization rate of textile industrial wastewater is faster with solar light than with UV light. The results indicated that solar energy could be effectively used for photocatalytic degradation of pollutants in wastewater.

Table. Effect of Irradiation Sources on Photocatalytic Decolorization of Textile Industrial Wastewater and Different Dyes

Process	Type of Treated Waste or Dye	Source of Irradiation	Type of Catalyst Hours	Time for Complete Mineralization/
Photocatalytic	Textile industrial wastewater	Solar	ZnO	2.7
Photocatalytic	Textile industrial wastewater	Solar	TiO_2	4
Photolysis	Murexide	Solar	–	7.5
Photocatalytic	Murexide	Solar	ZnO	3.8
Photocatalytic	Murexide	Solar	TiO_2	3.3
Photolysis	Thymol blue	Solar	–	7
Photocatalytic	Thymol blue	Solar	TiO_2	2.7
Photocatalytic	Thymol blue	Solar	ZnO	3.3
Photocatalytic	Textile industrial wastewater	Mercury lamp	TiO_2	2.6

Photocatalytic	Textile industrial wastewater	Mercury lamp	ZnO	3
Photocatalytic	Bismarck brown G	Mercury lamp	ZnO	1
Photocatalytic	Bismarck brown R	Mercury lamp	ZnO	0.8
Photocatalytic	Bismarck brown R	Mercury lamp	TiO_2	1.2
Photocatalytic	Textile industrial wastewater	Mercury lamp	TiO_2	3
Photocatalytic	Textile industrial wastewater	Mercury lamp	ZnO	1
Photocatalytic	Textile industrial wastewater	Solar	TiO_2	1.8
Photocatalytic	Textile industrial wastewater	Solar	ZnO	0.33

CONCLUSIONS

- Photocatalytic degradation techniques is the most efficient and clean technology.
- Textile industries have become worldwide. Thus, this method can be considered as a promising technique for providing formidable quantities of water especially for countries facing serious suffering from water shortage.
- The existence of catalyst and lights are essential for photocatalytic degradation of colored dyes.
- Photocatalytic degradation efficiency (PDE) of textile industrial wastewater is obviously affected by illumination time, pH, initial dye concentration and photocatalyst loading.
- Solar photocatalytic treatment has been proved to be an efficient technique for decolorization of industrial wastewater through a photocatalytic process and the transformation is practically complete in a reasonable irradiation time.
- In the countries where, intense sunlight is available throughout the year, solar energy could be effectively used for photocatalytic degradation of pollutants in industrial wastewater.
- The zero point charge is 6.4 and 9.0 for TiO_2 and ZnO, respectively above which the surface of photocatalyst is negatively charged by means of adsorbed hydroxyl ions; this favours the formation of hydroxyl radical, and as a result, the photocatalytic degradation of industrial wastewater increases due to inhibition of the photoholes and photoelectrons recombination.

13

Determining Fibre Properties and Yarn

AN OVERVIEW

Fibre, also spelled fibre, is a class of materials that are continuous filaments or are in discrete elongated pieces, similar to lengths of thread. They are very important in the biology of both plants and animals, for holding tissues together.

Human uses for fibres are diverse. They can be spun into filaments, string or rope, used as a component of composite materials, or matted into sheets to make products such as paper or felt.

Fibres are often used in the manufacture of other materials. Synthetic fibres can be produced very cheaply and in large amounts compared to natural fibres, but natural fibres enjoy some benefits, such as comfort, over their man-made counterparts.

TEXTILE FIBRE

A unit in which many complicated textile structures are built up is said to be textile fibre. Textile Fibre is the raw material required for the textile industry.

Natural Fibres

Natural fibres include those produced by plants, animals, and geological processes. They are biodegradable over time.

They can be classified just as to their origin:

- Vegetable fibres are generally based on arrange-ments of cellulose, often with lignin: examples include cotton, hemp, jute, flax, ramie, and sisal. Plant fibres are employed in the manufacture of paper and textile (cloth), and dietary fibre is an important component of human nutrition.
- Wood fibre, distinguished from vegetable fibre, is from tree sources. Forms include groundwood, thermomechanical pulp (TMP) and bleached or unbleached kraft or sulfite pulps. Kraft and sulfite, also called sulphite, refer to the type of pulping process used to remove the lignin bonding the original wood structure, thus freeing the fibres for use in paper and engineered wood products such as fiberboard.

- Animal fibres consist largely of particular proteins. Instances are spider silk, sinew, catgut, wool and hair such as cashmere, mohair and angora, fur such as sheepskin, rabbit, mink, fox, beaver, etc.
- Mineral fibres comprise asbestos. Asbestos is the only naturally occurring long mineral fibre. Short, fibre-like minerals include wollastonite, attapulgite and halloysite.

MAN-MADE FIBRES

Synthetic or man-made fibres generally come from synthetic materials such as petrochemicals. But some types of synthetic fibres are manufactured from natural cellulose, including rayon, modal, and the more recently developed Lyocell. Cellulose-based fibres are of two types, regenerated or pure cellulose such as from the cupro-ammonium process and modified cellulose such as the cellulose acetates.

Fibre classification in reinforced plastics falls into two classes:

1. Short fibres, also known as discontinuous fibres, with a general aspect ratio (defined as the ratio of fibre length to diameter) between 20 to 60, and
2. Long fibres, also known as continuous fibres, the general aspect ratio is between 200 to 500.

Cellulose Fibres

- Cellulose fibres are a subset of man-made fibres, regenerated from natural cellulose. The cellulose comes from various sources. Modal is made from beech trees, bamboo fibre is a cellulose fibre made from bamboo, seacell is made from seaweed, etc.

Mineral Fibres

- Fiberglass, made from specific glass, and optical fibre, made from purified natural quartz, are also man-made fibres that come from natural raw materials.
- Metallic fibres can be drawn from ductile metals such as copper, gold or silver and extruded or deposited from more brittle ones, such as nickel, aluminum or iron.
- Carbon fibres are often based on carbonised polymers, but the end product is pure carbon.

Polymer Fibres

- Polymer fibres are a subset of man-made fibres, which are based on synthetic chemicals (often from petrochemical sources) rather than arising from natural materials by a purely physical process.

  These fibres are made from:
  - Polyamide nylon,

  - PET or PBT polyester
  - Phenol-formaldehyde (PF)
  - Polyvinyl alcohol fibre (PVA)
  - Polyvinyl chloride fibre (PVC)
  - Polyolefins (PP and PE)
  - Acrylic polyesters, pure polyester PAN fibres are used to make carbon fibre by roasting them in a low oxygen environment. Traditional acrylic fibre is used more often as a synthetic replacement for wool. Carbon fibres and PF fibres are noted as two resin-based fibres that are not thermoplastic, most others can be melted.
  - Aromatic polyamids (aramids) such as Twaron, Kevlar and Nomex thermally degrade at high temperatures and do not melt. These fibres have strong bonding between polymer chains
  - Polyethylene (PE), eventually with extremely long chains / HMPE (*e.g.* Dyneema or Spectra).
  - Elastomers can even be used, *e.g.* spandex although urethane fibres are starting to replace spandex technology.
  - Polyurethane fibre
- Coextruded fibres have two distinct polymers forming the fibre, usually as a core-sheath or side-by-side. Coated fibres exist such as nickel-coated to provide static elimination, silver-coated to provide anti-bacterial properties and aluminum-coated to provide RF deflection for radar chaff. Radar chaff is actually a spool of continuous glass tow that has been aluminum coated. An aircraft-mounted high speed cutter chops it up as it spews from a moving aircraft to confuse radar signals.

Microfibers

Microfibers in textiles refer to sub-denier fibre (such as polyester drawn to 0.5 dn). Denier and Detex are two measurements of fibre yield based on weight and length. If the fibre density is known you also have a fibre diameter, otherwise it is simpler to measure diameters in micrometers. Microfibers in technical fibres refer to ultra fine fibres (glass or meltblown thermoplastics) often used in filtration.

Newer fibre designs include extruding fibre that splits into multiple finer fibres. Most synthetic fibres are round in cross-section, but special designs can be hollow, oval, star-shaped or trilobal. The latter design provides more optically reflective properties. Synthetic textile fibres are often crimped to provide bulk in a woven, non-woven or knitted structure. Fibre surfaces can also be dull or bright. Dull surfaces reflect more light while bright tends to transmit light and make the fibre more transparent. Very short and/or irregular fibres have been called fibrils. Natural cellulose, such as cotton or bleached kraft, show smaller fibrils jutting out and away from the main fibre structure.

ANIMAL FIBRES

ALPACA FIBRE

Alpaca fleece is the natural fibre harvested from an alpaca. It is light weight or heavy weight, depending on how it is spun. It is soft, durable, luxurious and silky natural fibre. While similar to sheep's wool, it is warmer, not prickly, and has no lanolin which makes it hypoallergenic.

Alpaca is naturally water-repellent. Huacaya, an alpaca that grows soft spongy fibre has natural crimp, thus making a naturally elastic yarn, perfect for knits. Suri has far less crimp and thus is best suited for woven goods, but is wonderfully luxurious as well. The designer Armani has used Suri alpaca to fashion Men's and Women's suits.

Alpaca fleece is made into various products, from very simple and inexpensive garments made by the aboriginal communities to sophisticated, industrially made and expensive products such as suits. In the United States, groups of smaller alpaca breeders have banded together to create "fibre co-ops," in order to make the manufacture of alpaca fibre products less expensive. The preparing, carding, spinning, weaving and finishing process of alpaca is very similar to the process used for wool.

Alpacas

Types

There are two types of alpaca: Huacaya (which produce a dense, soft, crimpy sheep-like fibre), and the mop-like Suri (with silky pencil-like locks, resembling dread-locks but not actually matted fibres). Suris are prized for their longer and silkier fibres, and estimated to make up between 19-20 per cent of the Alpaca population. Since its import into the United States, the number of Suri alpacas has grown substantially and become more colour diverse. The Suri is thought to be rarer, possibly because it is less hardy in the harsh South American mountain climates, as its fleece offers less insulation against the cold.

History of Fibre Industry

The Amerindians of Peru used this fibre in the manufacture of many styles of fabrics for thousands of years before its introduction into Europe as a commercial product. The alpaca was a crucial component of ancient life in the Andes, as it provided not only warm clothing but also meat. Many rituals revolved around the alpaca, perhaps most notably the method of killing it: An alpaca was restrained by one or more people, and a specially-trained person plunged his bare hand into the chest cavity of the animal, ripping out its heart. Today, this ritual is viewed by most as barbaric, but there are still some tribes in the Andes which practice it.

The first European importations of alpaca fibre were into Spain. Spain transferred that fibre to Germany and France. Apparently alpaca yarn was spun in England for the first time about the year 1808 but the fibre was condemned as an unworkable material. In 1830 Benjamin Outram, of Greetland, near Halifax, appears to have reattempted spinning it, and again it was condemned.

These two attempts failed due to the style of fabric into which the yarn was woven — a type of camlet. It was not until the introduction of cotton warps into Bradford trade about 1836 that the true qualities of alpaca could be developed into fabric. It is not known where the cotton warp and mohair or alpaca weft plain-cloth came from, but it was this simple and ingenious structure which enabled Titus Salt, then a young Bradford manufacturer, to use alpaca successfully. Bradford is still the great spinning and manufacturing centre for alpaca. Large quantities of yarns and cloths are exported annually to the European continent and the US, although the quantities vary with the fashions in vogue. The typical "alpaca-fabric" is a very characteristic "dress-fabric." Due to the successful manufacture of various alpaca cloths by Sir Titus Salt and other Bradford manufacturers, a great demand for alpaca wool arose which could not be met by the native product.

Apparently, the number of alpacas available never increased appreciably. Unsuccessful attempts were made to acclimatize alpaca in England, on the European continent and in Australia, and even to cross English breeds of sheep with alpaca.

There is a cross between alpaca and llama — a true hybrid in every sense — producing a material placed upon the Liverpool market under the name "Huarizo". Crosses between the alpaca and vicuña have not proved satisfactory. Current attempts to cross these two breeds are underway at farms in the US. The Alpaca Owners and Breeders Association, alpacas are now being bred in the US, Canada, Australia, New Zealand, UK, and numerous other places.

In recent years, interest in alpaca fibre clothing has surged, perhaps partly because alpaca ranching has a reasonably low impact on the environment. Individual U.S. farms are producing finished alpaca products like hats, scarves, and footwarmers.

Outdoor sports enthusiasts recognize that its lighter weight and better warmth provides them more comfort in colder weather, so outfitters such as R.E.I. and others are beginning to stock more alpaca products. Using an alpaca and wool blend such as merino is common to the alpaca fibre industry in order to improve processing and the qualities of the final product. In December 2006 the General Assembly of the United Nations proclaimed 2009 to be the International Year of Natural Fibres, so as to raise the profile of alpaca and other natural fibres.

Structure of Alpaca Fibre

In physical structure, alpaca fibre is somewhat akin to hair, being very glossy. Alpaca fibre is similar to that of merino wool fibre, and alpaca yarns tend to be stronger than wool yarns. The heel hole that appears in wool socks or in elbows of wool sweaters is non-existent in similar alpaca garments. In processing, slivers lack fibre cohesion and single alpaca rovings lack strength.

Blend these together and the durability is increased several times over. More twisting is necessary, especially in Suri, and this can reduce a yarn's softness. The alpaca has a very fine and light fleece. It does not retain water, is thermal even when wet and can resist the solar radiation effectively. These characteristics guarantee the animals a permanent and appropriate coat to fight against the extreme changes of temperature. This fibre offers the same protection to humans. Alpaca is sustainable as a fibre, and is naturally organic. Alpacas as animals are soft on the environment, making alpaca a truly green textile.

Alpaca fibre contains also microscopic airbags that make possible the manufacture of light textiles as well as different kinds of clothing. The cells of the central core may contract or disappear, forming air pockets which assist insulation. Fleeces vary from alpaca to alpaca and in some alpacas there may be a higher incidence of medullated fibres, compared to wool and mohair.

This can be an objectionable trait. Medullated fibres can take less dye, standing out in the finished garment, and are weaker. The proportion of medullated fibres is higher in the coarser, unwanted guard hairs: there is less or no medullation in the finer, lower micrometre fibres... These undesirable fibres are easy to see and give a garment a hairy appearance. Quality alpaca products should be free from these medullated fibres.

Quality of Alpaca Fibre

Good quality alpaca fibre is approximately 18 to 25 micrometres in diameter. Whilst breeders report fibre can sell for 2 to 4 dollars per ounce, the world wholesale price for processed pre-spun alpaca "tops" is only between about $10 to $24 US/kg, *i.e.* about $0.28 to $0.68 per oz. Finer fleeces, ones with a smaller diameter, are preferred, and thus are more expensive. As an alpaca gets older the width of the fibres gets thicker, at between 1 μm and 5 μm per year. This is often caused by over nutrition; if fed too much nutritious food the animal doesn't get fat, instead the fibre gets thicker.

As with all fleece-producing animals, quality varies from animal to animal, and some alpacas produce fibre which is less than ideal. Fibre and conformation are the two most important factors in determining an alpaca's value. Alpacas come in many shades from a true-blue black through browns-black, browns, fawns, white, silver-greys, and rose-greys. However, white is predominant, because of selective breeding: the white fibre can be dyed in the largest ranges of colours. In South America, the preference is for white as they generally have better fleece than the darker-coloured animals.

This is because the dark colours had been all but bred out of the animals. The demand for darker fibre sprung up in the United States and elsewhere, however in order to reintroduce the colours, the quality of the darker fibre has decreased slightly. Breeders have been diligently working on breeding dark animals with exceptional fibre, and much progress has been made in these areas over the last 5–7 years.

Dyeing Alpaca Fibre

Before dyeing the alpaca fibre must go through other stages:

- Selection of wool, according to colour, size and quality of fibre
- "Escarminado", Removal of grass, dirt, thorns, and other impurities
- Washing, to remove all the dirt and grease.
- Spinning

Once the fibre is clean then it is possible to begin with the process of dyeing. Natural dyeing: *(recipe used by Andean artisans)*: To dye 1 kg of alpaca wool with cochinilla (natural dye).

- Boil 5 litres of water in an aluminum can with 100 g of cochinilla for an hour.
- Sift and put the wool in the water.
- Boil again for an hour and ad 50 lemons cut in halves.
- Then take out the wool and hang for drying.

Note: For dyeing with another natural dye (native plants) add 2 kg, of the products to the water and boil.

ANGORA WOOL

Angora wool is an extraordinarily soft knitting fibre produced from the fur of the angora rabbit. It can only be spun from the hair of angora rabbits, while the product of angora goats is called mohair. Angora fibre has a very soft, silky texture and is usually 10 to 13 microns thick. The fibres are also hollow, which gives them loft and a characteristic floating feel. In addition, angora fibres are very short, which can effect stitch density when the knitter is using angora wool with a high percentage of angora.

Pure angora wool is impossible to make, because the fibres are too fine and the wool will simply unravel. Angora fibres are usually mixed with other soft fibres, such as cashmere and lambswool. Angora wool tends to be very warm and is frequently used to trim sweaters or to knit hats and scarves. Angora is generally viewed as a luxury fibre, and most angora wool products are very expensive, reflecting the laborious harvesting process and the small number of producers.

Angora rabbits can be combed or gently sheared for their fur, and they have been used for that purpose in Turkey for centuries. Angora first became popular in Europe in the late 18th century, when it was popularized by the French. In the Americas, angora wool didn't touch the popular imagination until the 20th century, when small cottage breeders began raising angora rabbits and spinning their fibre.

There are four breeds of angora rabbits, beginning with the English breed, which weighs 5 to 7.5 pounds (2 to 3.5 kilograms). The English breed produces a very large amount of fur and must be combed regularly to keep the hair free of tangles and debris. The good natured rabbits are popular in show, and their fibre wraps very tightly when spun, making an even and strong wool. The French breed ranges from 7.5 to 10.5 pounds (3.5 to 4.5 kilograms) and has a higher proportion of guard hair to wool. Guard hairs take colour better, and many coloured angora wool products come from the French breed. The French breed requires less grooming than other breeds and is recommended for novice breeders.

The Satin breed has very shiny, soft fur with a satiny appearance. It is easy to collect and spin, and the 6.5 to 9.5 pound (3 to 4 kilogram) rabbits are favoured by fibre collectors because their fur spins quickly and easily. The breed also tends to be richly coloured, producing wool with high colour saturation, although the all white Satins produce fur that takes dye very well.

The Giant breed is significantly larger than the other breeds, often weighing more than 20 pounds (9 kilos). The breed most often appears in white, and its wool must be harvested by shearing, because the rabbits do not naturally molt or shed. For sheer volume of collection, the giant breed is an excellent choice for breeders attempting to produce angora fibre for commercial use.

Angora wool can be harvested year round, although it requires careful handwork, as most rabbits are combed. Most angora rabbits are amenable, but nervous, and it is important to be gentle with them. The wool produced can be used in a wide variety of applications in which insulation, comfort, and warmth are needed, and the fibres are pleasurable to work with because of their luxurious look and feel.

CAMEL HAIR

Camel hair is, variously, the hair of a camel; a type of cloth made from camel hair; or a substitute for authentic camel hair; and is classified as a specialty hair fibre. When woven into haircloth, using the outer protective fur called guard hair, camel hair is coarse and inflexible. However, other varieties of camel hair cloth—especially those that blend camel hair with wool— or from the pure under coat are soft and plush. Pure camel hair, frequently used for coats, is gathered when camels molt in warmer seasons. This undercoat is very soft, and is separated from the dense, coarse guard hair for cloth use.

The fine fur of the camel hair are often blended with fine wool to create fabrics for men's and women's coats, jackets and blazers, skirts, hosiery, sweaters, gloves, scarves, mufflers, and caps and robes. The long coarser hair removed in the dehairing process is also used which can be made into carpet backing as well as waterproof coats that are very warm for colder climates.

Although most camel hair is left as its natural tones, the hair can be dyed in a multiple range of colours and accepts the dye equally compared to that of wool fibres. The best blends of camel hair in textiles are pure camel hair or blended with wool only. It is also commonly blended with nylon to make hosiery and other knitted products. Products containing camel hair should be dry cleaned or handwashed.

Paintbrush bristles are derived either from the hairs of horses, squirrels, goats, sheep, bears, or some combination of these. As a result, the texture, absorbency, and other properties of "camel hair" brushes vary markedly. Wooly camel hair is unsuitable for brush bristles.

CASHMERE WOOL

Cashmere is a luxuriant wool that many a fashion-conscious woman has dreamed of wearing against her skin. Its silken feel, feather-light weight, and appreciable status make it highly desirable.

Despite the glamour associated with cashmere, it hails from humble beginnings. Cashmere is the wool or fur of the Kashmir goat. Kashmir goats are primarily raised in Mongolia, but many are bred in Iran, Tibet, India and China. American herders have also joined the international cashmere production market in recent years.

Cashmere is harvested from the goats during their annual molting season through the shedding or the shearing of their down. In the frigid high desert climates where most of the goats are raised, the dense inner coat guards against harsh winter weather, but once seasons change, goats begin to lose the protective layer of down.

The finest cashmere comes from the underbelly and throat of the goats, but a lesser grade is also taken from the goats' legs and backs. Longer fibres from the belly and throat area make the wool especially soft and cause less "pilling" when the fibres are woven into garments such as sweaters, shawls, capes, dresses, and coats for both men and women.

The shorter fibres from the backs and legs are heavier and less expensive, making it easier to afford a luxury garment. Cashmere comes naturally in white, gray and brown, but the wool is easily dyed.

Garments made of cashmere were once only available to royalty because the rarity of the wool increased its value. Napoleon is said to have popularized the use of cashmere as shawls or wraps when he gave his second wife, Empress Eugenie, seventeen of them.

In more recent years, Old-Hollywood glamour girls graced the silver screen, bringing cashmere to the hearts of people everywhere. The "original sweater girl," Lana Turner, created a phenomenon when she wore a tight cashmere sweater in a 1937 film called, *They Won't Forget.*

Cashmere sweaters of all description soon became haute coutre; evening sweaters with heavily encrusted jewels and embroidery became popular during the 1940s, and the famed *sweater set* of best-dressed college coeds ruled the 50s. Avid collectors are now frantic to snatch up those fine examples of vintage cashmere sweaters.

Woven garments made of cashmere must be dry cleaned, but knitted articles may be hand washed. Home weavers and knitters cherish cashmere for its soft hand and practical warmth; cashmere wool is available for home projects at yarn shops or online via Internet craft and knitting sites.

The quality and feel of cashmere will leave you longing for more. Owning a garment made of cashmere is a fashion treat to be truly treasured—after all, it takes one little goat four years to produce enough wool to make just one cashmere sweater.

CHIENGORA

Chiengora is yarn spun from dog hair. The word is a combination of the french word for dog "chien", and angora. Chiengora yarns can be as varied as the many types of dogs. Usually, they are very soft, light, and warm. Chiengora can be spun in a variety of weights from laceweight through worsted.

Chiengora products are a treasure. Not only are they beautiful, one-of-a-kind, custom art works, they are very personal to their owners. The fibre used is brush out from their own sweet pets.

LLAMA

Llamas also have a fine undercoat which can be used for handicrafts and garments. The coarser outer guard hair is used for rugs, wall-hangings and lead ropes. The fibre comes in many different colours ranging from white, grey, redish brown, brown, dark brown and black.

The individual shafts of the wool can be measured in micrometres. 1 micrometre = 1/1000 millimetre.

Table. Average diameter of some of the finest, natural fibres

Animal	Fibre diameter(micrometres)
Vicuña	6 – 10
Alpaca (Suri)	10 - 15
Muskox (Qivlut)	11 - 13
Merino	12 - 20
Angora Rabbit	13
Cashmere	15 - 19
Yak Down	15 - 19
Camel Down	16 - 25
Guanaco	16 - 18
Llama (Tapada)	20 - 30

Chinchilla	21
Mohair	25 - 45
Alpaca (Huacaya)	27.7
Llama (Ccara)	30 - 40

MOHAIR

Mohair is a silky textile produced from the hair of the angora goat. It is most often spun into a thread that can be woven, knitted, or crocheted depending upon the application. Mohair is durable, warm, insulating, and light. It also has moisture wicking properties that carry moisture away from the skin of the wearer. The hair of the angora goat has been used to produce textiles for centuries, and the term *mohair* entered English usage from the Arabic *mukhayyar*, referring to a type of woven head cloth.

Despite being very soft, mohair can be irritating to the skin, like many animal products. As a result, many mohair garments are lined with silk or cotton, and mohair is frequently mixed with other fibres for comfort. The fibre has been used in the West since the 16th century, when Charles V first brought angora goats to Europe, although the delicacy of the species made mohair difficult to obtain and prohibitively expensive for many.

During the Second World War, the United States became concerned about wool production, as military uniforms were made from wool. The decision was made to blend wool and mohair, and the government began to heavily subsidize wool and mohair farmers in 1954 for their products in an attempt to keep the supply lines open. As late as 2005, mohair and wool producers could still receive financial assistance from the United States government.

Angora goats originate from Turkey, in the region now called Ankara. Bred for their special wool and milk products, the goats were used extensively throughout the Middle East for centuries before being introduced to the West. Angora goats are very delicate and highly susceptible to infestations of external parasites due to their shaggy coats. Angora goats also have high and very specific nutritional requirements, and they will begin to suffer from health problems if these are not met. Goats are sheared for their mohair one or two times a year, depending upon production requirements, the health of the goats, and other factors.

They also need to be combed regularly, to prevent clumping and the collection of debris. One angora goat can produce 11 to 17 pounds (5 to 8 kilograms) of mohair in one year. Mohair takes and holds dye well, making it a common choice for novelty knitted products.

PASHMINA

Pashmina refers to a type of fine cashmere wool and the textiles made from it. The name comes from *Pashmineh*, made from sajan Persian *pashm* ("wool"). The wool comes from changthangi or pashmina goat, which is a special breed of goat indigenous to high altitudes of the Himalayas. Pashmina shawls are hand spun, woven and embroidered in Kashmir, and made from fine cashmere fibre.

The goat sheds its winter coat every spring. One goat sheds approximately 80-170g (3-6 ounces) of the fibre. To meet the demand, the goats are now commercially reared in the Gobi Desert area in Inner and Outer Mongolia.

The region has identical harsh weather conditions to those of the Himalayan region, and is thereby apt for the goats to grow this inner wool, but also has acres of grazing ground to produce cashmere economically and commercially. During spring (the moulting season) the goats shed this inner wool, which regrows in winter.

The inner wool is collected and spun to produce cashmere. The quality of the cashmere produced in the Gobi Desert is just as high as that produced in the Himalayas, while the costs are less. A softening process is used by manufacturers of 100 per cent pashmina products, which gives the pashmina a soft, almost silken quality. Sometimes Pashmina is a blend of pure pashmina wool and silk. This gives strength and durability to the pashmina. As a general rule though, the higher content of pashmina wool, the more expensive the textile.

VICUÑA FIBRE

The fibre is popular due to its warmth. Its warming properties come from the tiny scales that are on the hollow air-filled fibres. It causes them to interlock and trap insulating air. At the same time, it is finer than any other wool in the world, but since it is sensitive to chemical treatment, the wool is usually left in its natural colour.

However, the vicuña will only produce about one pound of wool a year and gathering it requires a certain process. During the time of the Incas, vicuña fibres were gathered by means of communal efforts called chacu, in which multitudes of people herded hundreds of thousands of vicuña into previously laid funnel traps.

The animals were sheared and then released; this was only done once every four years. The vicuña was believed to be the reincarnation of a beautiful young maiden who received a coat of pure gold once she consented to the advances of an old, ugly king. Because of this, it was against the law for anyone to kill a vicuña or wear its fleece, except for Inca royalty. At present, the Peruvian government has a labeling system that identifies all garments that have been created through a government sanctioned chacu.

This guarantees that the animal was captured, sheared alive, returned to the wild, and cannot be sheared again for another two years. The programme also ensures that a large portion of the profits return to the villagers. However, annually up to 50,000 pounds of vicuña wool are exported as a result of illegal activities.

Because of this, some countries have banned the importation of the fibre to save the animal. And although it is possible to commercially produce wool from domesticated vicuñas, it is difficult because they tend to escape. Current prices for vicuña yarns and fabrics can range from $1,800 to $3,000 per yard. Vicuña fibre can be used for apparel (such as socks, sweaters, accessories, shawls, coats, and suits) and home fashion (such as blankets and throws). A scarf costs around $1500 while a man's coat can cost up to $20,000.

WOOL

Wool is the dense, warm coat of sheep, also called a fleece. The hair of sheep has many unique properties that make it well suited to textile production, something humans realised approximately 8000 BCE, when sheep first began to be domesticated. Wool is used in a variety of textiles and can be found woven or knitted.

Wool is highly flame resistant, and frequently used for mattresses and rugs for that reason. It is also highly durable, able to stretch up to 50 per cent when wet and 30 per cent when dry. In addition, wool has excellent moisture wicking properties, pulling moisture into the core of the fibre so that it doesn't feel wet or soggy to the wearer. Wool pulls moisture away from the skin, as well, and is worn by people in a wide variety of situations who prefer the feeling of dry air next to the skin to the clammy sense of perspiration.

Wool is favoured for textile production because it is easy to work with and takes dye very well. The springy fibres remember shapes when well cared for. Furthermore, wool takes to felting, a process in which fibres interlock into a tight mat, very well. Felt is used as insulation, for arts and crafts projects, and for decorative accents.

Wool production starts with the shearing of the sheep, which usually happens once a year. A skillful shearer can remove the entire fleece at once with long flowing strokes, keeping the fibres long. After shearing, the wool is washed to remove impurities. One of these impurities is lanolin, which is used in many cosmetics.

After washing, many producers combine wools for a specific blend and dye them together, so that the dye will take evenly, before carding the wool through a set of teethed rollers. Carding pulls the fibres straight, while removing any remaining dirt or vegetable matter from the wool. The wool is pulled into slivers, long strips of fibres loosely pulled together and running in the same direction. If, after carding, the wool is under three inches (almost eight centimeters), it is twisted into rovings, rope like strands that can be spun into wool for knitting. If the strands are longer, they must be combed and drawn before spinning.

Spinning pulls the fibres tightly together and twists them so that they retain a long yarn shape. Yarn can be spun in all sorts of thicknesses and gauges, depending upon the intended use. For wool weaving, the yarn tends to be very fine. For knitting, it may be quite chunky. There are myriad uses for wool, a highly versatile textile. For example, shorter, coarse wool will turn into carpeting, while medium length wool will be used in suits.

After knitting or weaving, the wool is often shrunk through a controlled process so that it won't shrink excessively for the end user. Most knits are also blocked on forms to set a shape, and many wool products are also brushed for a specific finish.

Proper care for wool begins with following the label directions. In general, wool should be allowed to rest between wearings, to retain its shape. It should never be compressed or stored on hangers, which will stretch it. Brushing wool will remove surface soil and stains before they are ground in, and a slightly damp cloth will remove deeper stains. Wool should be dried flat at room temperature, not exposed directly to heat.

SPIDER SILK

Spider silk is one of the 7 great wonders of the animal kingdom. That a small animal, often less than a millimetre across can make a substance that we humans with all our technology are unable to reproduce, a substance that is tough, stronger and more flexible than anything else we can make is surely a humble reminder of the fact that nature created us and not the other way around. One of the most astounding thing about spiders is their ability to make silk, their webs amaze and fascinate us, but they are also used by us in some cultures. All spiders possess spinning glands and make silk, they use it for a safety dragline, to fly, for making a highly durable cocoon for their eggs, to line their homes, to trap their prey and to immobilise their victims.

In some species males use silk to ritually immobilise the females before mating, and all male spiders make a special sperm-web or sperm-line to allow them to transfer the sperm from their genitals to their copulatory organs. Not all species use silk for all these reasons, but all species make egg cocoons. Amongst the most amazing and commonly admired webs are the various orb-webs.

The largest orb-webs in the world are spun spiders in the genus *Nephila* which may be 2 metres (6 ft) in diametre and are capable of catching small birds and bats. However the largest webs of all are built by communal spiders. *Ixeuticus socialis* in Australia builds webs that may be 1.2 metres wide and 3.7 metres long (12 ft by 4 ft) and is claimed by some authorities to build the largest nest.

However G. F. Masterman, once the British Ambassador to Paraguay, described social spiders there building webs 9 metres (30 ft) long and 2.4 metres (8 ft) wide in his book *Seven Eventful Years in Paraguay* (1888). Whatever the truth here, it is a fact that there are billions of kilometres of spider silk spread across the globe and it is all amazing.

The Nature of Silk

Spider silk is both light and strong. Hillyard "A typical strand of garden spider silk has a diametre of about 0.003 mm (0.00012 in) in diametre, compare this with silkworm silk which is 0.03 mm in diametre" or ten times as thick.

Spiders can produce different kinds of silks, at least 7 or 8 different kinds. However generally individual spider species use 5 to 6 types or less, in other words not all spiders produce all the possible types of silk. When assessing the strengths of silk scientists usually refer to dragline silk. It is also generally accepted that spiders in the genus *Nephila* produce the strongest draglines, however this may reflect their large size. Knowing this we should accept that any measurements we come across on the strength of spider silk will not apply to all the silks produced by one spider, and definitely not to the silk of all spiders. The data we do have applies to the toughest spider silk we know of, just how much variation there is between different silks and species one have not been able to find out.

Never-the -less the strength and resilience of spider silk has amazed people for thousands of years, reading around, on and off the web, we get a lot of facts about the strength of spider silk that sometimes seem contradictory, although they all agree that spider silk is way stronger than any other silk or natural material. Here are some quotes.

If all that is a little confusing, is it as strong as steel, stronger than steel or less strong, is it stronger than Kevlar or not, perhaps we can find some facts somewhere. The truth is it depends on how you measure it and what condition the silk is in. Silk absorbs moisture (a property given it by the large amount of the amino acid Alanine it contains).

The more water silk contains the less brittle and the more elastic it is. The range is from about 30 per cent to 300 per cent elasticity depending on the amount of water it contains. Thus we can see that changing the humidity of the environment the silk is spun in, and the length of time it exists in that environment before it is tested will effect the results of the tests. With all these things considered we learn that the tensile strength of spider silk under normal laboratory conditions is slightly less than steel if you compare it in terms of the threads diametre, but it is far greater than that of steel if you compare it in terms of the weight of the thread.

Obviously spider silk is lighter per unit area than steel. When we make the comparison with Kevlar we find that it is three times harder to break than spider silk, however spider silk is five times as elastic. Ok enough about the strength of spider silk, where does it come from and what is it. It is an unusual substance in that it is quite acidic and is not attacked by bacteria or fungi, which is why cobwebs remain around for so long. Silk is created as a proteinaceous liquid within special silk glands within the spider's abdomen. The proteins have a molecular weight of 30,000 daltons (the units that the weights of atoms and molecules are measured in). However when the silk is released from the spiders spinnerets it instantly solidifies and changes into a much larger molecule with a weight of between 200,000 and 300,000 daltons. Quite how this happens is not yet understood.

The proteins that make up dragline silk are secreted in long, sac-like glands, from which they are funnelled as a watery solution into a long, looping duct. This duct leads to a small spigot or nozzle on the end of a spinneret. There are tens or even hundreds of these spigots, and therefore of their accompanying glands, for each spinneret, and most spiders have 6 (3 pairs) of spinnerets.

Examining the complex structure of silk we see that its proteins come in two different forms called *á*-chains and *â*-pleated sheets. The *â*-pleated sheets are about 10 nm by 15 nm and are interspersed in a matrix of the *á*-chains. Spiders possess different glands that produce different silks, scientists distinguish the following glands and silks. Not all spiders have all these glands, in fact the last two are only found in orb-web weavers.

Table. Of Spider Silk Glands

Gland Name	Spinnerets Used	Type of Silk
Piriform	Anterior	Disk Attachment
Ampullate	Anterior/Median	Dragline and Web Frame
Aciniform	Median/Posterior	Wrapping silk, Sperm-web, Egg-cocoon (outer wall)
Tubuliform	Median/Posterior	Egg-cocoon
Aggregate	Posterior	Spiral parts of Sticky-web
Flagelliform	Posterior	Axial thread of Sticky-web

As if this wasn't enough some spiders have an additional silk spinning organ called a cribellum. The cribellum is situated in front of the three pairs of spinnerets and consists of between one and four plates covered in spigots.

The spider *Stegodyphus pacificus* has 40,000 spigots on its cribellum, but other species have much fewer. Spiders that possess a cribellum are called 'Cribellate', as compared with the remaining spiders which are sometimes referred to as 'Ecribellate'.

The silk is combed out of the spigots with a set of special hairs located on the 4th metatarsi called a 'Calamistrum'. The silk produced by the cribellum is especially fine and is used to help entangle prey.

Spiders can control the thickness of a strand of silk using muscles and valves in the necks of the spigots, larger spiders for instance need thicker draglines. Most strands of spider web are only a few micrometres in diametre, for instance the dragline of an adult orb-web has a diametre of about 2μm. In comparison with this hackle band silk is normally only 0.01μm to 0.02μm in diametre.

How Spiders Use Silk

So now that we know something of what silk really is, lets have a look at how spiders actually use it. Naturally enough different spiders use silk in different ways. However there are three nearly universal usages. The first is the dragline, all spiders use a dragline at sometime in their life.

In small species it is also safety line that will catch them if they jump into space in order to avoid a predator. Many spiders do not use silk to make a catching web, but do use it to make themselves a nice cosy cocoon or cell to live in when they aren't out hunting.

Such a neatly made hideaway helps them avoid predators and offers some protection from both excessive moisture and lack of moisture. The other two uses are gender specific, males spin a sperm web before mating to allow them to charge their pedipalps and females spin a web cocoon around their eggs, often it is composed of more than one kind of silk, you can read more about this in the page on reproduction as well. An unusual use of silk is flying, when small spiderlings stand head down and abdomen up to release a strand of silk which is caught by the breeze and lifted up, as the silk strand lengthens it eventually creates enough lift to raise the spiderling and it turns and grabs the thread and away it goes. Flying like this is a risky business as there is no way the spider can control where it ends, many must die as a result of landing in inappropriate places such as rivers and seas. However for those that land somewhere more accessible a new life is offered far away from all those brothers and sisters. Spiders can travel huge distances like this and their ability to live without food and water for long periods of time undoubtedly help.

In May 1984 the first scientist to visit the island of Krakatoa, which had exploded violently 9 months earlier exterminating all life on the island, reported that the only, and there fore the first, animal he found on the planet was a spider. Krakatoa is 37 km (25 miles) from the nearest other land. However spiders fly far further than this, scientific research has recorded spiders over 4.500 metres (14,000 ft) above sea level and also in the middle of the ocean 1,500 km (1,000 miles) from the nearest land. In late summer over grasslands in temperate climates you can experience huge numbers of these aeronautic spiders, the silk they deposit across the landscape as they descend towards evenings as the air cools and descends can also become very noticeable if you are out and about.

This is commonly called gossamer. In the past huge falls of gossamer were well known, but while they can be experienced anywhere in the world on warm days they appear to be less intense these days. In France gossamer is known as *fils de la Vierge* and in Germany as *Marienfaden*.

SILK

Silk is a filament fibre formed from proteins secreted by *Bombyx mori*, or silkworms. Silkworms are not actually worms, but caterpillars, despite their common name. Humans have practiced silk production, which originated in China, for thousands of years.

Highly prized for its softness, insulating properties, and strength, silk is a natural animal product and therefore quite expensive. Making silk requires monitoring and feeding the silkworms constantly, and a great deal of effort results in a surprisingly small amount of thread.

China managed to keep the secret of silk for thousands of years, exporting the rare textile to Europe over trade routes. Eventually, silkworm eggs were smuggled out, and in the 13th century, Western production of silk began in Italy. This by no means brought the cost down, as the extensive amount of work required to make silk remained the same. Thirty thousand eggs can end up eating one ton of mulberry leaves and producing 12 pounds (5.5 kilograms) of silk.

Silk manufacture begins when female silkworms lay their eggs. A single silkworm may produce hundreds of tiny eggs, which are incubated until they hatch into larva. The larva must be kept warm and fed on mulberry leaves frequently throughout their brief lives. In four to six weeks, the larva have reached their maximum size and have mustered enough energy to pupate.

It is the pupation stage that produces silk, as the larva attach to branches and spin a shell around themselves. The shell is unique to insect pupation, formed by two spinnerets on the larva. The fluid they secrete is high in protein and forms a continuous thread, which is repeatedly wound to form a pod to mature in. Unfortunately for the silkworm pupa, it is maturity that silk producers want to avoid. A small number of the larva are allowed to gestate into adults to carry on the lineage, while the rest are subjected to heat to kill the larva before they can begin to eat through the valuable silk thread. The pods are dipped into hot water to loosen the thread, which is then wound onto wheels. The dead larva are discarded. Once raw silk has been wound onto wheels, it can be spun into a variety of different types of thread, depending upon the intended use. Crepe is made by twisting multiple strands of silk together in different directions, while tram is made by twisting one to two threads in the same direction. Organzine is formed by twisting multiple threads together, switching direction, and repeating the process until the end of the thread.

Most wearers are familiar with all forms of silk thread, with single threads being used for fine and sheer garments and crepe being used to create textured and wrinkly silks. Organzine is used for warp threads in weaving, and tram creates the weft, or filling. Silk can also be used in knit fabrics. Silk can be treated to remove roughness or left raw, depending on the demand.

Silk takes dye well and is available in a dazzling array of colours from subtle to bright. The textile appears in scarves, sweaters, underwear, shirts, and everything in between. Despite its low yield, production of the sought after textile is rising globally. Many consumers prefer silk for its comfort, insulating qualities, look, and feel.

QIVIUT

Qiviut is an extremely soft fibre harvested from the coats of musk oxen, arctic mammals named for the distinctive musky odour carried by males. Qiviut is prized by knitters, because unlike wool, it does not shrink when exposed to water and temperature changes.

In addition, qiviut is very light, strong, insulating, and warm. Pure qiviut can be used directly next to the skin because of its extreme softness, although the fibre does not hold a knitted shape well. As a result, qiviut is often blended with cashmere or like wools to create sweaters and other similar garments.

A single musk ox will shed between five to seven pounds (two to three kilograms) of qiviut, which is the soft under layer of wool, in a year. Musk oxen can exceed 440 pounds (200 kilograms) and stand approximately five feet (4.7 meters) at the shoulder. They are bulky, sturdy beasts and can be found roaming the arctic in herds ranging in size between ten and 200. Musk oxen can also provide meat to enterprising hunters, although they are not used for milk, as most are not domesticated. Musk oxen have existed since the Pleistocene era and can be found in Alaska, Russia, Sweden, and Norway.

Many musk oxen are wild, and their qiviut is collected in the spring as they shed it in large sheets. Some farms raise musk oxen, and the qiviut is collected by combing in the spring as the oxen shed. Qiviut is never collected by shearing, as the coats of the musk oxen take a long time to grow back, and shearing would expose them to the winter elements.

Several collectives have been established in arctic regions to steward musk oxen and collect their qiviut. Musk oxen tend not to do well in more southern areas, because they have adapted to the extreme weather conditions in the arctic, and therefore, qiviut is generally sourced from the arctic. Qiviut has been used by the Inuit and other arctic peoples for centuries, because of its insulating properties, warmth, and versatility. It is often worn as an under layer, because it keeps the wearer warm while wicking moisture away from the body.

Qiviut makes light, floating scarves and shawls, and it is often spun very fine to make lacy yet insulating over garments. Integrated with other fibres, it's knitted into sweaters, hats, and socks as well. Many collectives that harvest raw qiviut also sell knitted qiviut goods made from their product. Most qiviut collectives have been established to support native peoples, and the majority of their earnings are returned to the village of origin for improvements.

VEGETABLES FIBRES

ABACÁ

Abaca is a relative of the banana plant that grows in hot, humid climates. One of the main producers of abaca is the Philippines, which is why the plant has been mistakenly called Manila hemp. Unlike banana plants, however, which are valued for their fruit, abaca's primary benefit is in its long leaf sheaths, which grow to be 12 to 20 feet high. All parts of the stalk—from the outer dark layer to the inner most layers—can be extracted, stripped and processed.

Abaca fibre has been used for centuries to make strong, breathable textiles that are comfortable to wear and long lasting. Abaca is popular for clothing, hats, shoes and slippers. In the Philippines there are at least 150 different traditional weaves. Because of the fiber's tensile strength, abaca clothing has been embroidered, hand-painted, dyed and beaded without any loss of luster and shape.

BAMBOO

Bamboo fibre resembles cotton in its unspun form, a puffball of light, airy fibres. Many companies use extensive bleaching processes to turn bamboo fibre white, although companies producing organic bamboo fabric leave the bamboo fibre unbleached. To make bamboo fibre, bamboo is heavily pulped until it separates into thin component threads of fibre, which can be spun and dyed for weaving into cloth.

Bamboo fabric is very soft and can be worn directly next to the skin. Many people who experience allergic reactions to other natural fibres, such as wool or hemp, do not complain of this issue with bamboo. The fibre is naturally smooth and round without chemical treatment, meaning that there are no sharp spurs to irritate the skin.

Bamboo fabric is favoured by companies trying to use sustainable textiles, because the bamboo plant is very quick growing and does not usually require the use of pesticides and herbicides to thrive. As a result, plantations can easily be kept organic and replanted yearly to replenish stocks. The process of making unbleached bamboo fibre is very light on chemicals that could potentially harm the environment.

In textile form, bamboo retains many of the properties it has as a plant. Bamboo is highly water absorbent, able to take up three times its weight in water. In bamboo fabric, this translates to an excellent wicking ability that will pull moisture away from the skin so that it can evaporate. For this reason, clothing made of bamboo fibre is often worn next to the skin. Bamboo also has many antibacterial qualities, which bamboo fabric is apparently able to retain, even through multiple washings.

This helps to reduce bacteria that thrive on clothing and cause unpleasant odours. It can also kill odour causing bacteria that live on human skin, making the wearer and his or her clothing smell more sweet.

In addition, bamboo fabric has insulating properties and will keep the wearer cooler in summer and warmer in winter. The versatility of bamboo fabric makes it an excellent choice for clothing designers exploring alternative textiles, and in addition, the fabric is able to take bright dye colours well, drape smoothly, and star in a variety of roles from knit shirts to woven skirts.

COIR

Coir is a product of the coconut tree, *Cocos nucifera*, and it is sometimes known as coco fibre. The substance is extracted from the hairy husk of coconuts, and used to create a variety of products such as mats, carpets, upholstery stuffing, and brushes.

Coir, pronounced KOY-er, is a very coarse, stiff fibre, and it is also extremely resistant to rot and salt water, making it an ideal material for situations in which other fibres would decay. The word comes from a Malayalam word, *kayar*, which is derived from *kayaru*, which means "to be twisted." The Malayalam language is spoken in Southern India, particularly in the state of Kerala.

The language is related to Tamil, Kota, and Tulu, among many other languages spoken in that region. For English speakers interested in trivia, "Malayalam" is among the longest palindromes in the English language, meaning that it reads the same backwards and forwards. India and Sri Lanka are the two biggest exporters of coir, accounting for most of the world's supply of the substance. To process coir, coconuts are split so that the stiff fibres are accessible. The outer husk is soaked to separate the fibres, which are sorted out into long fibres suitable for use as brush bristles, and shorter fibres which are used to make things like the padding inside inner coil mattresses. After soaking, the fibres are cleaned and sorted into hanks which may later be spun into twine, matted into padding, or used as individual bristles. Coir takes dye well, and many producers dye coir fibre before export.

There are two primary types of coir. The first type, known as white fibre, is extracted from young coconuts. The white fibre is somewhat finer and more pale in colour than coir from mature coconuts. This type of coir is often woven into yarn which is used to make mats, sacking, rope, and twine. In rope manufacturing, white coir fibre can sometimes be cheaper than other rope ingredients. Brown fibre, the other type of coir, is from mature coconuts, and it tends to be more stiff and unyielding.

Many Westerners are familiar with coir in the form of stiff doormats, which can be left out in all weathers because of coir's rot resistance. Coir matting can also be found inside some upholstered products, and a close inspection of an inner spring mattress may also yield coir matting, depending on when and where the mattress was made.

Coir matting is also sometimes made available in the form of a substrate for plants, since it can be impregnated with water and seeds to sprout flowers and groundcover. The matting will keep weeds back while the plants establish themselves.

COTTON

Cotton is a natural fibre harvested from the cotton plant. Cotton is one of the oldest fibres under human cultivation, with traces of cotton over 7,000 years old recovered from archaeological sites. Cotton is also one of the most used natural fibres in existence today, with consumers from all classes and nations wearing and using cotton in a variety of applications. Thousands of acres globally are devoted to the production of cotton, whether it be new world cotton, with longer, smoother fibres, or the shorter and coarser old world varieties. Cotton is in the mallow family and produces delicate, lovely flowers. Other members of the mallow family include hollyhocks and hibiscus, used to brighten gardens all over the world.

The cotton fibre forms around the seeds of the cotton plant and is designed to help carry the seeds long distances on the wind so that the plant can distribute itself. Early humans realised that the soft, fluffy fibres might be suitable for textile use and began to breed the plant, selecting for fluffy, easily spun varieties.

After harvesting, cotton must be combed to remove the seeds. This used to be a laborious process until the invention of the cotton gin, which quickly separates the seeds from the fibre and combs them for spinning. While a single cotton fibre is not terribly strong, when multiple curling fibres are straightened and twisted together, they form a strong, smooth thread that can be knitted or woven, as well as dyed.

Cotton is somewhat flammable, especially lighter cottons that hold a lot of air. Some cotton is chemically treated to reduce flammability. Many cottons are also blended with other natural fibres, such as linen, for particular properties, or to add texture and strength to the fibre.

Cotton can be woven or knitted. It can also be turned into flannel, corduroy, muslin, and a variety of other fabrics used so universally that the American Cotton Council uses "the fabric of our lives" as a tag line.

Cotton also carries environmental controversy, particularly in the developing world, where dangerous pesticides are heavily employed. Cotton is subject to infestation, and therefore many growers heavily douse the plant in pesticides that are harmful to human and animal health, as well as herbicides to eliminate competition for resources.

A number of producers also genetically modify the plant, which many outside the industry view as a questionable practice. Cotton also has very large water requirements, which may place stress on nations with limited water resources. In the late 20th century, there was a push for organic, sustainable cotton grown and harvested without the use of pesticides and human exploitation. This cotton is significantly more expensive than conventionally farmed cotton, however, and may not be practical for most consumers.

FLAX

Flax is a member of the genus *Linum* in the family Linaceae. It is native to the region extending from the eastern Mediterranean to India and was probably first domesticated in the Fertile Crescent. This is called as Agasi/Akshi in Kannada, Jawas/Javas or Alashi in Marathi. Flax was extensively cultivated in ancient Ethiopia and ancient Egypt. In a prehistoric cave in the Republic of Georgia dyed flax fibres have been found that date to 34,000 BC. New Zealand flax is not related to flax, but was named after it as both plants are used to produce fibres.

Flax is an erect annual plant growing to 1.2 m (3 ft 11 in) tall, with slender stems. The leaves are glaucous green, slender lanceolate, 20–40 mm long and 3 mm broad. The flowers are pure pale blue, 15–25 mm diameter, with five petals; they can also be bright red. The fruit is a round, dry capsule 5–9 mm diameter, containing several glossy brown seeds shaped like an apple pip, 4–7 mm long. In addition to referring to the plant itself, the word "flax" may refer to the unspun fibres of the flax plant.

FLAX FIBRES

Flax fibres are amongst the oldest fibre crops in the world. The use of flax for the production of linen goes back at least to ancient Egyptian times. Dyed flax fibres found in a cave in Dzudzuana (prehistoric Georgia) have been dated to 30,000 years ago. Pictures on tombs and temple walls at Thebes depict flowering flax plants.

The use of flax fibre in the manufacturing of cloth in northern Europe dates back to Neo-lithic times. In North America, flax was introduced by the Puritans. Currently most flax produced in the USA and Canada are seed flax types for the production of linseed oil or flax seeds for human nutrition. Flax fibre is extracted from the bast or skin of the stem of the flax plant. Flax fibre is soft, lustrous and flexible; bundles of fibre have the appearance of blonde hair, hence the description "flaxen".

It is stronger than cotton fibre but less elastic. The best grades are used for linen fabrics such as damasks, lace and sheeting. Coarser grades are used for the manufacturing of twine and rope. Flax fibre is also a raw material for the high-quality paper industry for the use of printed banknotes and rolling paper for cigarettes. Flax mills for spinning flaxen yarn were invented by John Kendrew and Thomas Porthouse of Darlington in 1787.

HEMP

Hemp is a natural fibre product of the *Cannabis sativa* plant. Astute readers may be aware of other byproducts of this plant, but hemp is produced from a type of *Cannabis sativa* specifically bred to yield long fibres. Cultivation of hemp for industrial purposes has been undertaken for thousands of years, and hemp was used to manufacture rope, canvas, paper, and clothing until alternative textiles for these purposes were discovered.

Traditionally, hemp has been a very coarse fibre, which made it well suited to rope but less than ideal for clothing designed to be worn against delicate human skin. Advances in breeding of the plants and treatment of the fibres have resulted in a much finer, softer fibre, which is ideal for weaving into clothing.

While hemp clothing in the late 20th century came to be associated with fringe movements, it was once widely utilized as a textile: the word *canvas*, for example, is related to *Cannabis*, one of the original components of canvas. As of 2006, it was still very difficult to grow industrial hemp in the United States, due to the plant's confusion with marijuana. Though the two plants are members of the same species, they have been bred to achieve different ends, and industrial hemp does not contain enough tetrahydro-cannabinol to make it a psychoactive substance. The growth of industrial hemp in the United States is heavily regulated, although the neighbouring nation of Canada grows commercial amounts of the plant product.

In addition to providing useful fibres, hemp seed also has high nutritional value, and the plant can be used to make biodegradable plastics, some fuels, and a variety of other things. While hemp is unlikely to save the world, as many proponents are fond of saying, it is an underutilized vegetable resource.

Hemp is rich in healthy fats and some vitamins, depending on how it is grown. As a result, it is frequently used in skin salves and balms, as well as in nutritional supplements. Hemp clothing tends to be strong, insulating, absorbent, and durable.

This durability makes it well suited to garments that will see hard wear, because hemp fibres can last up to three times longer than cotton fibres. Most frequently, hemp clothing is woven, although the fibres tend to form chunkier threads than other natural textile components like cotton. Hemp can also be used in knits.

Untreated hemp fibre is pale blonde in colour and takes dye well. Many hemp textile products are coloured with plant dyes, which gives hemp an undeserved reputation for being dull in colour. In fact, hemp can be dyed as vividly as other textiles like cotton.

JUTE

Jute is a type of plant fibre used to make common items such as rope, twine, chair coverings, curtains, sacks, hessian cloth, carpets, and even the backing used on linoleum.

This is accomplished by spinning the fibre into a coarse thread. Despite the fact that jute tends to be rough in texture, fine threads of it are sometimes used to create imitation silk. In addition, jute is increasingly being looked at as an alternative source for making paper, rather than cutting down trees for pulp.

The thread created from jute is quite strong, yet it is among the cheapest of natural fibres available. It also has exceptional insulating properties, low thermal conductivity, and antistatic characteristics. Non-etheless, synthetic materials are replacing jute in many applications, because they are still less costly to create and more efficient to use.

This is partly because jute has a tendency to become brittle and to yellow in sunlight. It also tends to lose its strength when wet and can become infested with microbes when used in humid regions. There are several applications for which jute is still used instead of synthetic fibres. These applications are mostly limited to those that require the use of a material capable of biodegrading.

Pots for plants that are planted directly into the ground with the plant, for example, are often made of jute. Jute cloth is also used in landscaping projects, in order to prevent erosion while still permitting natural vegetation to grow. Jute is also considered to be a possible alternative to wood. This is because its stem contains a woody inner core. Taking just four to six months to grow to maturity, jute can be harvested much more quickly than trees. Many hope to be able to use jute in order to slow down or prevent deforestation.

The majority of the jute used today is grown in the Ganges delta. This is because the plant prefers climates that are both warm and humid, with temperature ranging from 68 to 104°F (20 to 40°C) and a relative humidity of 70-80 per cent. It also requires about two to three inches (5 to 8 cm) of rainfall per week. China is the next largest producer of jute.

KAPOK

Kapok is a tropical tree of the order *Malvales* and the family *Malvaceae*, native to Mexico, Central America and the Caribbean, northern South America, and to tropical west Africa.

The word is also used for the fibre obtained from its seed pods. The tree is also known as the Java cotton, Java kapok, or ceiba. It is a sacred symbol in Maya mythology. The tree grows to 60-70 m (200-230 ft) tall and has a very substantial trunk up to 3 m (10 ft) in diameter with buttresses. The trunk and many of the larger branches are often (but not always) crowded with very large, robust simple thorns.

The leaves are compound of 5 to 9 leaflets, each up to 20 cm (8 in) and palm like. Adult trees produce several hundred 15 cm (6 in) seed pods. The pods contain seeds surrounded by a fluffy, yellowish fibre that is a mix of lignin and cellulose.

KENAF

Kenaf, *Hibiscus cannabinus*, is a plant in the Malvaceae family. *Hibiscus cannabinus* is in the genus *Hibiscus* and is probably native to southern Asia, though its exact natural origin is unknown. The name also applies to the fibre obtained from this plant. Kenaf is one of the allied fibres of jute and shows similar characteristics. Other names include Bimli, Ambary, Ambari Hemp, Deccan Hemp, and Bimlipatum Jute.

It is an annual or biennial herbaceous plant (rarely a short-lived perennial) growing to 1.5-3.5 m tall with a woody base. The stems are 1–2 cm diameter, often but not always branched. The leaves are 10–15 cm long, variable in shape, with leaves near the base of the stems being deeply lobed with 3-7 lobes, while leaves near the top of the stem are shallowly lobed or unlobed lanceolate. The flowers are 8–15 cm diameter, white, yellow, or purple; when white or yellow, the centre is still dark purple. The fruit is a capsule 2 cm diameter, containing several seeds.

PIÑA

Piña is a fibre made from the leaves of a pineapple and is commonly used in the Philippines. It is sometimes combined with silk or polyester to create a textile fabric. The end fabric is lightweight, easy to care for and has an elegant appearance similar to linen. Pina is also related to a drink from Puerto Rico known as Pina Colada. Piña comes from the leaves of the pineapple plant. "Each strand of the hand scraped Piña fibre is knotted one by one to form a continuous filament for hand weaving into the Piña cloth". The piña fibre is softer, and has a high luster, and is usually white or ivory in colour. It is also called Pina in Spanish.

RAFFIA PALM

The Raffia palms (*Raphia*) are a genus of twenty species of palms native to tropical regions of Africa, especially Madagascar, with one species (*R. taedigera*) also occurring in Central and South America.

They grow up to 16 m tall and are remarkable for their compound pinnate leaves, the longest in the plant kingdom; leaves of *R. regalis* up to 25.11 m long and 3 m wide are known. The plants are either monocarpic, flowering once and then dying after the seeds are mature, or hapaxanthic, with individual stems dying after fruiting but the root system remaining alive and sending up new stems.

RAMIE

Ramie is a flowering plant which is native to Asia. It is harvested and processed to yield strong fibres, also called ramie, which are used in the production of textiles, twine, upholstery, filters, and sacking. Like flax, jute, and hemp, ramie is considered a bast fibre crop, meaning that the usable portion of the plant is found in its connective tissue structures. The plant is widely cultivated in several Asian nations, which export ramie around the world.

The scientific name for the plant is *Boehmeria nivea*, and it is also sometimes called Chinese Grass. The plant grows in the form of stalks with heart shaped leaves which sprout up from an extensive underground root system. Ramie is in the nettle family, and it has the characteristic small silvery hairs associated with nettles, although the hairs do not sting.

The ramie stalk can be harvested up to six times each year in favourable cultivating conditions, although three to four crops of ramie annually are much more common. The plants must be extensively processed to yield ramie fibre. A series of beatings, washings, and chemical treatments extracts the usable part of the plant and de-gums the fibre so that it will be usable. Once processed, ramie can be spun into thread or yarn, and it is sometimes also blended with other textile materials to make it more versatile.

Pure ramie is very strong, resistant to mold and bacteria, lustrous, and it holds its shape very well. However, ramie is also stiff, not terribly elastic, and sometimes difficult to work with because it can be very brittle. In addition, ramie does not take dye very well.

All of these shortcomings make ramie more expensive than similar plant fibres, such as linen. Although the material has been used in the production of textiles for thousands of years, many producers prefer to produce it in blends rather than using it plain.

Like linen and other textiles made from woody fibre, ramie requires special care. Ideally, the fabric should be hand washed cold and not wrung or heavily pressed before being laid flat to dry. Once dried, ramie should be stored flat, and it should not be sharply creased or folded. In some cases, ramie can also be dry cleaned or machine washed on a gentle setting, and the care directions on the fabric should always be carefully followed to avoid damaging it.

SISAL

Sisal (*Agave sisalana*) is an agave that yields a stiff fibre traditionally used in making twine, rope and also dartboards. The term may refer either to the plant or the fibre, depending on context. It is sometimes incorrectly referred to as *sisal hemp* because hemp was for centuries a major source for fibre, so other fibres were sometimes named after it.

The plant's origin is uncertain; while traditionally it was deemed to be a native of Yucatan, there are no records of botanical collections from there. Gentry hypothesized a Chiapas origin, on the strength of traditional local usage. In the 19th century, sisal cultivation spread to Florida, the Caribbean islands and Brazil, as well as to countries in Africa, notably Tanzania and Kenya, and Asia. The first commercial plantings in Brazil were made in the late 1930s and the first sisal firer exports from there were made in 1948. It was not until the 1960s that Brazilian production accelerated and the first of many spinning mills was established. Today Brazil is the major world producer of sisal.

There are both positive and negative environmental impacts from sisal growing. Traditionally used for rope and twine, sisal has many uses, including paper, cloth, wall coverings and carpets.

WOOD FIBRE

Wood fibres are usually cellulosic elements that are extracted from trees, straw, bamboo, cotton seed, hemp, sugarcane and other sources. The end paper product (paper, paperboard, tissue, cardboard and etc.) dictates the species, or species blend, that is best suited to provide the desirable sheet characteristics, and also dictates the required fibre processing (chemical treatment, heat treatment, mechanical 'brushing' or refining etc.).

In North America, virgin (non-recycled) wood fibre is primarily extracted from hardwood (deciduous) trees and softwood (coniferous) trees. Wood fibres can also be recycled from used paper materials. Wood fibres are treated by combining them with other additives. They are then processed into a network of wood fibres, which constitutes the sheet of paper.

MINERAL FIBRES

MINERAL WOOL

Mineral wool has a unique range of properties combining high thermal resistance with long-term stability. It is made from molten glass, stone or slag that is spun into a fibre-like structure which creates a combination of properties that no other insulation material can match. It has the ability to save energy, minimize pollution, combat noise, reduce the risk of fire and protect life and property in the event of fire.

Mineral wool: Production and properties describes the technological process of mineral wool production and the physical characteristics of the melt and theoretical bases of multiregression and dimensionless theory. This is followed by the introduction of the fibre cooling model in the blow-away flow and the influence of temperature in the melt film (on the rotating centrifuge wheels) on the thickness of forming fibres.

The second part predominantly focuses on the use of computer-aided visualisation: tools for the diagnostics of fibre and primary layer formation. Special attention is given to the study of aerodynamic characteristics of the airflow which significantly influences the quality of the final product.

Mineral wool: Production and properties is suitable for engineers, researchers and for graduate and postgraduate students who want to broaden their knowledge of experimental methods in this field.

ASBESTOS

Asbestos is a mineral fibre that has been used commonly in a variety of building construction materials for insulation and as a fire-retardant. Because of its fibre strength and heat resistant properties, asbestos has been used for a wide range of manufactured goods, mostly in building materials (roofing shingles, ceiling and floor tiles, paper products, and asbestos cement products), friction products (automobile clutch, brake, and transmission parts), heat-resistant fabrics, packaging, gaskets, and coatings. When asbestos-containing materials are damaged or disturbed by repair, remodeling or demolition activities, microscopic fibres become airborne and can be inhaled into the lungs, where they can cause significant health problems.

Most Common Sources of Asbestos Exposure:

- Workplace exposure to people that work in industries that mine, make or use asbestos products and those living near these industries, including:
  - The construction industry (particularly building demolition and renovation activities),
  - The manufacture of asbestos products (such as textiles, friction products, insulation, and other building materials), and
  - During automotive brake and clutch repair work
- Deteriorating, damaged, or disturbed asbestos-containing products such as insulation, fireproofing, acoustical materials, and floor tiles.

Types and Associated Fibres

Six minerals are defined by the United States Environmental Protection Agency as "asbestos" including that belonging to the serpentine class chrysotile and that belonging to the amphibole class amosite, crocidolite, tremolite, anthophyllite and actinolite.

There is an important distinction to be made between serpentine and amphibole asbestos due to differences in their chemical composition and their degree of potency as a health hazard when inhaled .

Serpentine

White

Chrysotile, CAS No. 12001-29-5, is obtained from serpentinite rocks which are common throughout the world. Its idealized chemical formula is $Mg_3(Si_2O_5)(OH)_4$. Chrysotile fibres are curly as opposed to fibres from amosite, crocidolite, tremolite, actinolite, and anthophyllite which are needlelike. Chrysotile, along with other types of asbestos, has been banned in dozens of countries and is only allowed in the United States and Europe in very limited circumstances.

Chrysotile has been used more than any other type and accounts for about 95 per cent of the asbestos found in buildings in America. Applications where chrysotile might be used include the use of joint compound.

It is more flexible than amphibole types of asbestos; it can be spun and woven into fabric. The most common use is within corrugated asbestos cement roof sheets typically used for outbuildings, warehouses and garages. It is also found as flat sheets used for ceilings and sometimes for walls and floors. Numerous other items have been made containing chrysotile including brake linings, cloth behind fuses (for fire protection), pipe insulation, floor tiles, and rope seals for boilers.

Amphibole

Brown

Amosite, CAS No. 12172-73-5, is a trade name for the amphiboles belonging to the *Cummingtonite - Grunerite* solid solution series, commonly from Africa, named as an acronym from Asbestos Mines of South Africa. One formula given for amosite is $Fe_7Si_8O_{22}(OH)_2$. It is found most frequently as a fire retardant in thermal insulation products and ceiling tiles.

Blue

Crocidolite, CAS No. 12001-28-4 is an amphibole found primarily in southern Africa, but also in Australia. It is the fibrous form of the amphibole riebeckite. One formula given for crocidolite is $Na_2Fe^{2+}{}_3Fe^{3+}{}_2Si_8O_{22}(OH)_2$.

Notes: chrysotile commonly occurs as soft friable fibres. Asbestiform amphibole may also occur as soft friable fibres but some varieties such as amosite are commonly straighter. All forms of asbestos are fibrillar in that they are composed of fibres with widths less than 1 micrometer that occur in bundles and have very long lengths. Asbestos with particularly fine fibres is also referred to as "amianthus". Amphiboles such as tremolite have a crystal structure containing strongly bonded ribbonlike silicate anion polymers that extend the length of the crystal. Serpentine (chrysotile) has a sheetlike silicate anion which is curved and which rolls up like a carpet to form the fibre.

Other Materials

Other regulated asbestos minerals, such as tremolite asbestos, CAS No. 77536-68-6, $Ca_2Mg_5Si_8O_{22}(OH)_2$; actinolite asbestos, CAS No. 77536-66-4, $Ca_2(Mg, Fe)_5(Si_8O_{22})(OH)_2$; and anthophyllite asbestos, CAS No. 77536-67-5, $(Mg, Fe)_7Si_8O_{22}(OH)_2$; are less commonly used industrially but can still be found in a variety of construction materials and insulation materials and have been reported in the past to occur in a few consumer products.

Other natural and not currently regulated asbestiform minerals, such as richterite, $Na(CaNa)(Mg, Fe^{++})_5(Si_8O_{22})(OH)_2$, and winchite, $(CaNa)Mg_4(Al, Fe^{3+})(Si_8O_{22})(OH)_2$, may be found as a contaminant in products such as the vermiculite containing zonolite insulation manufactured by W.R. Grace and Company. These minerals are thought to be no less harmful than tremolite, amosite, or crocidolite, but since they are not regulated, they are referred to as "asbestiform" rather than asbestos although may still be related to diseases and hazardous.

BASALT FIBRE

Basalt fibre or fibre is a material made from extremely fine fibres of basalt, which is composed of the minerals plagioclase, pyroxene, and olivine. It is similar to carbon fibre and fiberglass, having better physicomechanical properties than fiberglass, but being significantly cheaper than carbon fibre. It is used as a fireproof textile in the aerospace and automotive industries and can also be used as a composite to produce products such as camera tripods.

The manufacture of basalt fibre requires the melting of the quarried basalt rock to about 1,400 °C (2,550 °F). The molten rock is then extruded through small nozzles to produce continuous filaments of basalt fibre. There are three main manufacturing techniques, which are centrifugal-blowing, centrifugal-multiroll and die-blowing.

The fibres typically have a filament diameter of between 9 and 13 μm which is far enough above the respiratory limit of 5 μm to make basalt fibre a suitable replacement for asbestos. They also have a high elastic modulus, resulting in excellent specific tenacity - three times that of steel.

GLASS WOOL

Glass wool is an insulating material, made from fibre glass, arranged into a texture similar to wool. Glass wool is produced in rolls or in slabs, with different thermal and mechanical properties.After the fusion of a mixture of natural sand and recycled glass at 1,450 °C, the glass that is produced is converted into fibres. The cohesion and mechanical strength of the product is obtained by the presence of a binder that "cements" the fibres together. Ideally, a drop of bonder is placed at each fibre intersection. This fibre mat is then heated to around 200 °C to polymerize the resin and is calendered to give it strength and stability.

The final stage involves cutting the wool and packing it in rolls or panels under very high pressure before palletizing the finished product in order to facilitate transport and storage. Glass wool having better advantages compair to other insulation materials.But it is hazardous due to its duct and very small glass particles which can travel into the human body during inhalation.

Glass wool is a thermal insulation that consists of intertwined and flexible glass fibres, which causes it to "package" air, resulting in a low density that can be varied through compression and binder content. It can be a loose fill material, blown into attics, or, together with an active binder sprayed on the underside of structures, sheets and panels that can be used to insulate flat surfaces such as cavity wall insulation, ceiling tiles, curtain walls as well as ducting. It is also used to insulate piping and for soundproofing.

SYNTHETIC FIBRES

ACRYLIC FIBRE

Acrylic fibres are synthetic fibres made from a polymer (polyacrylonitrile) with an average molecular weight of ~100,000, about 1900 monomer units. To be called acrylic in the U.S, the polymer must contain at least 85 per cent acrylonitrile monomer. Typical comonomers are vinyl acetate or methyl acrylate. The Dupont Corporation created the first acrylic fibres in 1941 and trademarked them under the name "Orlon".The polymer is formed by free-radical polymerization in aqueous suspension.

The fibre is produced by dissolving the polymer in a solvent such as N,N-dimethylformamide or aqueous sodium thiocyanate, metering it through a multi-hole spinnerette and coagulating the resultant filaments in an aqueous solution of the same solvent (wet spinning) or evaporating the solvent in a stream of heated inert gas (dry spinning). Washing, stretching, drying and crimping complete the processing.

Acrylic fibres are produced in a range of deniers, typically from 2 to 15, as cut staple or as a 500,000 to 1 million filament tow. End uses include sweaters, hats, hand-knitting yarns, rugs, awnings, boat covers, and upholstery; the fibre is also used as a precursor for carbon fibre. Production of acrylic fibres is centreed in the Far East, declining in Europe and now shut down (except for precursor) in the U.S. Former U.S. brands of acrylic were Acrilan (Monsanto), Creslan (American Cyanamid), and Orlon (DuPont).

Acrylic is lightweight, soft, and warm, with a wool-like feel. Acrylic is coloured before it is turned into a fibre as it does not dye very well but has excellent colorfastness. Its fibres are not very resilient, and wrinkle easily, but most acrylic fabrics have good wrinkle resistance.

Acrylic has recently been used in clothing as a less expensive alternative to cashmere, due to the similar feeling of the materials. The disadvantages of acrylic are that it tends to fuzz or pill easily and that it does not insulate the wearer as well as wool or cashmere. Many products, like fake pashmina or cashmina, use this fibre to create the illusion of cashmere.

Acrylic is resistant to moths, oils, chemicals, and is very resistant to deterioration from sunlight exposure. However, static and pilling can be a problem. Acrylic has a bad reputation amongst many knitters - however cheap the yarn is, its performance does not come near natural fibres. Also, some knitters complain that the fibre "squeaks" when knitted. Acrylic can irritate the skin of people with eczema.

ARAMID

Aramid fibres are a class of heat-resistant and strong synthetic fibres. They are used in aerospace and military applications, for ballistic rated body armor fabric, in bicycle tires, and as an asbestos substitute. The name is a shortened form of "aromatic polyamide".

They are fibres in which the chain molecules are highly oriented along the fibre axis, so the strength of the chemical bond can be exploited. Aromatic polyamides were first introduced in commercial applications in the early 1960s, with a meta-aramid fibre produced by DuPont under the tradename Nomex.

This fibre, which handles similarly to normal textile apparel fibres, is characterized by its excellent resistance to heat, as it neither melts nor ignites in normal levels of oxygen. It is used extensively in the production of protective apparel, air filtration, thermal and electrical insulation as well as a substitute for asbestos.

Meta-aramid is also produced in the Netherlands and Japan by Teijin under the tradename Teijinconex, in China by SRO Group (China) under the trade name X-Fiper , Yantai under the tradename New Star and a variant of meta-aramid in France by Kermel under the tradename Kermel. Based on earlier research by Monsanto Company and Bayer, a fibre - para-aramid - with much higher tenacity and elastic modulus was also developed in the 1960s-1970s by DuPont and Akzo Nobel, both profiting from their knowledge of rayon, polyester and nylon processing.

Much work was done by Stephanie Kwolek in 1961 while working at DuPont, and that company was the first to introduce a *para-aramid* called Kevlar in 1973. A similar fibre called Twaron with roughly the same chemical structure was introduced by Akzo in 1978. Due to earlier patents on the production process, Akzo and DuPont had a patent war in the 1980s. Twaron is currently owned by the Teijin company. *Para-aramids* are used in many high-tech applications, such as aerospace and military applications, for "bullet-proof" body armor fabric.

The Federal Trade Commission definition for aramid fibre is: A manufactured fibre in which the fibre-forming substance is a long-chain synthetic polyamide in which at least 85 per cent of the amide linkages, (–CO–NH–) are attached directly to two aromatic rings.

Production

World capacity of para-aramid production is estimated at about 41,000 tons/yr in 2002 and increases each year by 5-10 per cent. In 2007 this means a total production capacity of around 55,000 tons/yr.

Polymer Preparation

Aramids are generally prepared by the reaction between an amine group and a carboxylic acid halide group.

Simple AB homopolymers may look like:

$$n\mathrm{NH_2{-}Ar{-}COCl} \rightarrow -(\mathrm{NH{-}Ar{-}CO})_n- + n\mathrm{HCl}$$

The most well-known aramids (Nomex, Kevlar, Twaron, X-fiper and New Star) are AABB polymers. Nomex, X-Fiper, New Star and Teijinconex contain predominantly the meta-linkage and are poly-*metaphenylene isophtalamide*s (MPIA). Kevlar and Twaron are both *p*-phenylene terephtalamides (PPTA), the simplest form of the AABB para-polyaramide. PPTA is a product of *p*-phenylene diamine (PPD) and terephtaloyl dichloride (TDC or TCl).

Production of PPTA relies on a co-solvent with an ionic component (calcium chloride ($CaCl_2$)) to occupy the hydrogen bonds of the amide groups, and an organic component (N-methyl pyrrolidone (NMP)) to dissolve the aromatic polymer. Prior to the invention of this process by Leo Vollbracht, who worked at the Dutch chemical firm Akzo, no practical means of dissolving the polymer was known. The use of this system led to a patent war between Akzo and DuPont.

Spinning

After production of the polymer, the aramid fibre is produced by spinning the solved polymer to a solid fibre from a liquid chemical blend. Polymer solvent for spinning PPTA is generally 100 per cent (water free) sulfuric acid (H_2SO_4).

Appearances

- Fibre
- Chopped fibre
- Powder
- Pulp

Aramid Fibre Characteristics

Aramids share a high degree of orientation with other fibres such as ultra high molecular weight polyethylene, a characteristic which dominates their properties.

General:

- Good resistance to abrasion
- Good resistance to organic solvents
- Non-conductive
- No melting point, degradation starts from 500°C
- Low flammability
- Good fabric integrity at elevated temperatures
- Sensitive to acids and salts
- Sensitive to ultraviolet radiation
- Prone to static build-up unless finished

Para-aramids:

- Para-aramid fibres such as Kevlar and Twaron, provide outstanding strength-to-weight properties
- High Young's modulus
- High tenacity
- Low creep
- Low elongation at break (~3.5 per cent)
- Difficult to dye - usually solution dyed

Major industrial uses:

- Flame-resistant clothing

- Heat protective clothing and helmets
- Body armor, competing with PE based fibre products such as Dyneema and Spectra
- Composite materials
- Asbestos replacement (*e.g.* brake linings)
- Hot air filtration fabrics
- Tires, newly as Sulfron (sulfur modified Twaron)
- Mechanical rubber goods reinforcement
- Ropes and cables
- Wicks for fire dancing
- Optical fibre cable systems
- Sail cloth (not necessarily racing boat sails)
- Sporting goods
- Drumheads
- Wind instrument reeds, such as the Fibracell brand
- Loudspeaker diaphragms
- Boathull material
- Fibre reinforced concrete
- Reinforced thermoplastic pipes
- Tennis strings (*e.g.* by Ashaway and Prince tennis companies)
- Hockey sticks (normally in composition with such materials as wood and carbon)

CARBON FIBRE

Carbon fibre (carbon fibre), alternatively graphite fibre, carbon graphite or CF, is a material consisting of extremely thin fibres about 0.005-0.010 mm in diameter and composed mostly of carbon atoms. The carbon atoms are bonded together in microscopic crystals that are more or less aligned parallel to the long axis of the fibre.

The crystal alignment makes the fibre very strong for its size. Several thousand carbon fibres are twisted together to form a yarn, which may be used by itself or woven into a fabric. Carbon fibre has many different weave patterns and can be combined with a plastic resin and wound or molded to form composite materials such as carbon fibre reinforced plastic (also referenced as carbon fibre) to provide a high strength-to-weight ratio material. The density of carbon fibre is also considerably lower than the density of steel, making it ideal for applications requiring low weight.

The properties of carbon fibre such as high tensile strength, low weight, and low thermal expansion make it very popular in aerospace, civil engineering, military, and motorsports, along with other competition sports. However, it is relatively expensive when compared to similar materials such as fiberglass or plastic.

Carbon fibre is very strong when stretched or bent, but weak when compressed or exposed to high shock (eg. a carbon fibre bar is extremely difficult to bend, but will crack easily if hit with a hammer). In 1958, Dr. Roger Bacon created high-performance carbon fibres at the Union Carbide Parma Technical Centre, located outside of Cleveland, Ohio.

Those fibres were manufactured by heating strands of rayon until they carbonized. This process proved to be inefficient, as the resulting fibres contained only about 20 per cent carbon and had low strength and stiffness properties. In the early 1960s, a process was developed using polyacrylonitrile (PAN) as a raw material.

This had produced a carbon fibre that contained about 55 per cent carbon and had much better properties. The polyacrylonitrile (PAN) conversion process quickly became the primary method for producing carbon fibres. The high potential strength of carbon fibre was realised in 1963 in a process developed at the Royal Aircraft Establishment at Farnborough, Hampshire.

The process was patented by the Ministry of Defence and then licensed by the NRDC to three British companies: Rolls-Royce, already making carbon fibre, Morganite and Courtaulds. They were able to establish industrial carbon fibre production facilities within a few years, and Rolls-Royce took advantage of the new material's properties to break into the American market with its RB-211 aero-engine. Even then, though, there was public concern over the ability of British industry to make the best of this breakthrough. In 1969 a House of Commons select committee inquiry into carbon fibre prophetically asked: "How then is the nation to reap the maximum benefit without it becoming yet another British invention to be exploited more successfully overseas?" Ultimately, this concern was justified.

One by one the licensees pulled out of carbon-fibre manufacture. Rolls-Royce's interest was in state-of-the-art aero-engine applications. Its own production process was to enable it to be leader in the use of carbon-fibre reinforced plastics. In-house production would typically cease once reliable commercial sources became available.

Unfortunately, Rolls-Royce pushed the state-of-the-art too far, too quickly, in using carbon fibre in the engine's compressor blades, which proved vulnerable to damage from bird impact.

What seemed a great British technological triumph in 1968 quickly became a disaster as Rolls-Royce's ambitious schedule for the RB-211 was endangered. Indeed, Rolls-Royce's problems became so great that the company was eventually nationalized by Edward Heath's Conservative government in 1971 and the carbon-fibre production plant sold off to form Bristol Composites.

Given the limited market for a very expensive product of variable quality, Morganite also decided that carbon-fibre production was peripheral to its core business, leaving Courtaulds as the only big UK manufacturer.

The company continued making carbon fibre, developing two main markets: Aerospace and sports equipment. The speed of production and the quality of the product were improved. Continuing collaboration with the staff at Farnborough proved helpful in the quest for higher quality, but, ironically, Courtaulds's big advantage as manufacturer of the "Courtelle" precursor now became a weakness.

Low cost and ready availability were potential advantages, but the water-based inorganic process used to produce Courtelle made it susceptible to impurities that did not affect the organic process used by other carbon-fibre manufacturers. Nevertheless, during the 1980s Courtaulds continued to be a major supplier of carbon fibre for the sports-goodsmarket, with Mitsubishi its main customer. But a move to expand, including building a production plant in California, turned out badly.

The investment did not generate the anticipated returns, leading to a decision to pull out of the area. Courtaulds ceased carbon-fibre production in 1991, though ironically the one surviving UK carbon-fibre manufacturer continued to thrive making fibre based on Courtaulds's precursor. Inverness-based RK Carbon Fibres Ltd has concentrated on producing carbon fibre for industrial applications, and thus does not need to compete at the quality levels reached by overseas manufacturers.

During the 1970s, experimental work to find alternative raw materials led to the introduction of carbon fibres made from a petroleum pitch derived from oil processing. These fibres contained about 85 per cent carbon and had excellent flexural strength. Carbon Fibre is used in many other sprts cars other than Pagani *e.g.* Ferrarilink title.

Structure and Properties

Carbon fibres are the closest to asbestos in a number of properties. Each carbon filament thread is a bundle of many thousand carbon filaments. A single such filament is a thin tube with a diameter of 5–8 micrometers and consists almost exclusively of carbon. The earliest generation of carbon fibres (*i.e.*, T300, and AS4) had diameters of 7-8 micrometers. Later fibres (*i.e.*, IM6) have diameters that are approximately 5 micrometers.

The atomic structure of carbon fibre is similar to that of graphite, consisting of sheets of carbon atoms (graphene sheets) arranged in a regular hexagonal pattern. The difference lies in the way these sheets interlock. Graphite is a crystalline material in which the sheets are stacked parallel to one another in regular fashion. The intermolecular forces between the sheets are relatively weak Van der Waals forces, giving graphite its soft and brittle characteristics. Depending upon the precursor to make the fibre, carbon fibre may be turbostratic or graphitic, or have a hybrid structure with both graphitic and turbostratic parts present.

In turbostratic carbon fibre the sheets of carbon atoms are haphazardly folded, or crumpled, together. Carbon fibres derived from Polyacrylonitrile (PAN) are turbostratic, whereas carbon fibres derived from mesophase pitch are graphitic after heat treatment at temperatures exceeding 2200 C. Turbostratic carbon fibres tend to have high tensile strength, whereas heat-treated mesophase-pitch-derived carbon fibres have high Young's modulus and high thermal conductivity.

Applications

Carbon fibre is most notably used to reinforce composite materials, particularly the class of materials known as Carbon fibre or graphite reinforced polymers. Non-polymer materials can also be used as the matrix for carbon fibres. Due to the formation of metal carbides and corrosion considerations, carbon has seen limited success in metal matrix composite applications.

Reinforced carbon-carbon (RCC) consists of carbon fibre-reinforced graphite, and is used structurally in high-temperature applications. The fibre also finds use in filtration of high-temperature gasses, as an electrode with high surface area and impeccable corrosion resistance, and as an anti-static component. Molding a thin layer of carbon fibres significantly improves fire resistance of polymers or thermoset composites because a dense, compact layer of carbon fibres efficiently reflects heat..

It has also been used in experimental medical procedures to treat severe burns. Brian Eno underwent extensive surgery in early 2009 which complemented a regular skin graft on his arm with carbon fibre threads. Being carbon based, doctors were able to fuse together his skin cells with the carbon fibre. Most recently, carbon fibre composites have been used in Helios braces, braces for persons with Charcot-Marie-Tooth disease and other Peripheral neuropathy disorders. Many racecars have carbon fibre reinforcing and panels incorporated in them. Many window cleaners are now using carbon fibre poles to get off ladders. The Tucker metal pole built in 1958 lead the way but now the SimPole Carbon Fibre 12 pound poles are being used by window cleaners.

Synthesis

Each carbon filament is produced from a precursor polymer. The precursor polymer is commonly rayon, polyacrylonitrile (PAN) or petroleum pitch. For synthetic polymers such as rayon or PAN, the precursor is first spun into filaments, using chemical and mechanical processes to initially align the polymer atoms in a way to enhance the final physical properties of the completed carbon fibre.

Precursor compositions and mechanical processes used during spinning may vary among manufacturers. After drawing or spinning, the polymer fibres are then heated to drive off non-carbon atoms (carbonization), producing the final carbon fibre. The carbon fibres may be further treated to improve handling qualities, then wound on to bobbins. Wound bobbins are then used to supply machines that produce carbon fibre threads or yarn.

A common method of manufacture involves heating the spun PAN filaments to approximately 300 °C in air, which breaks many of the hydrogen bonds and oxidizes the material. The oxidized PAN is then placed into a furnace having an inert atmosphere of a gas such as argon, and heated to approxi-mately 2000 °C, which induces graphitization of the material, changing the molecular bond structure. When heated in the correct conditions, these chains bond side-to-side (ladder polymers), forming narrow graphene sheets which eventually merge to form a single, columnar filament. The result is usually 93–95 per cent carbon. Lower-quality fibre can be manufactured using pitch or rayon as the precursor instead of PAN. The carbon can become further enhanced, as high modulus, or high strength carbon, by heat treatment processes. Carbon heated in the range of 1500-2000 °C (carbonization) exhibits the highest tensile strength (820,000 psi or 5,650 MPa or 5,650 N/mm^2), while carbon fibre heated from 2500 to 3000 °C (graphitizing) exhibits a higher modulus of elasticity (77,000,000 psi or 531 GPa or 531 kN/mm^2).

Textile

Precursors for carbon fibres are polyacrylonitrile (PAN), rayon and pitch. Carbon fibre filament yarns are used in several processing techniques: the direct uses are for prepregging, filament winding, pultrusion, weaving, braiding, etc. Carbon fibre yarn is rated by the linear density (weight per unit length, *i.e.* 1 g/ 1000 m = 1 tex) or by number of filaments per yarn count, in thousands. For example, 200 tex for 3,000 filaments of carbon fibre is three times as strong as 1,000 carbon fibres, but is also three times as heavy. This thread can then be used to weave a carbon fibre filament fabric or cloth. The appearance of this fabric generally depends on the linear density of the yarn and the weave chosen. Some commonly used types of weave are twill, satin and plain.

MICROFIBER

Microfiber or microfibre refers to synthetic fibres (fibre) that measure less than one denier. The most common types of microfibers are made from polyesters, polyamides (nylon), and or a conjugation of polyester and polyamide.

Microfiber is used to make non-woven, woven and knitted textiles. The shape, size and combinations of synthetic fibres are selected for specific characteristics, including: softness, durability, absorption, wicking abilities, water repellency, electrodynamics, and filtering capabilities. Microfiber is commonly used for apparel, upholstery, industrial filters and cleaning products.

Material

Microfiber is a fibre with less than 1 denier per filament. (Denier is a measure of linear density and is commonly used to describe of the size of a fibre or filament. Nine thousand meters of a 1-denier fibre weighs one gram.) Fibres are combined to create yarns. Yarns are knitted or woven in a variety of constructions. While many microfibers are made of polyester, they can also be composed of polyamide (nylon) or other polymers.

Clothing

Microfiber apparel is often used for athletic wear, such as cycling jerseys, because the microfiber material wicks moisture (sweat) away from the body, keeping the wearer cool and dry. Microfiber is also very elastic, making it suitable for women's undergarments. Microfibre is also now used in men's underwear for extra comfort and stretch, for example in some of the Calvin Klein underwear collections.

Microfibers were used by the US military and many federal agencies, such as in the so-called Future Force Warrior Programme in the United States as they dried rapidly and caused less skin irritation due to moisture, but the U.S. military has since banned the wearing of most synthetic clothing due to melting and burn risk.

Microfiber is also used to make tough, very soft-to-the-touch materials for general clothing use, often used in skirts and jackets. Microfiber can be made into an imitation suede not made from animal products which is cheaper and easier to clean and sew than real suede.

MODACRYLIC

A modacrylic is a synthetic copolymer. Modacrylics are soft, strong, resilient, and dimensionally stable. They can be easily dyed, show good press and shape retention, and are quick to dry. They have outstanding resistance to chemicals and solvents, are not attacked by moths or mildew, and are non-allergenic.

Among their uses are in apparel linings, furlike outerwear, paint-roller covers, scatter rugs, carpets, and work clothing and as hair in wigs. Commercial production of modacrylic fibre began in 1949 by Union Carbide Corporation in the United States . Modacrylic and acrylic fibres are similar in composition and at one time were in the same category. In 1960 the Federal Trade Commission decided to separate the two fibres and establish a category for each.

The Federal Trade Commission defines modacrylic fibres as manufactured fibres in which the fibre-forming substance is any long-chain synthetic polymer composed of less than 85 per cent, but at least 35 per cent weight acrylonitrile units except when the polymer qualifies as rubber.

Modacrylic fibres are modified acrylic fibres made from acrylonitriles, but larger amounts of other polymers are added to make the copolymers. The modacrylic fibres are produced by polymerizing the components, dissolving the copolymer in acetone, pumping the solution into the column of warm air (dry-spun), and stretching while hot.

Modacrylics are creamy or white and are produced in tow and staple form. If looked at in cross section views they have an irregular shape. Modacrylic fibres are also produced in many different lengths, crimp levels, deniers and they can have various shrinkage potentials. Current modacrylic fibre producers include Solutia Inc. in the U.S. and Kaneka Corporation in Japan.

A modacrylic has properties that are similar to an acrylic. However, modacrylics are flame retardant and do not combust. The fibres are difficult to ignite and will self-extinguish. In addition to a modacrylic's flame retardant properties it has a relatively high durability that is comparable to wool. Modacrylic fibres have a moderate resistance to abrasion and a very low tenacity.

Modacrylics are poor conductors of heat. The fabrics are soft, warm and resilient but are prone to pilling and matting. Modacrylics display high performance when it comes to appearance retention. The fibres are quite resilient and will not wrinkle. They also have great dimensional stability and high elastic recovery, which gives them the ability to hold their shape. Modacrylics are sensitive to loss of appearance due to improper care; therefore, it is important to know how to care for modacrylics. Modacrylics are resistant to acids, weak alkalis, and organic solvents. These fibres are also resistant to moths, mildew and sunlight.

Modacrylic fabrics can be machine washed using warm water and tumble dried on a low setting. Modacrylic pieces can also be dry-cleaned, however, they should not be steamed and should only be tumbled on cold. Some fabrics may also be cleaned using the furrier method (a special non-immersion cleaning process).

The fibres are heat sensitive and will shrink at 250 °F and will stiffen at temperatures over 300 °F. Modacrylics are used primarily in applications where environmental resistance or flame retardancy is necessary or required.

Modacrylics have the ability to combine flame retardancy with a relatively low density, meaning protective gear is not uncomfortably heavy (*i.e.* shirts and trousers worn by electrical linemen). The combination of flame retardancy and low density is also useful in furnishings, draperies, and outdoor fabrics.

Modacrylics are also commonly used in fake fur fabrics, toupées, wigs and fleece-type fabric. By mixing the various forms of fibres it is easy to create a realistic fur look. The fabrics can then be sheared or embossed to create a closer resemblance to fur. The heat-sensitivity of modacrylic also allows wigs and hairpiece to be curled and heat styled without damage.

Modacrylics are also used in fleece, knit-pile fabric backings, and non-woven fabrics. Other uses of modacrylics include paint rollers, industrial fabrics, stuffed toys and filters.

NYLON

Nylon is a synthetic fabric made from petroleum products. It was developed in the 1930s as an alternative to silk, although it quickly became unavailable to civilian consumers, because nylon was used extensively during the war. Nylon, like many synthetics, was developed by Wallace Carothers at the Dupont Chemical company, which continues to manufacture it today. Nylon is valued for its light weight, incredible tensile strength, durability, and resistance to damage. It also takes dye easily, making nylon fabrics available in a wide array of colours for consumers.

Today, nylon is among the many polymer products in common daily use throughout the world. It is the second most used fibre in the United States, since it is so versatile and relatively easy to make. Like most petroleum products, it has a very slow decay rate, which unfortunately results in the accumulation of exhausted nylon products in landfills around the world.

Nylon is made through a chemical process called *ring opening polymerization*, in which a molecule with a cyclic shape is opened and flattened. Other forms of nylon are made through the chemical reaction between two monomers: adipoyl chloride and hexamethylene diamine.

When stretched, nylon fibres even out, thin, and smooth until they reach a point at which they have no more give, yet are still very strong. Therefore, after nylon is extruded in a thread form, it is drawn or stretched after it cools to make long, even fibres. Before drawing, nylon has a tangled structure, which straightens out into parallel lines.

The strength of nylon comes from amide groups in its molecular chain, which bond together very well. Nylon also has a very regular shape, which makes it well suited to creating fabrics designed to stand up to intense forces. In fact, nylon was the primary material used in parachutes and ropes during the Second World War for this reason. It is also used for bulletproof vests and other hard wearing items.

Nylon is very sensitive to heat and should be washed and dried on cool settings. Nylon can also be hung dry, and it is favoured by campers because it dries very rapidly.

Nylon is a flexible textile, and as a result, it appears in a wide range of applications, from clothing to climbing equipment. Depending on how it is processed, nylon can be formed into the gossamer-like threads used in stockings or into thick toothbrush bristl,es.

OLEFIN FIBRE

Olefin fibre is a synthetic fibre made from alkenes. It is used in the manufacture of various textiles as well as clothing, upholstery, wallpaper, ropes, and vehicle interiors. Olefin is also referred to as polypropylene, polyethylene, or polyolefin. Olefin's advantages are its strength, colourfastness and comfort, stain, mildew, abrasion and sunlight resistance, and good bulk and cover.

Olefin fibres have great bulk and cover while having low specific gravity. This means "warmth without the weight." The fibres have low moisture absorption, but they can wick moisture and dry quickly. Olefin is abrasion, stain, sunlight, and chemical resistant.

It does not dye well, but has the advantage of being colorfast. Since Olefin has a low melting point, textiles can be thermally bonded. The fibres have the lowest static of all manufactured fibres and a medium luster.

One of the most important properties of olefin is its strength. It keeps its strength in wet or dry conditions and is very resilient. The fibre can be produced for strength of different properties.

Production Method

The Federal Trade Commission's official definition of olefin fibre is "A manufactured fibre in which the fibre forming substance is any long-chain synthetic polymer composed of at least 85 per cent by weight of ethylene, propylene, or other olefin units"

Polymerization of propylene and ethylene gases, controlled with special catalysts, creates olefin fibres. Dye is added directly to the polymer before melt spinning is applied. Additives, polymer variations and different process conditions can create a range of characteristics.

High pressure production, which uses ten tons per square inch, creates a film for molded materials. Low pressure production uses a low temperature with a catalyst and hydrocarbon solvent. This process is less expensive and produces a polyethylene polymer more suitable for textile use. The polymer is then melted, spun into water, or air cooled. The fibre is drawn out to six times the spun length. Gel spinning is a new method in which a gel form of polyethylene polymers is used.

POLYESTER

Polyester is a manufactured product made from synthesized polymers. It tends to be very resilient, quick drying, resistant to biological damage such as mold and mildew, easy to wash, and able to hold forms well. While polyester is often maligned as a textile, it has many useful applications. Polyester is, however, highly flammable, so care should be taken when wearing it. Most synthetic fabrics are subject to flammability by nature, because they are made from polymers.

Polyester is not just a textile. It is made from polyethylene terephthalate (PET), the same material used to make plastic drink bottles. Many drink bottles are recycled by being reheated and turned into polyester fibres, which in addition to being an efficient use, also helps keep polymers out of landfills.

Polyester is a plastic, invented in Britain in the early 1940s by two gentlemen working with synthetic polymers. In the 1950s, the new textile took off, becoming popular for its easy care, drape, and versatility. In the United States, Dupont is one of the major manufacturers of PET and has been making Dacron polyester since 1950.

Dupont has researched the properties of polyester extensively and devised a number of uses well beyond textiles, such as mylar and other polyester films. Dupont has also been driven to find ways to recycle polyester products to help keep costs down, as the base chemical substances that combine to form polyester are petroleum products.

To make polyester, ethylene glycol and dimethyl terephthalate are mixed together. The chemical reaction results in bisterephthalate. This substance is heated to 270° Fahrenheit (132° Celsius), where it reacts again to form polyethylene terephthalate. Like many chemical reactions that result in polymers, the polyester making process results in unhealthy off gassing, and protection should be worn while making PET. After synthesizing the polymers, the manufacturer decides what to do with them.

PET can be formed into plastics that can later be recycled. It is a highly malleable material and appears in all sorts of applications, from drink bottles to floppy disk liners. PET can also be used to make polyester fibre, which is used in auto upholstery, quilt batting, and clothing of all sorts.

To make polyester fibre, an extruder is used to produce very fine threads of PET. Polyester clothing tends to be slippery and silky in feel, although it can cause skin irritation for some wearers. The polyester fibres used to make clothing can be knitted or woven, although most are knit, to maximize the flexibility of polyester. Some polyester is blended with other fabrics to provide more loft or stretch, or to minimize skin irritation.

ZYLON

Zylon is a trademarked name for a range of thermoset liquid crystalline polyoxazole. This synthetic polymer material is manufactured by the Toyobo Corporation. Zylon was invented and developed by SRI International in the 1980s. Like Kevlar, Zylon is used in a number of applications that require very high strength with excellent thermal stability. Tennis racquets, table tennis blades, various medical applications, and some of the Martian rovers are some of the more well known instances. More common usage is in drive belts for snow mobiles manufactured by MBL USA Corporation. IUPAC name: poly(p-phenylene-2,6-benzobisoxazole)

VINALON

Vinalon is a synthetic fibre, produced from polyvinyl alcohol, using anthracite and limestone as raw materials. Vinalon was first developed in 1939 by Japanese scientist, Ichiro Sakurada, Yi Sung-ki and H. Kawakami Production of this fibre was delayed for WW2. The fibre was largely ignored in Korea until Ri defected to North Korea in 1950. Trial production began in 1954 and in 1961 the massive February 8 Vinalon Complex was built in Hamhung. Its success and widespread usage in North Korea is often pointed to in propaganda as an example of the success of the *juche* philosophy.

While Hamhung remains a major production centre for vinalon; in 1998, a vinalon factory was opened up in South Pyongan. In early 2010, Kim Jong-il himself attended a mass rally at the February 8 Vinalon Complex in Hamhung to celebrate it's reopening after 16 years of inactivity. While Kim is often seen at politial rallies or military parades, this is the first documented time he has ever attended a industrial mass rally, said an anonymous South Korean security official.

Vinalon, also known as "Juche fibre" in the DPRK, has become the national fibre of North Korea and is used for the majority of textiles, outstripping fibre such as cotton or Nylon, which are only produced in small amounts in North Korea. Other than clothing, Vinalon is also used for shoes, ropes and quilt wadding. Vinalon is resistant to heat and chemicals but has numerous disadvantages: it is stiff, uncomfortable, shiny, prone to shrinking and difficult to dye.

SPANDEX

Spandex, Lycra or elastane is a synthetic fibre known for its exceptional elasticity. It is stronger and more durable than rubber, its major non-synthetic competitor. It is a polyurethane-polyurea copolymer that was invented in 1959 by chemist Joseph Shivers at DuPont's Benger Laboratory in Waynesboro, Virginia. When first introduced, it revolutionized many areas of the clothing industry.

The name "Spandex" is an anagram of the word "expands". It is the preferred name in North America; in many European countries it is referred to as "elastane". Other brand names associated with Spandex include Lycra, Elaspan (also Invista's), Creora (Hyosung), ROICA and Dorlastan (Asahi Kasei), Linel (Fillattice), and ESPA (Toyobo).

POLYETHYLENE

Polyethylene is a type of polymer that is classified as a thermoplastic, meaning that it can be melted to a liquid and remolded as it returns to a solid state. As the name implies, polyethylene is chemically synthesized from molecules that contain long chains of ethylene, a monomer that provides the ability to double bond with other carbon-based monomers to form polymers. Polyethylene is known by other, non-official names, such as polythene in the United Kingdom. In addition, it is sometimes spelled as polyethylyne, or abbreviated to simply PE.

The first laboratory creation of polyethylene occurred in 1898 by accident at the hands of Hans von Pechmann while applying heat to another compound the German chemist previously discovered — diazomethane. Ironically, the synthesis of polyethylene via extreme heat and pressure in an industrial setting was again made by accident, but 35 years later.

A few years later, another chemist employed by the same England-based chemical company devised a method to consistently produce polyethylene under the same conditions. As a result, polyethylene became the primary source of low-density polyethylene (LDPE) production in 1939.

While polyethylenc is cssential to the economic health of the plastics industry, most consumers readily recognize the role it plays in everyday life. In fact, this substance is found in many ordinary household items, such as food wrap, shampoo bottles, milk containers, toys, and the common plastic bag used to tote groceries home from the store.

However, polyethylene is also present in numerous other products that contain plastic components. For instance, it is used to manufacture artificial knee and hip replacement parts, bulletproof vests, and even glassy flooring for ice skating rinks. Polyethylene may fall under one of several types. The distinction between them is determined by its molecular weight and branching, which is affected by its crystallization. LDPE is an example of branched polyethylene since its carbon molecules are attached to long chains of polyethylene instead of hydrogen.

Otherwise, a linear structure of carbon to hydrogen occurs, which is known as high-density polyethylene (HDPE). However, further variances in structure and molecular weight produce other forms, such as ultra high molecular weight polyethylene (UHMWPE), medium-density polyethylene (MDPE), or very low-density polyethylene (VLDPE). While polyethylene may help to make numerous useful and durable products possible, its environmental impact is cause for concern.

For one thing, it does not readily biodegrade and can reside in a landfill for hundreds of years. However, diligent recycling may significantly reduce this problem. In addition, scientists are exploring the possibility of employing *Sphingomonas*, an aerobic bacteria shown to shorten biodegrading of some forms of polyethylene to just a few months. Environmental preservation efforts have also led to the development of bioplastics, with the aim of synthesizing polyethylene from ethanol obtained from sugarcane.

Bibliography

Allen, Frederick James: *The shoe industry*, Boston: The Vocation bureau of Boston, 1916.

Allen, Nellie B: *Cotton and other useful fibers*, Boston: Ginn, 1929.

Allington, Sara May: *Practical sewing and dressmaking*, Boston: Page Co., 1913.

Allinson, May: *Dressmaking as a trade for women in Massachusetts*, Washington: Govt. print. off., 1916.

Archana Ruhela : *A Handbook of Livestock Management Techniques*, Oxford Book Company, 2008.

Arvind M. Nawale : *Anita Desai's Fiction : Themes and Techniques*, B.R. Publishing, 2011.

Atkinson, Edward: *Cotton and Cotton Manufactures in the United States*, Boston: Franklin Press, 1880.

Avtar Singh Bimbraw : *Agro-techniques for Umbelliferous Medicinal and Aromatic Plants of India*, International Book Distributing Co, 2006.

Baird, Robert H: *The American Cotton Spinner, and Managers' and Carders' Guide: A Practical Treatise on Cotton Spinning*, Philadelphia: H. C. Baird, 1869.

Batchelder, Samuel: *Introduction and Early Progress of the Cotton Manufacture in the United States*, Boston: Little, Brown and Comp

Brian K. Nunnally : *Analytical Techniques in DNA Sequencing*, Taylor and Francis Group, 2009.

C. Emmanuel, Fr. S. Ignacimuthu,s.j. and S. Vincent : *Applied Genetics : Recent Trends and Techniques*, MJP Pub, 2006.

Chinmoy Goswami and and Abhijit Paintal : *A Textbook of Laboratory Techniques*, Wisdom Press, 2011.

Ekwal Imam : *Applied Statistical Techniques*, New India Publishing Agency, 2015.

James Turley : *Advanced 80386 Programming Techniques*, Tata McGraw-Hill, 2008.

Khaja Mohtesha Muddin : *A Practical Manual of Veterinary Andrology and Reproductive Techniques*, Satish Serial Publishing House, 2015.

M.C. Nautiyal and B.P. Nautiyal : *Agrotechniques for High Altitude Medicinal and Aromatic Plants*, Bishen Singh Mahendra Pal Singh, Dehradun, 2004.

Mahinder Singh : *A Textbook of Analytical Chemistry: Instrumental Techniques*, Dominant, Publisher, 2005.

Michael Armstrong : *A Handbook of Management Techniques*, Viva Books, 2010.

Michael Ward : *50 Essential Management Techniques*, Jaico, Publisher, 2002.

Michael Wash : *54 Tools and Techniques for Business Excellence*, Westland, 2002.

O.P. Sharma, Narendra Singh and Ruchira Shukla : *Advanced Management Techniques*, Agrotech, Publication, 2012.

P. Venkatesan : *A Textbook of Biomedical Laboratory Techniques*, Atlantic, Publication, 2011.

R C Upadhyaya : *Agrotechniques of Medicinal Plants*, Anmol, Publication, 2008.

R.C. Mandal and P.T.N. Nambiar : *Agricultural Statistics Techniques and Procedures*, Agrobios, Publication, 2012.

Rajeev K. Prithiani : *Algorithms and Techniques for the Recognition of Various Protein Structures*, Vista International, 2010.

Richard J. Maturi : *100 World Famous Stock Market Techniques*, Vision Books, 2004.

S Felix : *Aquaculture Management Techniques*, Narendra Pub., 2007.

S. Jayapandian : *Accounting for Managers : Effective Techniques for Decision Making*, Ane Books, 2009.

S.P. Tiwari, S. Rajagopal and Usha Rani Mehra : *Analytical Techniques in Animal Nutrition*, Satish Serial Pub, 2012.

Sadhana Singh : *Agro-Techniques of Medicinal Plants*, Wisdom Press, 2012.

Samuel A. Malone : *A to Z of Training and Development: Tools and Techniques*, Jaico Publication, 2004.

Sangeeta Gupta : *Accounting and Statistical Techniques*, Pointer, Publisher, 2008.

Subha Gnaguly : *A Laboratory Manual on Virology and Tissue Culture Techniques*, Narendra Publishing House, 2014.

Sumit Narula and R.K. Jain : *An Introduction to Journalism : Principles and Techniques*, Regal Pub., 2012.

V.K. Agarwal and S.D. Upadhyaya : *Agrotechniques of Medicinal and Aromatic Plants*, Satish Serial, 2006.

Index